PSAM 12
Probabilistic Safety Assessment and Management
22–27 June 2014 • Sheraton Waikiki, Honolulu, Hawaii, USA

CONFERENCE ROCEEDINGS
Volume 3 – Monday PM II

PSAM 12

Probabilistic Safety Assessment and Management

22 - 27 June, 2014

Sheraton Waikiki, Honolulu, Hawaii USA

CONFERENCE PROCEEDINGS

Volume 3

Monday PM II

Foreword

It is was our honor to welcome you to Honolulu, Hawaii, for the twelfth rendition of the Probabilistic Safety Assessment and Management (PSAM) Conference. The planning for PSAM Honolulu began back in 2007 (before PSAM 9 in Hong Kong), when we looked at several locations around the United States, included Arizona, California, Boston, and even considered locations in Oceania. Based upon the feedback both during and after the conference, PSAM 12 proved to be a great success.

We would like to thank all of the volunteers, those that served before, during, and after the Conference. Members of the Technical Program Committee, the Organizing Committee, the session chairs, and the presenters have our gratitude for making PSAM 12 the most memorable PSAM yet.

This publication represents the technical proceedings for the Conference. Due to the large number of published papers (a total of 391), we have subdivided the technical content (papers) into five volumes, one for each day of the conference.

On behalf of the International Association for Probabilistic Safety Assessment and Management Board of Directors, we hope that this publication will provide a valuable technical resource in addition to a reminder of the memorable stay in the Hawaiian Islands.

Dr. Curtis Smith
Technical Program Chairs

Dr. Todd Paulos
General Chair

/ Sponsors

PSAM 12 - Probabilistic Safety Assessment and Management
JUNE 22-27, 2014

Sponsors

Technical Program Committee

Technical Program Chair: Curtis Smith, INL USA
Assistant Technical Program Chairs: Steve Epstein, Lloyd's Register Japan
Vinh Dang, PSI Switzerland
Ted Steinberg, QUT Australia

We would like to thank the members of the PSAM 12 Technical Program Committee. These individuals helped to make PSAM 12 a success by reviewing abstracts, technical papers, organizing sessions, and providing technical leadership for the conference.

Technical Committee Members:

- Roland Akselsson
- S. Massoud (Mike) Azizi
- Tito Bonano
- Ronald Boring
- Roger Boyer
- Mario Brito
- Kaushik Chatterjee
- Vinh Dang
- Claver Diallo
- Nsimah Ekanem
- Steve Epstein
- Fernando Ferrante
- Federico Gabriele
- Ray Gallucci
- S. Tina Ghosh
- David Grabaskas
- Katrina Groth
- Seth Guikema
- Steve Hess
- Christopher J. Jablonowski
- Moosung Jae
- Jeffrey Joe
- Vyacheslav S. Kharchenko
- James Knudsen
- Zoltan Kovacs
- Ping Li
- Harry Liao
- Francois van Loggerenberg
- Jerome Lonchampt
- Soliman A. Mahmoud
- Diego Mandelli
- Donoval Mathias
- Zahra Mohaghegh
- Thor Myklebust
- Cen Nan
- Mohammad Pourgolmohammad
- Marina Roewekamp
- Clayton Smith
- Shawn St. Germain
- Ted Steinberg
- Kurt Vedros
- Smain Yalaoui
- Robert Youngblood
- Enrico Zio

Organizing Committee

General Chair: Dr. Todd Paulos
General Vice Chair: Prof. Stephen Hora, USC
Technical Program Chair: Curtis Smith, INL USA
Webmaster, Registration, Support for Papers/Abstracts Submission and Review: Hanna Shapira, TICS

Table of Content

Page	Paper	
10	46	**Proof Testing of Safety-Instrumented Systems: New Testing Strategy Induced by Dangerous Detected Failures** Yiliu Liu, Marvin Rausand *Department of Production and Quality Engineering, Norwegian University of Science and Technology, Trondheim, Norway*
18	77	**A New Interfacing Approach between Level 1 and Level 2 PSA** Nicolas Duflot, Nadia Rahni, Thomas Durin, Yves Guigueno and Emmanuel Raimond *IRSN, Fontenay aux Roses, France*
30	78	**An Approach to Ensure the Availability of Complex Systems** Kaushik Chatterjee, Kumar Bhimavarapu, Robert Kasiski, and William Doerr *FM Global, Norwood, MA, USA*
37	101	**Reliability/Availability Methods for Subsea Risers and Deepwater Systems Design and Optimization** Annamaria Di Padova (a), Fabio Castello b), Fabrizio Tallone (a), Michele Piccini (b) *a) Saipem S.p.A., Fano, Italy, b) RAMS&E S.R.L., Turin, Italy*
49	56	**Statistical Analysis of Common Cause Failure Events Using ICDE Data** S. Yu, M.D. Pandey (a), S. Yalaoui and Y. Akl (b) *a) University of Waterloo, Waterloo, Canada, b) Canadian Nuclear Safety Commission, Ottawa, Canada*
58	133	**Extending the Alpha Factor Model for Cause Based Treatment of Common Cause Failure Events in PRA and Event Assessment** Andrew O'Connor, Ali Mosleh *Center for Risk and Reliability, University of Maryland, College Park, United States*
69	383	**Estimating Common Cause Failure Probabilities for a PRA Taking into Account Different Detection Methods** Kalle E. Jänkälä *Fortum Power and Heat Oy, Espoo, Finland*
81	473	**Time Dependent Analysis with Common Cause Failure Events in RiskSpectrum** Pavel Krcal (a,b) and Ola Bäckström (a) *a) Lloyd's Register Consulting, Stockholm, Sweden, b) Uppsala University, Uppsala, Sweden*
89	9	**A State of the Practice Investigation Guiding the Development of Visualizations for Minimal Cut Set Analysis** Yasmin I. Al-Zokari, Liliana Guzman (a), Barboros Can Conar (b), Dirk Zeckzer (c), Hans Hagen (a) *a) TU Kaiserslautern, Kaiserslautern, Germany, b) University of Applied Sciences, Kaiserslautern, Germany, c) Leipzig University, Leipzig, Germany*
108	71	**Risk Analysis and Decision Theory: An Extended Summary** E. Borgonovo, V. Cappelli, F. Maccheroni, M. Marinacci (a), and C. Smith (b) *a) Department of Decision Sciences and IGIER, Università Bocconi, Milan, Italy, b) Idaho National Laboratory, Idaho Falls, Idaho, USA.*
112	105	**Maritime Oil Spill Risk Assessment for Hanhikivi Nuclear Power Plant** Juho Helander *Fennovoima, Helsinki, Finland*
124	159	**Using Bond Graphs for Identifying and Analyzing Technical and Operational Hazards in Complex Systems** Ingrid Bouwer Utne, Eilif Pedersen and Ingrid Schjølberg *Department of Marine Technology, Norwegian University of Science and Technology, Trondheim, Norway*
236	33	**Estimating Farmer's Risk Aversion** Patrick Momal *IRSN, Fontenay-aux-Roses, France*
244	42	**Development of a Methodological Approach to Strategic Fire Service Planning Combining Concepts of Risk, Hazard and Scenario-based Design** Adrian Ridder, Uli Barth *University of Wuppertal, Wuppertal, Germany*

Table of Content

Page	Paper	
149	43	**Ambiguity in Risk Assessment** Inger Lise Johansen and Marvin Rausand *Norwegian University of Science and Technology, Trondheim, Norway*
161	157	**How is Capability Assessment Related to Risk Assessment? Evaluating Existing Research and Current Application from a Design Science Perspective** Hanna Palmqvist, Henrik Tehler, and Waleed Shoaib *Division of Risk Management and Societal Safety, Centre for Risk Assessment and Management, and Centre for Societal Resilience, Lund University, Lund, Sweden*
173	60	**Analyses of AP1000® Expanded Event Tree Sequences Based on Best-Estimate Calculations** J.Montero-Mayorga, C.Queral, J.Gonzalez-Cadelo and G. Jimenez *Universidad Politecnica de Madrid, Madrid, Spain*
185	88	**Application of Web-based Risk Monitor in Tianwan Nuclear Power Plant** Hao Zheng (a), Wei Wang (b), Xiaohui Gu, Yong Qu, Zhenli Bao (c), Xuhong He (b) *a) Lloyd's Register Consulting, Beijing, China, b) Lloyd's Register Consulting, Stockholm, Sweden, c) Jiangsu Nuclear Power Co., Lianyungang, China*
192	106	**Analyzing System Changes with Importance Measure Pairs: Risk Increase Factor and Fussell-Vesely Compared to Birnbaum and Failure Probability** Janne Laitonen, Ilkka Niemelä *Radiation and Nuclear Safety Authority (STUK), Helsinki, Finland*
202	15	**Energy Loss Optimization in Basic T-Shaped Water Supply Piping Networks for Probabilistic Demands** KW Mui, LT Wong, and CT Cheung *Department of Building Service Engineering, The Hong Kong Polytechnic University, Hong Kong, China*
214	502	**Insights and Improvements Based on Updates to Low Power and Shutdown PRAs** J. F. Grobbelaar, J. A. Julius, K. D. Kohlhepp, and M. D. Quilici *Scientech, a Curtiss-Wright Flow Control Company, Tukwila, WA, U.S.A.*
223	34	**When Is It Justified to Delay the Implementation of Safety Improvements After They Have Been Approved?** Patrick Momal *IRSN, Fontenay-aux-Roses, France*
229	65	**The Underlying Principles and Quantitative Values of Risk Limits** Dennis R. Damon *U. S. Nuclear Regulatory Commission, Washington DC*
234	86	**Development of A Framework for Establishment of Risk-informed Safety Goals for Nuclear Power Plants Operation in the UAE** Jun Su Ha (a), Sung-yeop Kim (b), Jamila Khamis Al Suwaidi (c), and Philip Beeley(a) *a) Khalifa Univ. of Science, Technology and Research, Abu Dhabi, UAE, b) Korea Advanced Institute of Science and Technology (KAIST), Republic of Korea, c) Federal Authority for Nuclear Regulation (FANR), Abu Dhabi, UAE*
245	254	**Insights from PSA Comparison in Evaluation of EPR Designs** Ari Julin, Matti Lehto (a), Patricia Dupuy, Gabriel Georgescu, Jeanne-Marie Lanore (b), Shane Turner, Paula Calle-Vives (c), Anne-Marie Grady, Hanh Phan (d) *a) Radiation and Nuclear Safety Authority (STUK), Finland, b) Institute of Radiological Protection and Nuclear Safety (IRSN), France, c) Office for Nuclear Regulation (ONR), United Kingdom, d) Nuclear Regulatory Commission (USNRC), United States of America*
259	237	**OECD WGRISK – Challenges and Recent Tasks** Marina Roewekamp (a), Jeanne-Marie Lanore (b), Kevin Coyne (c), Milan Patrik (d), Abdallah Amri, Neil Blundell (e) *a) Gesellschaft für Anlagen- und Reaktorsicherheit (GRS) mbH, Köln, Germany, b) Institut de Radioprotection et de Sûreté Nucléaire (IRSN), Fontenay-aux-Roses, France, c) U.S. Nuclear Regulatory Commission, Washington, DC USA, d) UJV Rez, Rez, Czech Republic, e) OECD Nuclear Energy Agency (NEA), Issy-les-Moulineaux, France*
268	146	**RAPP: Method for Risk Prognosis on Complex Failure Behaviour in Automobile Fleets Within the Use Phase** Stefan Bracke and Sebastian Sochacki *University of Wuppertal, Chair of Safety Engineering and Risk Management, Wuppertal, Germany*
280	168	**Stress-Dependent Weibull Shape Parameter Based on Field Data** Jochen Juskowiak and Bernd Bertsche *University of Stuttgart, Stuttgart, Germany*

Table of Content

Page   Paper

291   *314*   **APTA Approach: Analysis of Accelerated Prototype Test Data Based on Small Data Volumes Within a Car Door System Case Study**
Marcin Hinz, Philipp Temminghoff, and Stefan Bracke
University of Wuppertal, Chair of Safety Engineering and Risk Management, Wuppertal, Germany

Proof testing of safety-instrumented systems:
New testing strategy induced by dangerous detected failures

Yiliu Liu[*], Marvin Rausand
Department of Production and Quality Engineering, Norwegian University of Science and Technology, Trondheim, Norway

Abstract: Some dangerous failures of safety-instrumented systems (SISs) are detected almost immediately by diagnostic self-testing, whereas other dangerous failures can only be detected by proof-testing. The first type is called dangerous detected (DD) failures and the second type is called dangerous undetected (DU) failures. Proof tests are usually carried out at constant time intervals. DD-failures are repaired almost immediately whereas a DU-failure will persist until the item is proof-tested. Many items can have a DU- and a DD-failure at the same time. After the repair of a DD-failure is completed, the maintenance team has two options: to perform an "insert" proof test for DU-failure or not. If an insert proof test is performed, it is necessary to decide whether the next scheduled proof test should be postponed or performed at the scheduled time. This paper uses Petri nets to model the proof test strategies after DD-failures and to analyze the effects of the different strategies on the SIS performance. It is shown that insert proof tests reduce the unavailability of the system, whereas the adjustment (or not) of the test schedule does not have any significant long term effect.

Keywords: safety-instrumented system, proof test, dangerous detected-failure, dangerous undetected-failure

1. INTRODUCTION

Safety-instrumented systems (SISs) are widely used in many industries (e.g., process, nuclear, oil and gas industry) to prevent hazardous events and to mitigate the consequences of such events [1, 2]. A modern SIS has built-in facilities for diagnostic self-testing during operation. Such tests can detect many dangerous failures almost immediately such that a repair action can be initiated. These dangerous failures are called dangerous detected (DD) failures. On the other hand, dangerous failures that are not detected by diagnostic testing are called dangerous undetected (DU) failures and are only revealed in proof tests that are carried out at regular intervals (e.g., once per year).

The mean time from a DD-failure occurs until the function is restored, MTTR, is usually rather short (e.g., 5-8 hours), and DD-failures will therefore not be a main contributor to the unavailability of a SIS that is operated in low-demand mode (i.e., where demands for the safety function do not occur more often than once per year). For some channels, DD-failures can be repaired on-line, while the process is running as normal during the repair. In most cases, however, the process section has to be brought to a safe state (most often stopped) during the repair of the DD-failure. For some channels, DD- and DU-failures can be present at the same time and repairing a DD-failure does not guarantee that a DU-failure is not remaining in the channel. In some cases, it may be possible to proof-test for a DU-failure as part of the repair of the DD-failure.

Such proof tests can be regarded as "insert tests" between two scheduled tests, such that the number of proof tests in a certain time period will increase. This means that the average proof test interval will be reduced. Since the length of the proof test interval has a significant influence on the availability

---

[*] Corresponding author, yiliu.liu@ntnu.no

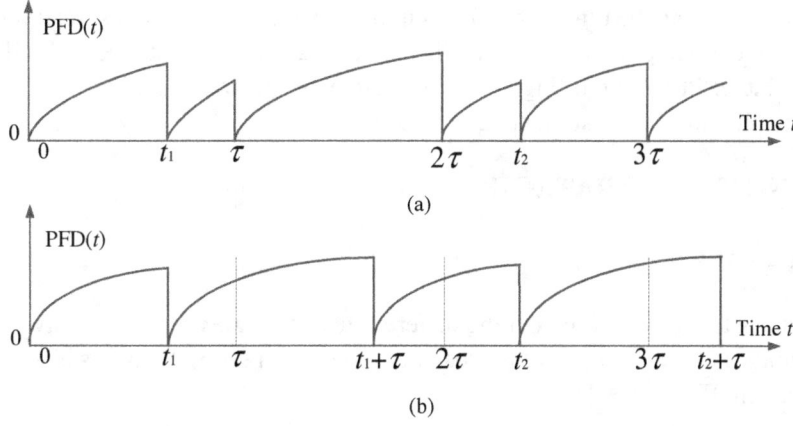

Figure 1: PFD for test strategies 2 (a) and 3 (b) as a function of time t

performance of a SIS [3], the new proof tests induced by DD-failures should also have influence. Thus, the objective of this paper is to model the relationship between such proof tests induced by DD-failures and SIS performance, as well to examine effects of these tests.

The remainder of the paper is organized as follows: Section 2 presents the possible follow-up test strategies after a DD-failure is revealed. Next, the modeling approach is briefly introduced, and some Petri net models for different strategies are studied in section 3. The effects of different test strategies on the SIS availability performance are analyzed in section 4. Finally, section 5 presents conclusions and research perspectives.

2. TEST STRATEGIES INDUCED BY DD-FAILURES

In this paper, we study a simple SIS subsystem with only one channel. When a DD-failure in this system is detected, the maintenance team can repair the SIS channel in a short time, and then they have three options for testing the SIS for DU-failures:

– Strategy 1: Finish the work, but do not perform any insert proof test for DU-failures.

– Strategy 2: Perform an insert proof test for DU-failures, but keep the proof test schedule unchanged.

– Strategy 3: Perform an insert proof test for DU-failures, and change the proof test schedule (mostly postpone the subsequent proof test).

To illustrate the difference between strategies 2 and 3, consider a pressure safety valve that is scheduled to be proof-tested each April and assume that a DD-failure occurs in September. If strategy 2 is applied, a proof test for DU-failures is initiated immediately after having repaired the DD-failure, and the next proof test is still carried out the next April. If, on the other hand, strategy 3 is applied, the next proof test is postponed till next September keeping the same interval between two proof tests.

We use the probability of failure on demand (PFD) as measure for the safety performance of the SIS. DU-failures are always the main contributor to the PFD of a SIS because they may put the SIS in an unavailable state for a long time until a proof test is carried out. Fig. 1 briefly illustrates the changing trends of the PFD(t) as a function of the time t when test strategies 2 and 3 are applied, respectively.

In Fig. 1, $t_0, t_1, t_2, \ldots$, denote the times when DD-failures occur, and τ is the test interval. It is shown in Fig. 1(a) that the predefined proof test schedule is kept unchanged under strategy 2, and each proof test can reduce the value of PFD(t) to 0. Fig. 1(b) for strategy 3, illustrates that the time to the next proof test is re-counted after an insert test induced by a DD-failure.

3. MODELING TEST STRATEGIES

3.1. Modeling approach

In this paper, Petri nets are used to model the different test strategies. Petri nets have been adapted to SIS reliability analysis [4, 5] especially for test strategies of SISs [6, 7], and is also a recommended modeling approach in IEC 61508 [1].

The international standard IEC 62551 [8] defines the terminology of Petri nets in dependability analysis. There are two basic elements in such models, places (shown as circles in Fig. 2) and transitions (shown as bars) are connected with directed arcs. Tokens are illustrated as bullets to express the movable resources in the system and reside in the places. For each arc, a multiplicity is assigned to denote the token delivering capacity of the arc. The distribution of tokens in the places is regarded as a marking, and each marking represents a system state.

When all input places to a transition have at least as many tokens as the multiplicities of the associate arcs to the transition, the transition is enabled. And then, the transition can be fired to absorb the tokens in these places according to the multiplicities of the associate arcs. In this paper, timed Petri nets are applied, where a firing time (delay from enabled to fired) can be assigned to each transition. Such Petri nets have two types of transitions: immediate transitions (zero firing time, shown as thin bars) and timed transitions (shown as thick bars). In IEC 62551 [8], a blank bar is used to denote a transition with exponential firing time, and a filled thick bar is for the transition with constant firing time.

In addition, an inhibitor arc (shown as a small circle at the end of an arc) is sometimes used to prevent a transition from being enabled. Such a special arc enables its output transition when there is no token in the associate place. More details for Petri nets can be found in IEC 62551 [8].

3.2. Modeling test strategy 1

In the following subsections, three Petri net models are established to illustrate the three test strategies in section 2 after the occurrence of a DD-failure.

Fig. 2 shows the Petri net model for test strategy 1. In this model, a DD-failure occurs when the token in p_{DW} is removed by the transition t_{DD} and deposited to p_{DF}. In the same way, the places p_{UW}, t_{UW} and p_{UF} are used to denote the occurrence of a DU-failure. Both t_{DD} and t_{DU} are blank, which means that the failure times are exponentially distributed. In addition, exponentially distributed transition t_{DR} is used to denote the repair times of the SIS from the fault state due to DD-failures, respectively.

Proof tests are reflected by firing t_T and depositing a token to p_T. The filled bar shows that proof tests are carried out at constant intervals.

Transitions t_1 and t_2 express the two situations in a proof test. If a DU-failure in the SIS is revealed, t_1 can be fired; otherwise, t_2 can be fired. The inhibitor here is for enabling t_2 when there is no token in p_{UF}. No matter whether t_1 or t_2 is fired, a token can be deposited to p_S, meaning that the test is finished and the test resources are restored and ready for the next proof test. In this figure, both t_1 and t_2 have exponential distributed firing times, reflecting the assumption that the time required to perform a proof

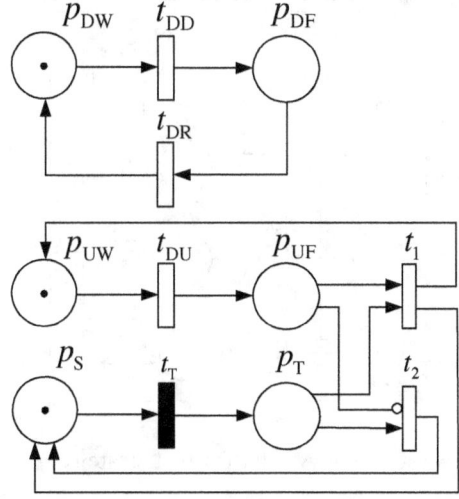

Figure 2: Model for test strategy 1

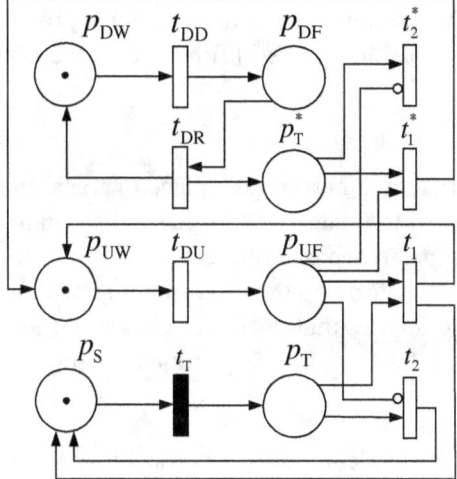

Figure 3: Model for test strategy 2

test is exponentially distributed. In fact, the firing time of t_1 includes the time for the proof test and for the repair of the DU-failure.

It should be noted that the part describing the DD-failure is separate from the parts describing the DU-failure and the proof test in this model. The moving of tokens between p_{DW} and p_{DF} has no influence on other tokens, since maintenance does not do anything except for repairing the DD-failure in strategy 1.

3.3. Modeling test strategy 2

Test strategy 2 is modeled in Fig. 3 by adding one place and two transitions to the model for strategy 1. When t_{DR} is fired in this model, a token is released to p_{DW} and another token can enter p_T^* to initiate a proof test for DU-failure. The functions of t_1^* and t_2^* are similar with those of t_1 or t_2, in absorbing the token in p_T^* to complete the test. In the case where a token resides in p_{UF}, it will be absorbed by t_1^*. Thereafter, a token is released to p_{UW}, such that a DU-failure is revealed by the proof test induced by

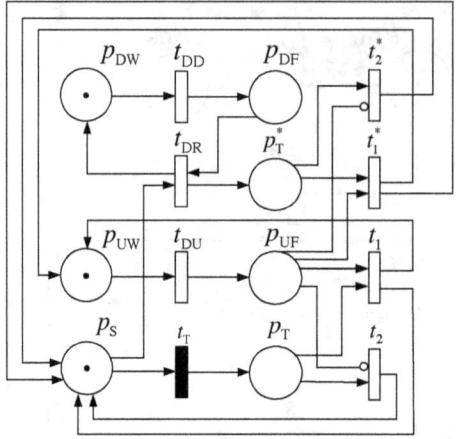

Figure 4: Model for test strategy 3

the DD-failure and repaired.

There is no relation between DD-failure part and the scheduled proof test part in this model. Even if there is an insert proof test, the transition t_T is still fired at the scheduled time.

3.4. Modeling test strategy 3

The model in Fig. 4 for test strategy 3 has no new place or transition compared with the model in Fig. 3, but some arcs are involved to relate DD-failures and scheduled proof tests. In Fig. 4, t_{DR} is fired by absorbing the token in p_S. When one of the transitions, t_1^* and t_2^* is fired, a token can come back to p_S. In other words, when performing the insert proof test in the model, the enabling condition of t_T is stopped. Such a change implies that once there is a proof test induced by a DD-failure, the resource for scheduled testing will be used, and the time counting process for the next proof test is re-initiated.

It should also be noted that such a model ignores the probability that a DD-failure and the scheduled proof test take place at the same time. This assumption has little influence on the following analyses since this probability is very low.

4. EFFECTS ON SIS PERFORMANCE

In the analysis of the influences of the three test strategies on SIS performance, the following assumptions are made:

- DU- and DD-failures occur independent of each other;

- The occurrences of DU- and DD-failures have constant failure rates;

- The system can have both DU-and DD-failures at a specific moment;

- Only one maintenance team is available for testing and repair of the SIS;

- A failed system can be restored to a fully functioning state (i.e., without any failure);

- The test and repair times are exponentially distributed.

The average, long term PFD (or PFD_{avg}) is used in this section as the measure to analyze the effects of different test strategies on the performance of a SIS. PFD_{avg} is determined by the contributions from DD- and DU-failures, and for all the three test strategies, PFD_{avg} can be calculated as:

$$\text{PFD}_{\text{avg}} = \Pr(\mathbf{m}(p_{\text{DF}}) = 1) + \Pr(\mathbf{m}(p_{\text{UF}}) = 1) - \Pr(\mathbf{m}(p_{\text{DF}}) = 1) \cdot \Pr(\mathbf{m}(p_{\text{UF}}) = 1) \quad (1)$$

where $\Pr(\mathbf{m}(p_{\text{DF}}) = 1)$ denotes the long term sojourn probability of the model where place p_{DF} holds one token, such that the SIS is unavailable due to a DD-failure. Similarly, $\Pr(\mathbf{m}(p_{\text{UF}}) = 1)$ is equal to the unavailability of the SIS due to a DU-failure.

Since DD- and DU failures are independent from each other, both $\Pr(\mathbf{m}(p_{\text{DF}}) = 1)$ and $\Pr(\mathbf{m}(p_{\text{UF}}) = 1)$ involve the probability of simultaneous DD- and DU-failures, and thus when we calculate PFD_{avg}, the probability $\Pr[(\mathbf{m}(p_{\text{DF}}) = 1) \cap (\mathbf{m}(p_{\text{UF}}) = 1)]$ should be subtracted to remove this overlap. Of course, for highly reliable SISs, $\Pr[(\mathbf{m}(p_{\text{DF}}) = 1) \cap (\mathbf{m}(p_{\text{UF}}) = 1)]$ is very small and without a significant influence on the calculated result.

The efficiency of the diagnostic test for a SIS is measured by the diagnostic coverage (DC), which is the fraction of the rate of dangerous failures that are revealed by the diagnostic testing relative to the rate of all dangerous failures. Suppose that the total dangerous (D) failure rate of a SIS is 2×10^{-5} per hour, and DC is 0.9. Thus, the rate of DD-failures (λ_{DD}) is $0.9 \times 2 \times 10^{-5} = 1.8 \times 10^{-5}$ per hour, whereas the rate of DU-failures (λ_{DU}) is $(1 - \text{DC}) \times 2 \times 10^{-5} = 2 \times 10^{-6}$ per hour. In this case, we can regard the scheduled test interval (τ) as 1 year (8760 hour), and all test and repair rates as 0.125 per hour (meaning that the average time of testing and repair is 8 hours). Therefore, the transition rates of $t_{\text{DR}}, t_{\text{UR}}, t_1, t_2, t_1^*$, and t_2^* are 0.125 in all the three models. Here, the testing and potential repair processes are partly simplified: Repair time is included in the test duration. Since such time is rather short compared with the test interval, the simplification has little impact on the calculations.

The firing time of t_T in the model in Fig. 3 is set to be 8752 hours, which is equal to the test interval minus the average test time. While the corresponding firing time t_T in Fig. 4 is 8744 (8760-8-8) hours, since the maintenance resource is used in the test for DD-failures and then in the testing for DU-failures. If these transitions are approximated as exponential distributed ones, the firing rate also should be determined in the consideration of testing and repair time.

Monte Carlo simulations for the three models in Fig. 2-4 are carried out by using the software GRIF.[1] The results are shown in Table 1.

Table 1: PFD_{avg} of the SIS under different test strategies

	Strategy 1	Strategy 2	Strategy 3
$\text{PFD}_{\text{avg(DD)}}$	1.4389×10^{-4}	1.4403×10^{-4}	1.4455×10^{-4}
$\text{PFD}_{\text{avg(DU)}}$	8.6930×10^{-3}	8.0501×10^{-3}	8.0511×10^{-3}
PFD_{avg}	8.8356×10^{-3}	8.1930×10^{-3}	8.1945×10^{-3}

Strategies 2 and 3 change the proof test schedules since the maintenance team performs follow-up actions induced DD-failures, and thus the rates of DD-failures can be simply assumed to influence the effects of test strategies 2 and 3. In order to measure this effects, we check four DD-failure rates from 2×10^{-5} to 2×10^{-4} per hour, while keeping the DU-failure rate at 2×10^{-6} per hour. The corresponding DCs vary from 0.91 to 0.99.

---

[1] http://grif-workshop.com

In Table 2, we only present the PFD$_{\text{avg}}$ values due to the DU-failures, because they are the dominant parts of the unavailability of the SIS.

Table 2: PFD$_{\text{avg}}$ of the SIS with different DD-failure rates

λ_{DD}	PFD$_{\text{avg(DU)}}$		
	Strategy 1	Strategy 2	Strategy 3
2×10^{-5}	8.6995×10^{-3}	7.9740×10^{-4}	7.9792×10^{-3}
6×10^{-5}	8.6970×10^{-3}	6.8910×10^{-3}	6.8792×10^{-3}
1×10^{-4}	8.6902×10^{-3}	6.0541×10^{-3}	6.0513×10^{-3}
2×10^{-4}	8.6937×10^{-3}	4.6436×10^{-3}	4.6440×10^{-3}

The results illustrate the reduction of PFD$_{\text{avg(DU)}}$ obtained by using test strategies 2 and 3. Since DD-failures have no relation with proof testing, the values of PFD$_{\text{avg(DU)}}$ have a constant level. Meanwhile, with test strategies 2 and 3, PFD$_{\text{avg(DU)}}$ decreases with more frequent DD-failures and the follow-up proof testing for DU-failures.

If only DU-failures are taken into account, PFD$_{\text{avg}}$ is often approximated as (e.g., in [4])

$$\text{PFD}_{\text{avg(DU)}} \approx \frac{\lambda_{\text{DU}} \tau}{2} \qquad (2)$$

The idea behind the formula is that the unavailability of the SIS is the DU-failure rate multiplied by the mean time until the DU-failure is revealed ($\tau/2$). This can also be formulated as the DU-failure rate divided by the mean revealing rate of the DU-failure (i.e., $2/\tau$). Results of PFD$_{\text{avg}}$ based on such an approximation is close to those by the model under test strategy 1, where DD-failures do not induce more tests for DU-failures. However, two types of proof tests exist under test strategy 2. With the same philosophy as used to develop Eq. 2, the unavailability of the SIS due to DU-failures also can be approximated as the DU-failure rate divided by the mean revealing rate of the DU-failure.

$$\text{PFD}_{\text{avg(DU)}} \approx \frac{\lambda_{\text{DU}}}{\frac{2}{\tau} + \lambda_{\text{DD}}} = \frac{\lambda_{\text{DU}} \tau}{2 + \lambda_{\text{DD}} \tau} \qquad (3)$$

Based on Eq. 3, it can be found that if the rate of DD-failure is high, the reduction of PFD$_{\text{avg}}$ with test strategy 2 is more obvious. The data in Table 2 shows the same trend.

Test strategies 2 and 3 have approximately the same effects on the SIS performance. Even if the two strategies after some few DD-failures result in different PFD profiles, their average, long term PFD are very close. When the rate of DD-failures is low compared with the scheduled test frequency, the number of proof tests induced by DD-failures is very low, and the influence of postponing one proof test is therefore low. If the rate of DD-failures is relatively high (e.g., equal or higher than the scheduled test interval), the two strategies are still similar because the changed proof test schedule will often be changed again by another DD-failure before the next proof test. For example, when the rate of DD-failure is high enough, a DD-failure often occurs between two scheduled proof tests, and thus the next scheduled proof test may be postponed again and again. PFD$_{\text{avg}}$ due to DU-failures under test strategy 3 can therefore be approximated by Eq. 3.

The managerial implications based on the findings include: Proof testing for DU-failures induced by DD-failures can improve the performance of a SIS, especially when DD-failures occur rather frequently. After the insert tests, it is not important for the long term SIS performance whether the original proof test schedule is changed or not. Selection of test strategies 2 or 3 should involve other factors in consideration, such as the cost of each proof test, and cost due to the adjustment of the testing schedule.

5. CONCLUSION

This paper studies follow-up activities after a DD-failure has been detected. Insert proof testings to reveal DU-failures are found to be able to reduce the unavailability of the SIS, especially when DD-failures occur frequently. Based on the presented Petri net models, the average values of the PFD_{avg} of the SIS in the long term are similar no matter if the original proof test schedule is changed or not. Such findings are helpful for operators to select their appropriate strategies on the basis of the requirements for safety and cost.

Studies of SISs with more complex structures are needed to verify the conclusion of this study, since only the simplest one-out-of-one system is considered here. Common cause failures (CCFs) are always dominating factors for the unavailability of complex SISs with redundancies, and thus it is necessary to consider such kind failures in prospective studies. In addition, since the sizes of Petri net models in this paper will greatly increase when involving more elements in the system, suitable modeling methods should also be developed.

References

[1] IEC 61508, *Functional Safety of Electrical/Electronic/Programmable Electronic Safety-Related Systems. Part 1-7*. Geneva: International Electrotechnical Commission, 2010.

[2] IEC 61511, *Functional safety: safety instrumented systems for the process industry sector, part 1-3*. Geneva: International Electrotechnical Commission, 2003.

[3] M. Rausand and A. Høyland, *System reliability theory; models, statistical Methods, and applications*. Hoboken, NJ.: Wiley, 2nd ed., 2004.

[4] M. Rausand, *Reliability of Safety-Critical Systems: Theory and Applications*. Hoboken, NJ: Wiley, 2014.

[5] ISO/TR 12489, *Petroleum, petrochemical and natural gas industries – reliability modeling and calculation of safetyt systems*. Geneva: International Organization for Standardization, 2013.

[6] Y. Liu and M. Rausand, "Reliability effects of test strategies on safety-instrumented systems in different demand modes," *Reliability Engineering & System Safety*, vol. 119, pp. 235–243, 2013.

[7] Y. Liu, "Optimal staggered testing strategies for heterogeneously redundant safety systems," *Reliability Engineering & System Safety*, vol. 126, pp. 65–71, 2014.

[8] IEC 62551, *Analysis techniques for dependability Petri net techniques*. Geneva: International Electrotechnical Commission, 2012.

A new interfacing approach between level 1 and level 2 PSA

Nicolas Duflot[a], Nadia Rahni[a], Thomas Durin[a], Yves Guigueno[a] and Emmanuel Raimond[a]

[a] IRSN, Fontenay aux Roses, France

Abstract: IRSN (TSO of the French Nuclear Safety Authority) has been developing L2 PSAs for many years, using its own probabilistic tool, KANT (probabilistic event trees software) associated to a very fast-running source term code (MER). Since the IRSN L1PSAs event trees are developed with one other dedicated software, the L1-L2 PSA interface methodology is a key and difficult point of the IRSN PSA methodology.

In the previous versions of the IRSN PSAs, L1-L2 PSA interface was a mostly manual process, resulting in significant resources allocation. To cope with such a difficulty, a new interfacing approach, allowing computerized generation of plant damage states (PDSs), has been developed. This approach is based on the introduction of flag events (basic events with a probability of one) into the L1PSA minimal cut sets (MCSs) in order to transfer information related to front lines systems (needed for accident management) status and operators actions. Afterwards, the MCSs are filtered to identify automatically the different PDSs of the L1-L2 PSA interface using a new dedicated tool.

The automatic PDS generation allows implementing a very detailed L1-L2PSA interface easy to update. Since this new IRSN interfacing approach is based on fault trees only, it can be implemented with most of the level 1 PSA tools.

Keywords: Level 2 PSA, Interface, Plant Damage States (PDS), bridge trees.

1. INTRODUCTION

The purpose of this paper is to present the new interfacing approach used at IRSN (the French TSO) to interface the level 1 PSA (L1PSA), which assesses the probability of core damage, with the level 2 PSA (L2PSA), which aims to assess the risk of radiological release of a Nuclear Power Plant. This paper presents the motivation for the development of a new automatized interfacing method, the principles which sustain this method illustrated by examples and some results.

1.1. Context

As presented in the reference [1], in a L1PSA, *"the design and operation of the plant are analysed in order to identify the sequences of events that can lead to core damage and the core damage frequency is estimated. Level 1 PSA provides insights into the strengths and weaknesses of the safety related systems and procedures in place or envisaged as preventing core damage.*

In Level 2 PSA, the chronological progression of core damage sequences identified in Level 1 PSA are evaluated, including a quantitative assessment of phenomena arising from severe damage to reactor fuel. Level 2 PSA identifies ways in which associated releases of radioactive material from fuel can result in releases to the environment. It also estimates the frequency, magnitude, and other relevant characteristics of the release of radioactive material to the environment. This analysis provides additional insights into the relative importance of accident prevention, mitigation measures, and the physical barriers to the release of radioactive material to the environment (e.g. a containment building)".

Since the L2PSA is the prolongation of the L1PSA's sequences after the beginning of core degradation, an interface between the L1PSA and the L2PSA is required to transfer, from L1PSA to L2PSA, all information needed for L2PSA. As presented in [2], the interface is defined by plant damage states (PDS). The PDSs are the equivalent of initiating events in the level 1 event trees, i.e. they are the initiating events in the level 2 event trees. Accident sequences from L1PSA are grouped together into PDSs in such a manner that all accidents within a given PDS can be treated in the same way. Each PDS represents a group of level 1 accident sequences that have similar characteristics for severe accident scenarios, e.g. accident timelines, potential for generation of loads on the containment or systems availability, thereby resulting in similar severe accident progression and radiological source terms.

The plant damage states have to characterize the parameters that are needed to describe the sequences in the L2PSA analysis and those that influence the accident progression and source term. The PDSs definition is based on the definition of interface variables which can have different values. Each PDS corresponds to a specific combination of values of the different interface variables (an example of interface variables and their values is given in section 2.1).

As presented in [2], there are two approaches in developing the probabilistic model of a L2PSA:
- the integrated models,
- the separated models.

With the integrated model approach, the same computer tool is applied for L1 and L2PSA and the model database contains all level 1 and level 2 information. The advantage of such an approach consists in the possibility to use, in the level 2 event trees, the same fault trees as the ones used in the level 1. Consequently, the interface between level 1 and level 2 event trees is simplified since it is not required to codify in the PDSs the status of each system at the core damage time to make available this information in the level 2. On the other hand, the L2PSA has to be developed with the same event tree formalism as the L1PSA (i.e. event tree and fault trees, with Boolean fusion algorithms), which has limitation regarding L2PSA needs (especially for the modeling of severe accident phenomenology, dependencies between events and uncertainties). In addition, the quantification of such integrated models, if detailed, may be very time consuming since L1PSA and L2PSA have to be quantified simultaneously.

With the separated models approach, two different probabilistic event tree softwares can be used: one appropriate to the L1PSA and another one appropriate to L2PSA. It is thus required to implement a very detailed interface since the only link between L1PSA and L2PSA is the PDSs list. The list of interface variables has to be developed to inform about:
- the initiating events that start the accidental sequences leading to core damage,
- the status of the front lines systems[a] (e.g. fully available / partly available / not operable),
- the status of the key operator actions (initiated or not) like the initiation of the Reactor Coolant System (RCS) feed and bleed on LWR.

Furthermore, if some systems are modeled in L1PSA and some other in L2PSA, it may be complicated to guaranty the effective correlation between these systems failure (due to support systems, shared components...).

This disadvantage, intrinsic to the separated models, is counterbalanced by the possibility to use a L2PSA dedicated tool specifically designed to take care of the modeling of the level 2 part (accident progression after core damage). For example, IRSN, the French safety authority's TSO, has developed its own L2PSA tool package (see ref. [3]) which consists in KANT (probabilistic event trees software), MER (for source term calculations) and MERCOR (for standardized radiological consequences assessment). Since these codes are specifically designed for L2PSA, KANT allows, for example, the development of simplified (fast running) physical models to simulate each phenomenon during accident progression and allows also the transmission of physical values (time, pressures, temperatures...), allowing a precise description of the NPP status, through the Accident Progression Event Tree (APET). In the same manner, MER allows detail releases assessment for each severe accident sequence generated by the APET quantification.

At IRSN, the L1PSA is developed with a commercial tool: RiskSpectrum. The L2PSA is developed with the tool package KANT / MER / MERCOR. These codes, which deal with probabilistic

---

[a] A front line system is a system whose operating influences directly the accident progress. For example, the Safety Injection System (SIS) is a front line system whereas the Component Cooling Water System (CCWS) - even if it is a key system regarding safety - is not a front line system, but a support system.

quantification of severe accident phenomena, systems and human failures, allow fast-running consequences/frequencies severe accident calculations for a very large number of accidental sequences (thousands of release categories). The severe accident phenomena analyses are mostly performed with the reference code ASTEC[a] (Accident Source Term Evaluation Code). Each IRSN L2PSA (one for each French type of NPP) is supported by large set of ASTEC accident scenario calculations (between 100 and 200) to consider in detail and in a "best estimate" manner (few conservative assumptions) the different possible accidental sequences on a reactor

1.2. IRSN motivation for a new interface development

Thanks to recent developments (see ref. [4]), the IRSN tools package allows consequences/frequencies calculations for a large amount of severe accidental sequences in a detailed and best estimate (plus uncertainties) manner. To take advantage of these possibilities, a detailed and precise interface has to be developed.

The generation of such an interface has historically been made by adding bridge trees in the L1PSA model. These bridge trees are event trees connected to L1PSA core damage sequences and are used to specify the state of systems that are not considered in L1PSA sequences (for additional information regarding bridge trees construction, see section 2.2.4 from ref. [2]).

Even if this approach allows precise definition of PDS directly from an "extended" L1PSA, it requires a manual development of a large amount of sequences and the attribution of the PDSs to these sequences is also manual (at least in RiskSpectrum, the L1PSA software tool used by IRSN). In addition, the creation or the deletion of an interface variable involves a manual modification of all these sequences and consequences.

Thus, the objective of the new IRSN interface method development was:
- to offer the same level of precision as the bridge trees approach,
- to avoid deep modification of L1PSA event trees,
- to avoid any manual post processing of the L1PSA results,
- to guaranty a reasonable computation time when updating the PDSs frequencies.

2. IMPLEMENTATION OF A NEW INTERFACE APPROACH

2.1. Preamble

The construction of the interface between L1PSA and L2PSA requires, first, that all the information from L1PSA that should be used in the L2PSA, is identified. Then, this information has to be attributed to the different interface variables, which have to be precisely defined for this purpose.

For example, on a PWR, the "*manual start of the high pressure safety injection*" is identified as one key operator action that has to be considered in the L2PSA. In parallel, the *High Pressure Safety Injection System* (HPSI) is identified as a front line system whose availability has to be known in the interface. It is, then, possible to combine both pieces of information in a variable as shown in Table 1.

Table 1: Example of interface variable attributes (HPSI)

Values for HP variable	Description
1	The HPSI is available and started by the operators
2	The HPSI is started by the operators and available until the switch in recirculation mode
3	The HPSI is available but the operators have failed to start it (human error)
4	The HPSI is available until the switch in recirculation mode but the operators have failed to start it (human error)
5	The HPSI is not available

---

[a] ASTEC is an integral code which is able to simulate the plant behaviour from the initiating event to the possible release of radioactive products (the 'source term') outside the containment. It covers all the severe accident phenomena except steam explosion and containment mechanical integrity.

As an example, the list of interface variable considered in the on-going update of IRSN's 900MWe PWR L2PSA is given in the Table 2.

Table 2: Example of interface variable list (900 MWe PWR)

Variable name	Description
AC	Status of the safety injection accumulators
AE	Status of the Emergency Steam Generators Feed Water tank makeup water means
AG	Status of gravitational reactor pool makeup
AP	Status of the ultimate makeup from the neighbor plant
AP-RC	Status of the automatic makeup when operating at mid-loop level
AR	Status of normal steam generator feedwater system
AS	Status of the containment spray system
AT	Status of the diversified RCS pumps seals water injection pump
BP	Status of the low pressure safety injection system (LPSI)
CR	Status of the reactor criticality after initiating event occurrence (i.e. ATWS)
DL	Presence of a coolant boron dilution
EF	Status of the ultimate containment venting system
EG	Possible presence of a boron plugging in the core (i.e. failure of simultaneous injection in hot leg and cold leg if required)
ET	Initial plant state (at the time of the initiating event)
GM	Status of the reactor coolant pumps (operating or not)
GV	Status of the secondary cooling (merging the Emergency Feed Water System (EFWS) and the steam discharge (in atmosphere) system)
HP	Status of the High Pressure Safety Injection System (HPSI)
IE	Status of the containment isolation system
IG	Capability to perform Steam Generators isolation
LC	Status of the relaying 48V DC power
LH	Staus of the 6,6 kV AC emergency supplied distribution system
PL	RCS break location (if any)
PT	RCS break size (if any)
RA	Status of the residual heat removal system (CHRS)
RC	Status of the chemical and volume control system
RT	Type of Steam Generator Tube Rupture (if any)
SE	Type of steam line break (if any)
SO	Status of the pressurizer safety valves
SR	Status of the residual heat removal system's safety valves
VL	Type of interfacing LOCA (V-LOCA), if any
IC	Availability of the main control room
KT	Status of the automatic information transmission system from the main control room (MCR) to the national emergency teams
MP	Status of the containment pressure measurement (for severe accident)
MG	Status of the steam generators water level measurement
DD	Status of the in containment dose measurement
TR	Status of the core outlet temperature measurement

2.2. Principle of the method

The three steps of this new interface method consist in:
- firstly, expressing the different values of each interface variable thanks to binary questions (i.e. with answer yes / no). Each value of a given interface variable can be expressed as a combination of the answers to these questions,
- secondly, adding, in the Minimal Cut Set (MCS) from the L1PSA, flag events (i.e. basic events with a probability of one) which indicate the answer to these binary questions,
- finally, identifying automatically the PDS corresponding to each MCS based on a flag events filtering thanks to a dedicated tool named OIPK.

In the following paragraphs, these three steps are presented in more details and illustrated by an example issued from table 1.

First step:
This step consists in expressing the different values of a given interface variable thanks to binary questions. These questions are formulated in such a manner that the answer *yes* corresponds to the normal situation when systems are operational (and *no* corresponds to their unavailability). If the example of table 1 is used, the *HP* variable can be expressed based on the following questions:
a) Have the operators started manually the HPSI? (yes / no)
b) Is the HPSI available for direct injection mode (before switching in recirculation mode) ? (yes / no)
c) Is the HPSI available after the switch in recirculation mode? (yes / no)

As a result, the value 2 (*The HPSI is started by the operators and available until the switch in recirculation mode*) of variable *HP* corresponds to the following combination of binary questions:
$$\text{question a} = yes * \text{question b} = yes * \text{question c} = no$$

This step has to be done for all interface variables to be able to express all their values in term of combination of binary questions.

Second step:
This step consists in adding, if needed, flag events in L1PSA MCSs. A flag event is a basic event with a probability equal to one (i.e. when a flag event is added in a MCS, the MCS frequency is not modified). There is one flag event for each binary questions from step one. For a given binary question, there are two possibilities when a L1PSA is considered:
- either the MCS involves an unique answer to the binary question,
- or the MCS does not allow to identify an unambiguous answer to the binary question.

If there is a unique answer and if the answer to the binary question is *yes*, the flag event is added in the MCS. It is not added if the answer is *no*. No additional MCS is created.
If the answer to the binary question is unknown, both cases are considered. In a first case, the answer is supposed to be *yes* and the flag event is added in the MCS. In a second time, the answer *no* is supposed, the flag event is not added and one or several MCSs are created to consider the initial MCS plus the adverse event corresponding to the answer *no* (in our example, the MCS is combined with the failure of HPSI in injection mode).

For example, let us consider that the MCS from the L1PSA corresponds to a large break LOCA with a common cause failure of all emergency busbars:
$$LB\text{-}LOCA * CCF\_busbars$$

It is then obvious that the answer to the question *Is the HPSI available for direct injection?* is *no* since the HPSI pumps are not powered. Consequently, no flag event is added, the MCS is not modified.
Let us consider, now, a MCS which corresponds to a large break LOCA with a total failure of the containment spray system:
$$LB\text{-}LOCA * CCF\_Cont.spray$$

It is not possible, based on this MCS, to know if the HPSI is available for direct injection or not. Consequently, both cases are considered:

initial MCS plus HPSI success → $LB\text{-}LOCA * CCF\_Cont.spray * \underbrace{\S HP\_INJ\_OK}_{\text{Flag event which indicates that direct injection is available}}$

initial MCS plus HPSI failure → $LB\text{-}LOCA * CCF\_Cont.spray * \underbrace{(Failure\ of\ HPSI\ in\ direct\ injection)}_{\text{All the MCS corresponding to HPSI failure in injection mode}}$

As a result,
- the initial MCS is preserved with the additional information direct injection available. Its frequency is unchanged (flag event has a probability of one) and
- a sub set of MCSs is created to consider the occurrence of the initial MCS and the failure of the direct injection. The frequency of this sub set of modified MCS is consistent with the frequency of the initial MCS considered in conjunction with the HPSI direct injection failure.

This second step **has to be performed for all the binary questions of all the interface variables**. As a result each initial MCS, corresponding to core damage, may contain many flag events and/or may have been modified to consider additional failures not considered in L1PSA. The new MCSs set obtained does not correspond anymore to the core damage but to the core damage and to the success or failure of all systems relevant for the PDSs construction. This new MCSs set is named extended MCSs set. Each MCS corresponding to core damage is extended to consider relevant information for PDSs construction.

The effective implementation of this second step in a fault trees / event trees context is presented in section 2.3.

Third step:

This third step consists in identifying, for each extended MCS, the corresponding PDS. This step is supported by a dedicated tool developed by IRSN and named OIPK. This tool has, as an input, a file containing all the extended MCSs. Before identifying automatically the existing PDSs and defining their frequencies, the user has to define, in OIPK, the interface variables and their values. Then, each value of each interface variable has to be expressed as a combination of flag events. The combination can used AND, OR, NOT and brackets.

Let us consider that
- the flag event which indicates that the HPSI direct injection is available is §HP_INJ_OK,
- the flag event which indicates that the HPSI recirculation is available is §HP_RECR_OK,
- all the basic events name, corresponding to the operator failure to start HPSI, start with H_SIS

It is then possible to define, in OIPK, the interface variable *HP* which contains five values. Based on the flag events name given above, the different values of HP are expressed as:

Table 1: Example of filters defined in OIPK for the variable HP (HPSI)

Num.	Values of HP variable	Filter
1	The HPSI is available and started by the operators	§HP_INJ_OK and §HP_RECR_OK and not H_SIS*
2	The HPSI is started by the operators and available until the switch in recirculation mode	§HP_INJ_OK and not §HP_RECR_OK and not H_SIS*
3	The HPSI is available but the operators have failed to start it (human error)	§HP_INJ_OK and §HP_RECR_OK and H_SIS*
4	The HPSI is available until the switch in recirculation mode but the operators have failed to start it (human error)	§HP_INJ_OK and not §HP_RECR_OK and H_SIS*
5	The HPSI is not available	not §HP_INJ_OK

Once the interface variables and their values are defined and expressed in terms of combinations of flag events, OIPK is able to
- identify automatically the existing PDS (no need to define the potentially existing PDS manually),
- built the sub MCSs set corresponding to each PDS and,
- based on this sub MCSs set, define the frequency of each PDS.

These inputs (list of PDS and their frequencies) are then transmitted to the L2PSA software (the software in charge of the Accident Progression Event Tree (APET) modeling).
Further information about OIPK software is presented in section 2.4.

2.3. Implementation of this method in L1PSA

Step one: construction of the prolongation fault tree

The construction of the extended MCSs set is obtained with a unique fault tree in charge of adding flag events. This fault tree contains, under an AND gate, one sub fault tree for each interface variable. This unique fault tree is named the "prolongation fault tree". The sub fault trees implement the flag events corresponding to each interface variable. The figure 1 gives an example of a prolongation fault tree.

Figure 1: Example of a prolongation fault tree

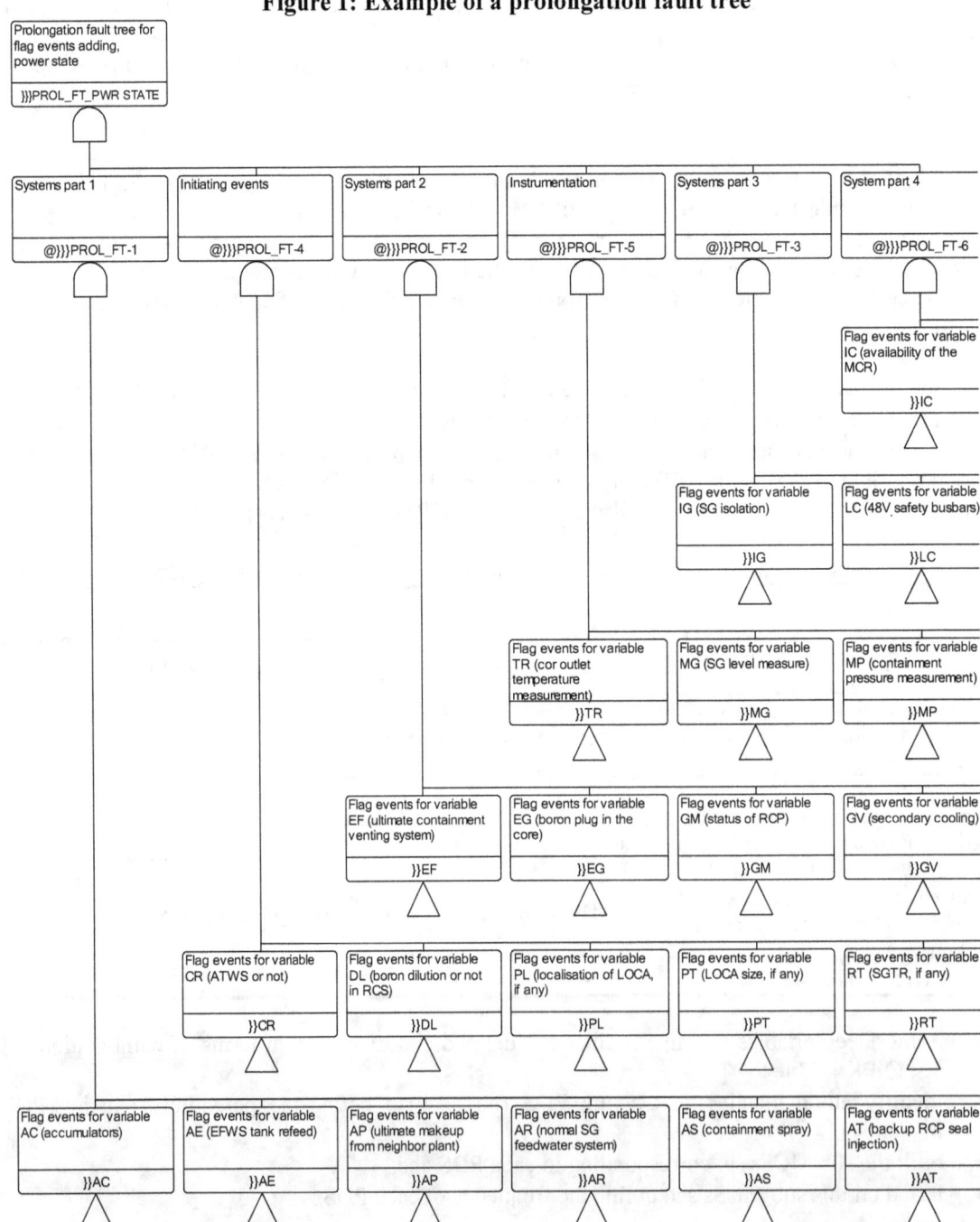

Each sub fault tree, corresponding to an interface variable, is made of one top AND gate and sub OR gates as shown in figure 2.

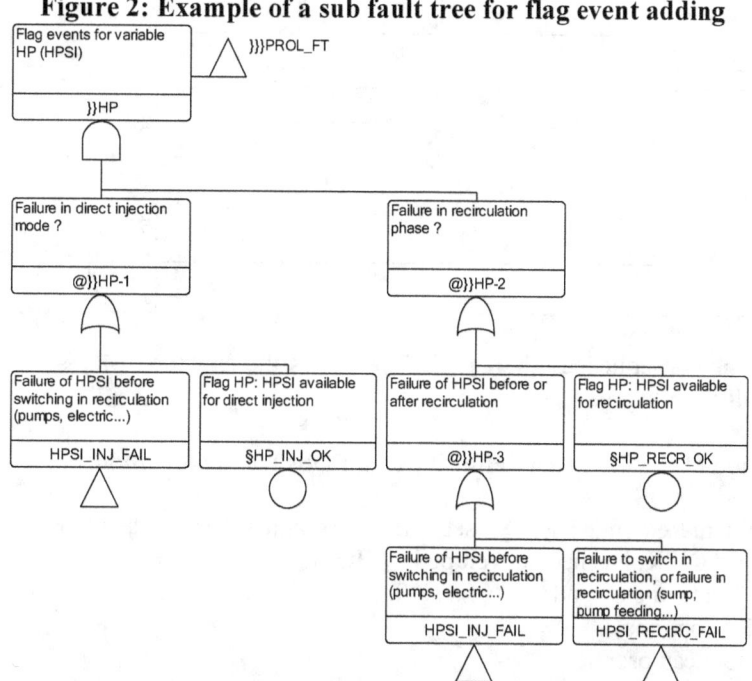

Figure 2: Example of a sub fault tree for flag event adding

As shown on the above figure, if the failure of HPSI in direct injection mode is given in the MCS from L1PSA (due to CCF on the safety busbars, for example), the top gate of the fault tree HPSI_INJ_FAIL is true and the flag events are not added into the extended MCS.

In these sub fault trees, the same systems modeling as the one used in L1PSA fault trees is used to reduce the work load and to ensure consistency[a]. For example, the system fault tree HPSI_INJ_FAIL is also used in L1PSA modeling (for example, it is included in the function event LHSI of figure 4).

Step two: integration of the prolongation fault tree in the event trees
The "failure" coming from the prolongation fault tree has to be considered in all core damage sequences of the L1PSA event trees. To do so, it is possible, either to put the prolongation fault tree in a unique bridge tree connected to all the L1PSA sequences leading to core damage (see figure 3) or, if the PSA software does not allow event trees linking, to add it in all core damage sequences (see figure 4).

Figure 3: Utilization of a bridge tree

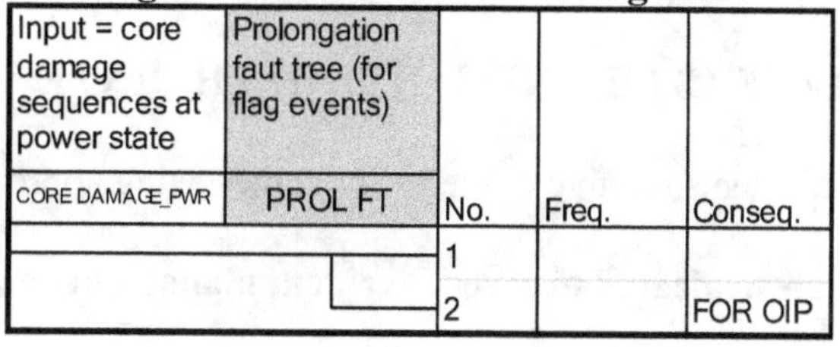

Input = core damage sequences at power state	Prolongation faut tree (for flag events)			
CORE DAMAGE_PWR	PROL FT	No.	Freq.	Conseq.
		1		
		2		FOR OIP

---

[a] A careful attention has to be paid on the consistency between the initial L1PSA system fault trees and the added flag events systems fault trees: the failure of a system in the L1PSA must not be missed in the interface.

Figure 4: Prolongation fault tree directly in L1PSA event trees

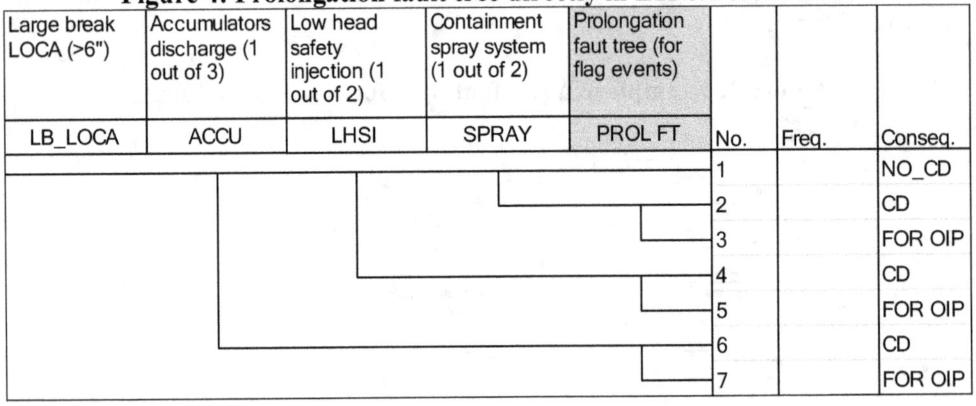

Large break LOCA (>6")	Accumulators discharge (1 out of 3)	Low head safety injection (1 out of 2)	Containment spray system (1 out of 2)	Prolongation faut tree (for flag events)	No.	Freq.	Conseq.
LB_LOCA	ACCU	LHSI	SPRAY	PROL FT			
					1		NO_CD
					2		CD
					3		FOR OIP
					4		CD
					5		FOR OIP
					6		CD
					7		FOR OIP

When quantified, the consequence "FOR OIPK" produces the extended MCSs set which constitutes the input for the OIPK software.

2.4. Automation of the PDSs list definition and PDSs frequency updating with OIPK

To identified, from the extended MCSs set, the PDSs and to calculate their total frequencies, an interface tool, named OIPK, has been developed by IRSN.

This tool uses, as inputs, the extended MCSs set in a text format (required) and the list of basic events of the extended PSA (optional).

As an output, this tool can produce:
- the list of PDSs and their frequency in a text format adapted to KANT,
- the list of PDS and their frequency in an Excel spreadsheet for results analysis,
- the MCSs set corresponding, for example, to a given PDS or to all the PDS sharing a given value for a given interface variable.

To produce such results, as presented in the step 3 of section 2.2, the OIPK user needs to define the interface variables (name + description). Then, for each interface variable, its possible values have to be defined and expressed in terms of combination of flag events through a graphical interface. As a result, tables similar to Table 1 are produced in OIPK.

One of the advantages of OIPK is that the user does not have to define, a priori, the PDSs that have to be quantified. Based on the interface variables definition made in this software, the existing PDSs are automatically identified and quantified from L1PSA results. The number of PDSs generated depends:
- on the number of interface variables and the number of their values and
- on the cut-off frequency applied when the extended MCSs set has been generated with L1PSA tool.

The computation time to produce PDSs set, once the extended MCSs set has been generated, is limited (about half an hour in the worst case).

3. OTHER IRSN'S SPECIFICITIES REGARDING THE INTERFACE APPROACH

In this section, some other elements relative to the IRSN's interfacing approach are presented.

3.1. Modeling of systems dedicated to severe accident management

To guaranty a good consistency between the systems modeled in the L1PSA and some active systems useful for severe accident management (containment isolation system, severe accident instrumentation in MCR, reactor building venting …) IRSN has chosen to model these systems in the extended L1PSA and to transfer their status through dedicated interface variables. For example, the filtered containment venting system (FCVS) is modeled in the extended L1PSA (i.e. in the prolongation fault tree) to consider, consistently with other systems, the availability of its electric heaters.

As a result, all the active systems are modeled in the extended L1PSA[a]. Their state is defined for L2PSA event trees, thanks to the interface, in the PDSs. However, a system which is considered as available in the interface may be considered later unavailable in the Accident Progression Event Tree (APET) if its environmental conditions are too degraded (temperature, pressure, radiations…) or in case of energetic phenomena during the severe accident (like steam explosion). These induced failures are easy to consider in KANT since these physical values (temperature, pressure, time…) are transmitted and possibly modified from one node of the APET to another.

3.2. Definition of ASTEC accident simulations to support L2PSA development

As introduced in section 1.1, each L2PSA developed by IRSN is supported by a large set of ASTEC accident simulations (typically between 100 and 200 scenarios of accidents are calculated). These ASTEC simulations are defined to fit with the PDSs. However, the PDSs attributes contain information not useful to define an ASTEC simulation (for example, the status of the containment isolation system, which is only considered in KANT and MER). Thus, different PDS may lead to the same ASTEC run.

To facilitate the definition of needed ASTEC accident simulations, a graphical interface has been implemented in OIPK to allow an additional PDS merging by a "merging" tree formalism. The "head events" considered in these trees are the interface variables. Each sequence corresponds to the definition of an ASTEC run. In the nodes of a merging tree, several values for a given interface variable can be grouped together (if their frequency is to low or if the two values lead to similar plant behavior in the given context). OIPK automatically updates the frequency of each node. It is then easy to identify the branches which are not significant and to adapt the grouping accordingly.

In the figure 6, a (simplified) example is given to illustrate the "merging tree" used to define the ASTEC simulations in case of large break LOCA. In this definition, the interface variable SO (pressurizer safety valves), GV (secondary system availability: EFWS, MSB) and AE (EFWS tank water makeup) are not considered to differentiate the ASTEC simulations. Indeed, in this simplified example, it is assumed that, in case of large break LOCA, neither the pressurizer safety valves opening nor the secondary system cooling will modify significantly the accident progression. To neglect these interface variables in a given context (i.e. large break LOCA context), they can be either not mentioned (like SO) or not used to split sequences (like GV and AE).

As seen in this figure, some branches are stopped due to their low frequency whereas some others are grouped.

This "merging" tree functionality is a flexible tool to define and document the ASTEC runs definition based on PDS. It can also be used to reproduce some L1PSA event trees and, then, verify that the frequencies are consistent with the ones of the L1PSA sequences.

---

[a] Some passive systems, like the autocatalytic H2 recombineurs, are not modeled in the extended L1PSA since they have no link with the other systems (i.e. recombineurs do not require cooling or electric supply, they neither include shared components).

Figure 6: Example of an ASTEC runs definition tree, from OIPK

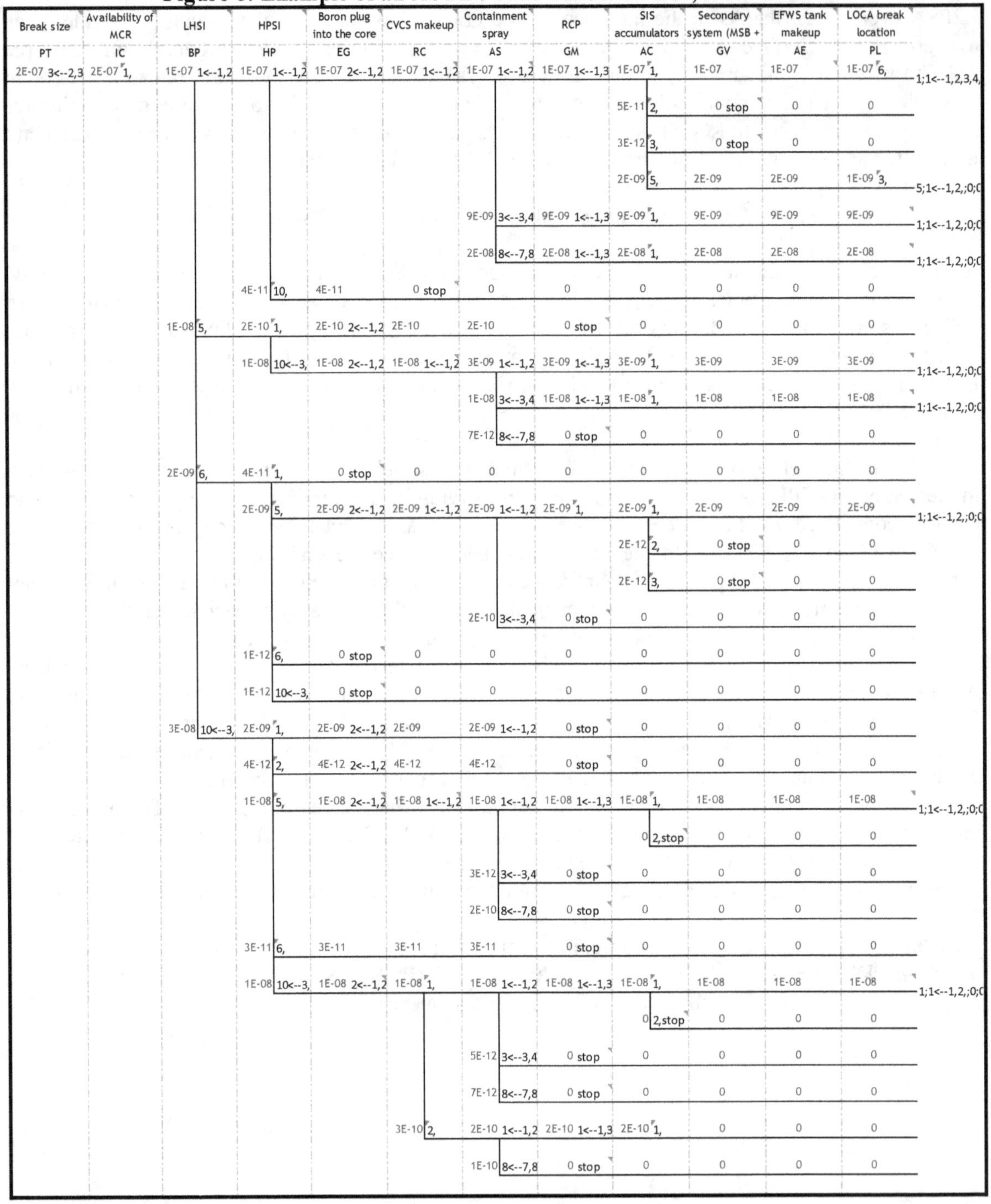

4. CONCLUSION

Results obtained:
Due to its automation and its flexibility, this new interfacing approach, developed by IRSN, allows a **transmission** of information between L1PSA and L2PSA **with the same precision as obtained with an integrated model** with an affordable cost. But, thanks to the use of separated and appropriate dedicated tools for L1PSA and L2PSA, parallel work on L1PSA and L2PSA is facilitated. L2PSA can be quantified separately from L1PSA. The L1PSA is not calculated again for the L2PSA quantifications.

After a L1PSA modification, the interface update is a **"button click" update** and the **development of a new interface is strongly facilitated**.

In the same manner, the **adjunction of additional information in the interface is very simple and efficient**. For example, adding a new interface variable is done by simply introducing the corresponding flag events in the prolongation fault tree and rerunning the extended L1PSA[a]. Eventually, a new variable in OIPK has to be declared and OIPK rerun.

If the L1PSA software does not allow event trees linking, this interfacing approach allows to preserve the L1PSA event trees readability[b] (only one additional function event is added in the event trees). In addition, since the extended L1PSA model remains almost as compact as before its extension, **the CPU time to run it is not strongly impacted** (at least with RiskSpectrum tool).

Regarding the PDS validation, **the results produced by OIPK can be easily verified** through the MCSs set presented for each PDS.

The interfacing approach presented in this paper can be **declined for any L1PSA software** using fault trees and event trees. This approach is especially interesting if the L1PSA software does not allow event tree linking since it is an alternative to the bridge trees construction.

This approach is particularly suitable **to implement internal and external hazards** in L1-L2PSA interface. Indeed, since the sub fault trees used in the prolongation fault tree are mainly based on a reuse of L1PSA fault trees, the interface is "automatically" updated when the L1PSA fault trees are adapted to the internal and external hazards modeling. In addition, due to the facility to add an interface variable and to modify the existing ones, it is simple and fast to add hazard specific variables in the interface (for internal fire localization by example).

Perspectives:

Coupled KANT-MER-MERCoR calculations provide a large amount of data, which allows various ways to analyze and present the L2PSA results. In particular, it is possible to assess contribution of each PDS regarding containment failure modes, radioactive release and radiological consequences.

With OIPK, it could be possible to identify the contribution of a given basic event of the L1PSA to the different PDSs. Consequently, it could be feasible to compute automatically the importance measures of the extended L1PSA's basic events, in regards of the dose consequences of their failure (and not only in regards of core damage). Such importance measures would allow building a better hierarchy of components importance, especially for the components dedicated to severe accident management and to those contributing to the core damage prevention and to the releases mitigation.

References

[1] AIEA, "*Development and Application of Level 2 Probabilistic Safety Assessment for Nuclear Power Plants*", Specific Safety Guide N° SSG-4.

[2] "*ASAMPSA2 best-practices guidelines for L2PSA development and applications - Volume 2 - Best practices for the Gen II PWR, Gen II BWR L2PSAs. Extension to Gen III reactors*", 2013; collaboration between IRSN, GRS, NUBIKI, TRACTEBEL, IBERINCO, UJV, VTT, ERSE, AREVA NP GmbH, AMEC NNC, CEA, FKA, CCA, ENEA, NRG, VGB, PSI, FORTUM, STUK, AREVA NP SAS, SCANDPOWER,
available on internet at http://www.asampsa.eu or http://www.asampsa2.eu

[3] N. Rahni et al., "*A methodology for the characterization of the severe accident consequences and the results presentation in level 2 probabilistic safety assessment*", conference PSA 2011, Wilmington

[4] T. Durin et al., "*Very fast running codes for the characterization of severe accident radiological consequences and the results presentation in level 2 PSA*", conference PSAM 11, Helsinki

---

[a] That is the L1PSA model with the prolongation fault tree added.
[b] This was not an issue for IRSN since RiskSpectrum allows event trees linking.

An Approach to Ensure the Availability of Complex Systems

Kaushik Chatterjee[*], Kumar Bhimavarapu, Robert Kasiski, and William Doerr
FM Global, Norwood, MA, USA

Abstract: Availability of a system depends on: (1) the components' reliabilities; and (2) the Inspection, Testing and Maintenance (ITM) characteristics (i.e., inspection/testing frequency, repair/replacement duration, and maintenance restoration factor). Complex systems typically have several sub-systems and components with complicated interactions and dependencies. In order to ensure a desired availability of a safety-critical complex system such as a fire protection system throughout its lifetime, it is necessary to: (1) ensure the needed reliability of the critical components through carefully planned durability/life tests; and (2) perform ITM actions at appropriate intervals (or frequencies).

This paper presents a comprehensive approach to: (1) establish the reliability targets and the ITM frequencies for the critical components based on the desired availability of the system; and (2) estimate durability/life test duration and sample size requirements based on the established reliability targets for these critical components. The steps of the comprehensive approach have been demonstrated using a typical foam-water sprinkler protection system. The comprehensive approach, when applied to a safety-critical complex system, would help achieve the desired availabilities of the critical components, which in turn would ensure the desired availability of the system throughout its lifetime.

Keywords: Availability; reliability targets; Inspection, Testing, and Maintenance (ITM); safety-critical complex systems.

1. INTRODUCTION

Safety-critical complex systems such as fire protection systems need to be available when called upon to prevent catastrophic losses. Complex systems typically have several sub-systems and components with complicated interactions and dependencies. Availability of a system depends on: (1) the components' reliabilities; and (2) the Inspection, Testing and Maintenance (ITM) characteristics (i.e., inspection/testing frequency, repair/replacement duration, and maintenance restoration factor). In order to ensure a desired availability of a safety-critical complex system throughout its lifetime, it is necessary to: (1) ensure the needed reliability of the critical components through carefully planned durability/life tests; and (2) perform ITM actions at appropriate intervals (or frequencies).

Durability/life tests are typically used to evaluate component failures from degradation mechanisms such as wear, fatigue or corrosion. Development of durability/life test methods includes two major steps: i) definition of test environmental requirements (e.g., temperature, humidity, vibration, and cycling rate) considering the physics-of-failure (failure mode and the causal degradation mechanisms); and ii) estimation of test duration and sample size requirements based on the established reliability targets. Reliability target is the reliability at a specified lifetime that a component is expected to demonstrate.

This paper presents a comprehensive approach to: (1) establish the reliability targets and the ITM frequencies for the critical components based on the desired availability of the system; and (2) estimate durability/life test duration and sample size requirements based on the established reliability targets for these critical components.

---

[*] Email: Kaushik.chatterjee@fmglobal.com

2. COMPREHENSIVE APPROACH

The objective of the comprehensive approach is to ensure that the critical components achieve the desired availabilities, which would ensure the desired availability of the system throughout its lifetime. The key steps of the approach include definition of the system (see Section 2.1); Failure Mode and Effect Analysis (see Section 2.2); availability model development for the system and its components (see Section 2.3); system availability analysis to establish the reliability targets and the ITM frequencies for the critical components (see Section 2.4); and estimation of the durability/life test duration and sample size requirements for the critical components based on the established reliability targets (see Section 2.5).

2.1. System Definition

A clear definition of a complex system is necessary to ensure that all the sub-systems/components and other relevant details are considered appropriately in the evaluation of the system. More specifically, the system definition, which forms the basis for the next steps, includes delineation of: (1) the sub-systems and components, (2) the system boundaries, and (3) the postulates where appropriate.

2.2. Failure Mode and Effect Analysis

Failure Mode and Effect Analysis (FMEA) identifies the credible failure modes of the components, the causes of failures, and their effect on the system. This analysis is necessary to build the system availability model, as well as to identify the causes of component failures. The causes include degradation through mechanisms such as wear or corrosion, and/or human errors. For a component failure mode, identification of the causal degradation mechanisms helps in defining the environmental requirements for the durability/life tests.

2.3. Availability Model Development for the System and its Components

The system availability model is developed in order to determine the logic (i.e., possible combinations of credible component failure modes) that can lead to the system failure. A system availability model can be developed using a fault tree or a reliability block diagram.

The system availability is estimated using the system availability model and the component availabilities. The instantaneous availability, $A(t)$, of a component at any random lifetime t is the sum of the probability of two mutually exclusive events, i.e., either the component is functioning properly at time t, the probability of which is its reliability $R(t)$; or the component is functioning properly since the last repair at time x $(0 < x < t)$, the probability of which is $\int_0^t R(t-x)m(x)dx$ [1]. These contributions are shown in Equation (1), where $m(x)$ is the renewal density function that incorporates maintainability information. The renewal density function is dependent on the probability density functions of the time to failure and the repair time distributions.

$$A(t) = R(t) + \int_0^t R(t-x)m(x)dx \qquad (1)$$

2.3.1. Reliability Distribution Parameters

The Weibull distribution is typically used for modelling reliability. The reliability $R(t)$ at a component lifetime t can be estimated using Equation (2), where, β is the shape parameter, and η is the scale parameter or characteristic life.

$$R(t) = \exp\left[-\left(\frac{t}{\eta}\right)^{\beta}\right] \tag{2}$$

The shape parameter determines the shape of the failure distribution curve. The value of the shape parameter for a component depends on the physics of failure, i.e., the failure mode and the causal degradation mechanisms. The characteristic life is the lifetime at which the component reliability equals 0.368 (or 63.2% of the components fail). For a given shape parameter, the component reliability varies as a function of the characteristic life. For example, corresponding to a shape parameter value of 3, the reliability values at 50 years would be 0.368, 0.89 and 0.98 respectively for characteristic life values of 50, 100 and 200 years.

2.3.2. ITM Parameters

The ITM parameters include the inspection/testing frequency, repair/replacement duration, and maintenance restoration factor. Restoration factor indicates the percentage (of new condition) to which a component will be restored after the performance of the maintenance action. While performing availability estimations, the restoration factor is assigned a value from 0 to 1. A restoration factor value of 1 indicates that the component will be "as good as new" after the maintenance action. Thus the age of the component will be reset to zero. A restoration factor value of 0 indicates that the component will not be improved at all by the maintenance action. Thus the age of the component will be the same as the age before the maintenance was performed[2].

The type of maintenance (i.e., corrective or preventive) also plays a key role in the system availability. Corrective maintenance is usually performed to restore a failed system to operational status by replacing or repairing the component that is responsible for the system failure[2]. Preventive maintenance is usually performed to replace components before they fail in order to maintain uninterrupted system operation[2]. Choice of corrective or preventive maintenance can affect the needed reliability targets. For example, if only corrective actions are considered, then relatively higher reliability values for components may be necessary to achieve the desired system availability, as compared to when preventive actions are also considered.

Another important factor is the repair/replacement time, which indicates whether the repair commences immediately after failure, or when a failure is detected during a periodic ITM. For example, in the case of revealed failures (e.g., a pipe rupture) the repair/replacement may commence immediately after failure, whereas in the case of unrevealed failures (e.g., a valve spring failure) the repair/replacement may commence after the failure is detected during a periodic ITM or a real demand.

2.4. System Availability Analysis to Establish Reliability Targets and ITM Frequencies

The reliability targets and the ITM frequencies for the critical components are established through an iterative process so as to achieve the desired system availability, as shown in Figure 1. The current system availability is estimated using: (1) the system availability model (e.g., fault tree); and (2) the current reliabilities of components (estimated based on the respective Weibull distribution parameters, i.e., shape parameter and characteristic life) and ITM parameters (i.e., inspection/testing frequency, repair/replacement duration, and maintenance restoration factor).

If the estimated current system availability is found to be unacceptable (less than the desired value), then the reliabilities and/or the ITM frequencies of the critical components (with focus on specific

---

[2] The descriptions have been adopted from the ReliaSoft BlockSim software manual.

failure modes) are increased to reasonable levels[3], and system availability is re-estimated. This is done iteratively until the desired system availability is achieved. This procedure establishes the reliability targets and the ITM frequencies for the critical components.

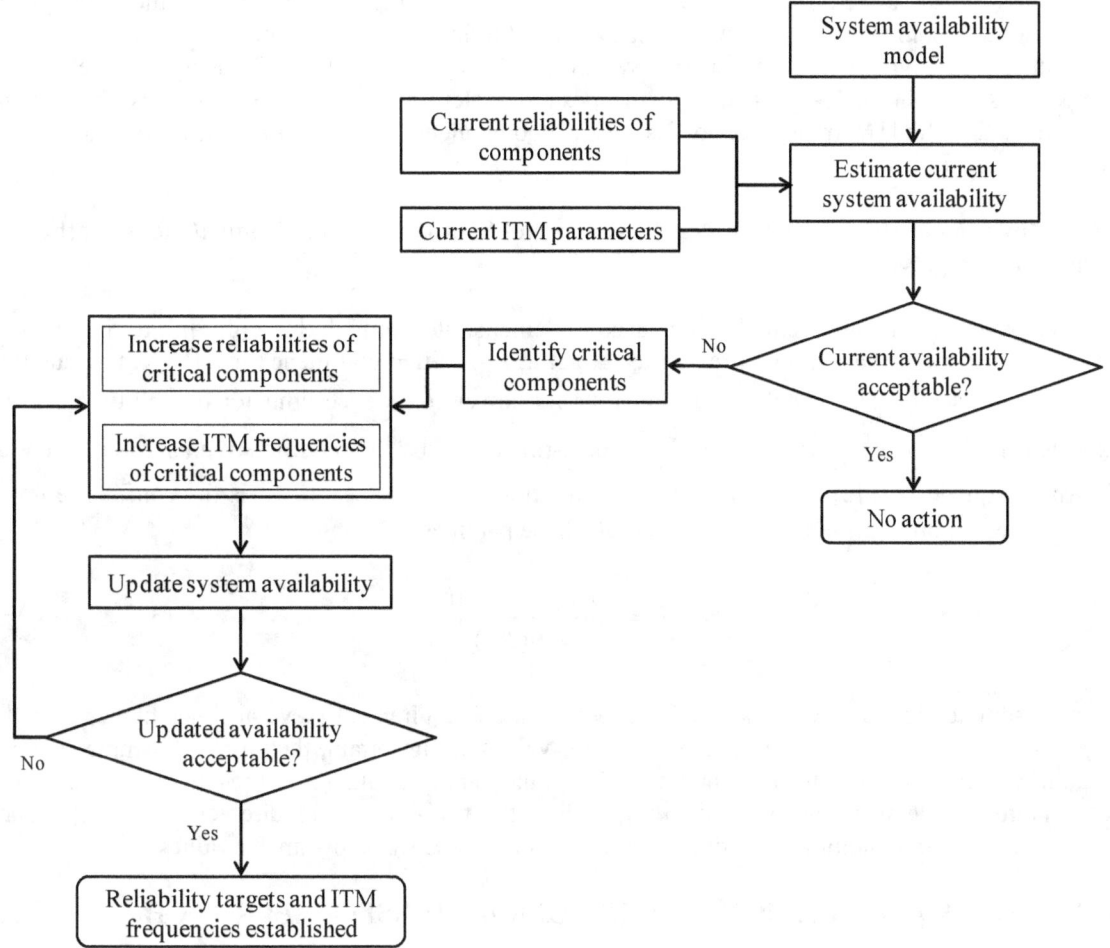

Figure 1: Iterative process for establishment of reliability targets and ITM frequencies

In order to increase the reliabilities of the critical components, the respective Weibull characteristic lives can be increased based on the relative contributions of each component to the system downtime. For example, the Weibull characteristic life of the most critical component (with highest contribution to the system downtime) can be increased by the highest factor, whereas the Weibull characteristic lives of the remaining critical components (with relatively smaller contributions to the system downtime) can be increased by relatively smaller factors.

Generally, the ITM frequencies are constant throughout the life of a system. Since the component reliabilities reduce with age, it can be useful to increase the ITM frequencies with age of the components to improve their availabilities. For example, for a component with specific reliability, prescribing a high ITM frequency only towards the end of life can be more useful and effective in achieving the desired minimum availability throughout the lifetime, when compared to prescribing a constantly high ITM frequency for the entire lifetime. Such an increase in the ITM frequency would also allow for less stringent reliability targets.

---

[3] A component's reliability or ITM frequency can only be increased to a certain level beyond which it may not be feasible owing to technology and cost constraints. Introducing component redundancy can also increase the system reliability, however, this method has not been considered in this study.

2.4.1. Identification of Critical Components

Critical component failure modes can be identified and ranked based on the Downtime Criticality Index (DTCI). DTCI is a relative index showing the contribution of each component to the system's downtime (i.e., the system downtime caused by a particular component divided by the total system downtime)[2]. A component with the highest DTCI value has the highest contribution to the system downtime, and is most critical for the system availability. Therefore, focusing on the critical components not only helps in achieving the desired system availability, but also ensures that the reliabilities and the ITM frequencies of less critical components are not unnecessarily increased from the current levels.

2.5. Estimation of Durability/Life Test Duration and Sample Size Requirements to Meet the Reliability Targets

The Weibayes zero-failure method can be used to demonstrate (through durability/life testing without observing any failure) a minimum reliability target at a specified confidence level. The test duration for a specific sample size needed to demonstrate the lower confidence limit for reliability R_d at the expected lifetime of t_d (cycles or years) can be estimated using the modified Weibayes equation, as shown in Equation (3) [2], where T is the test duration (cycles or years), C is the confidence level, N is the test sample size, and β is the Weibull shape parameter.

$$T = t_d \left[\frac{\ln(1-C)}{N \ln(R_d)} \right]^{\frac{1}{\beta}} \quad (3)$$

A small sample size can result in very high test duration requirement (several times higher than the expected lifetime). Further, if the shape parameter values are uncertain, then a small sample size can result in wide variability in the test durations. Depending on the availability of resources (i.e., time and cost), large sample sizes can be used to: (1) reduce the test duration requirements; and (2) reduce variability in the test durations resulting from uncertainty in the shape parameter values.

3. EXAMPLE APPLICATION OF THE COMPREHENSIVE APPROACH

An active foam-water sprinkler protection system has been used to demonstrate the key steps of the comprehensive approach. The foam-water protection system consists of two major sub-systems: (1) the water supply (consisting of components such as controller, motor, pump, and valves); and (2) the foam supply (consisting of components such as bladder tank, foam concentrate, and hydraulic concentrate valve). These two sub-systems supply water and foam at the desired flow rate and pressure to a proportioner, which mixes the water and foam in the right proportion to prepare the foam-water solution for discharge through automatic sprinklers. The foam-water protection system failure is defined as 'no or low' foam-water discharge from sprinklers when required in the event of a fire. The fault tree technique has been used to develop the system availability model to determine the logic (possible combinations of credible component failure modes) leading to the system failure.

Based on the system fault tree, availability analysis was performed for an assumed lifetime of 30 years using the software package ReliaSoft BlockSim. The current system availability (curve shown by the dotted line in Figure 2), estimated using the current values of the reliability distribution and the ITM parameters, is below the desired availability value for most of the system lifetime. In order to achieve the desired system availability, an iterative availability analysis was performed by increasing the reliabilities and the ITM frequencies of the critical components (identified using the DTCI values). The achieved availability curves (for systems updated – 1 & 2) are shown in Figure 2.

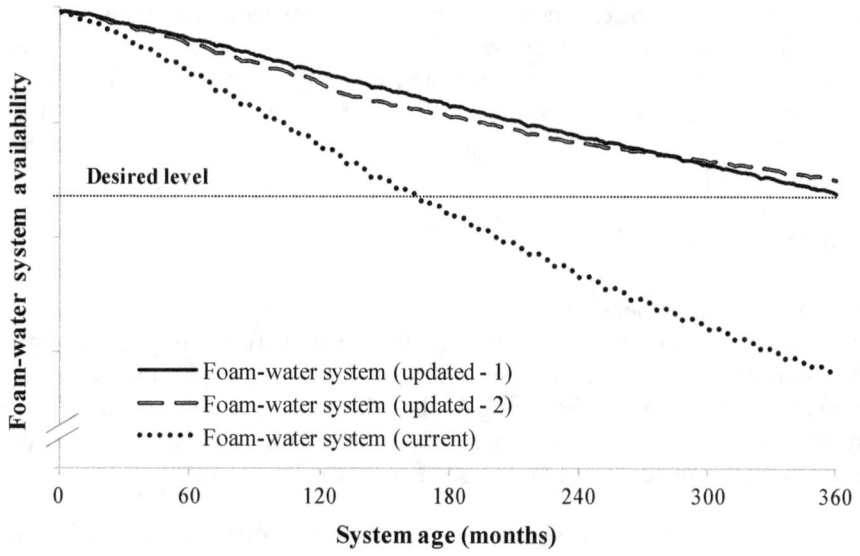

Figure 2: Foam-water system availabilities (current and updated)[4]

In Figure 2, the availability curves (updated – 1 & 2) were achieved by increasing the reliabilities of the critical components to the same level, but with different ITM frequencies. The solid line (updated – 1) represent the availability curve achieved by increasing the ITM frequencies by a factor of 1.33 throughout the lifetime. However, the dashed line (updated – 2) in Figure 2 represent the availability curve achieved by increasing the ITM frequencies with age. The current ITM frequencies were used for first 10 years of lifetime, increased by a factor of 1.33 from the current level for the next 10 years of lifetime, and increased by a factor of 2 from the current level for the last 10 years of lifetime.

When compared to the availability curve (updated – 1) shown by the solid line, the availability curve (updated – 2) shown by the dashed line has higher availability in the later part of the lifetime. Thus, for the system with the availability curve (updated – 2) shown by the dashed line, the reliability targets of components can be made relatively less stringent (i.e., lowered further) and still meet the desired availability level.

The durability/life test duration and sample size requirements were then estimated for the critical components based on the established reliability targets (used to achieve the updated curves in Figure 2) and at a specified confidence level. Figure 3 presents the durability/life test requirements for a component with an expected lifetime of 1000 cycles.

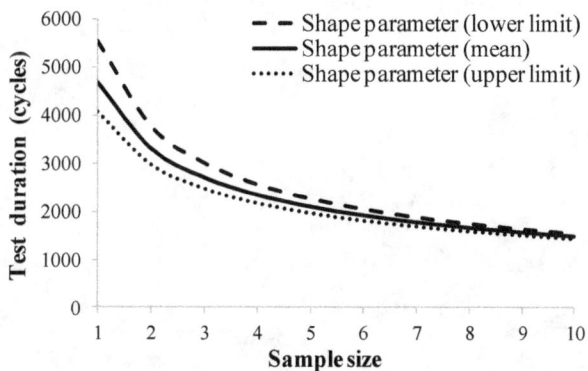

Figure 3: Test duration and sample size requirements for a component

---

[4] The absolute values of system availability are not provided in this figure, which is only intended to demonstrate the comprehensive approach.

The test duration requirements corresponding to a range of shape parameter values (lower limit, mean, upper limit) have been provided to demonstrate the benefit of using large sample sizes when shape parameter values are uncertain. As can be seen from Figure 3, for small sample sizes, there is higher variability in the test durations corresponding to the three shape parameter values. However, as the sample size is increased, the test durations (corresponding to the three shape parameter values) decrease exponentially and tend to converge.

4. CONCLUSION

This paper presented a comprehensive approach to: (1) establish the reliability targets and the ITM frequencies for the critical components based on the desired availability of the system; and (2) estimate durability/life test duration and sample size requirements based on the established reliability targets for these critical components. Focusing on the critical components not only helps in achieving the desired system availability, but also ensures that the reliabilities and the ITM frequencies of less critical components are not unnecessarily increased from the current levels.

Since the component reliabilities reduce with age, it can be useful to increase the ITM frequencies with age of the components to improve their availabilities. For example, for a component with specific reliability, prescribing a high ITM frequency only towards the end of life can be more useful and effective in achieving the desired minimum availability throughout the lifetime, when compared to prescribing a constantly high ITM frequency for the entire lifetime. Such an increase in the ITM frequency would also allow for less stringent reliability targets.

A small sample size can result in very high test duration requirements (several times higher than the expected lifetime). Further, if the shape parameter values are uncertain, then a small sample size can result in wide variability in the test durations. Depending on the availability of resources (i.e., time and cost), larger sample sizes can be used to: (1) reduce the test duration requirements; and (2) reduce variability in the test durations resulting from uncertainty in the shape parameter values.

The steps of the developed comprehensive approach have been demonstrated using a typical foam-water sprinkler protection system. The comprehensive approach, when applied to a safety-critical complex system, would help achieve the desired availabilities of the critical components, which in turn would ensure the desired availability of the system throughout its lifetime.

References

[1] E.A., Elsayed, *"Reliability Engineering"*, John Wiley & Sons, 2012, Hoboken, New Jersey.
[2] J.C., Wang, *"Sample size determination of bogey tests without failures"*, Quality and Reliability Engineering International, Vol. 7, pp: 35-38, 1991.

Reliability/availability methods for subsea risers and deepwater systems design and optimization

Annamaria Di Padova[a], Fabio Castello[b] Fabrizio Tallone[a]*, Michele Piccini[b],

[a] Saipem S.p.A., Fano, Italy
[b] RAMS&E S.R.L., Turin, Italy

Abstract: The restriction of construction licenses for onshore oil/gas treatment plants and regasification units along with energy demand growth has increased the development of offshore installations. Furthermore the discover of new offshore deep water fields enhance the engineering efforts towards the development of engineering of submarine systems and plants. Due to the complexity of these submarine systems, the severe environment where they operate and the difficulty or the impossibility to repair a component, a high system availability is becoming a key requirement. In this framework, to have a system architecture verified also from the reliability and availability point of view, the RAM analysis are becoming an essential part of the design. This paper describes the application of reliability/availability methods (RBD, Montecarlo method, FMEA risk assessment) to support the design of subsea deep water systems. In particular, two case studies are presented, the first aiming at the definition of the optimum configuration of retrievable and permanent deep water modules, the second addressing the verification of design configurations and the suggestion of tests and inspection plans to guarantee system integrity along operating life. Moreover the paper summarizes also difficulties to find subsea equipment reliability data and proposes solutions for reliability components characterization.

Keywords: Reliability, Availability, Deep water systems, Subsea systems, Risers.

1. INTRODUCTION

In the frame of the oil and gas industry, the improvement of offshore installation, more and more competitive, complex and challenging is becoming the rule. The reasons for the development of new offshore technologies, as subsea plants, are mainly two: the optimization of the resources and the discover of new offshore fields in harsh and severe environment and/or in deep and ultra-deep water.
Due to the nature and the location of these new subsea installations, an high system availability is a mandatory requirement for the remunerativeness of the investment. The causes of this strict requirement are various: the complex system architecture, the severe environment where they operate, the difficulty or the impossibility to repair the components.

In this framework the reliability methods are an essential tool to verify and validate the design. They also represent an effective method to individuate the system weakness and criticality, aiming at the design optimization. Notwithstanding, differently from the standard oil and gas industry, the subsea installations are often characterized by Non Reparable components (due to the onerous operations needed for repair or substitute an item) and by the difficulty to retrieve reliability data, mainly due to the new techniques employed.

In this paper two RAM analysis approaches are presented and applied to two case studies to support the design of subsea deep water systems.

*fabrizio.tallone@saipem.com

2. CASE-STUDY 1

2.1. Description

Case-study 1 concerns a subsea system for deep-water application.
This subsea system (Figure 1) consists of subsea separation of the well fluids, boosting of the liquid phase to topside and dumping of the gas phase at sea or at topside.

RAM analysis has been applied to this subsea system in order to verify and validate the design and to provide, if necessary and applicable, input and recommendations to the engineering design team in order to improve and enhance the availability of facilities and to reach the availability target.

Figure 1: Case-study 1 - simplified scheme.

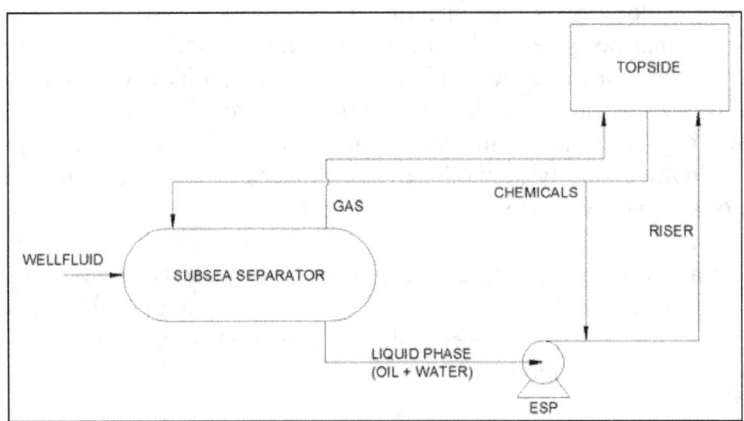

The design life of this subsea system is 30 years but the operating life is very short, 6 months; in this period the system availability shall be at least 90%.
The maintenance and repair activities shall be minimized or, if possible, eliminated.

2.2. Methodology

The analysis has been developed as for the scheme in Figure 2.

Figure 2: Flow diagram adopted for the Case-study 1 assessment.

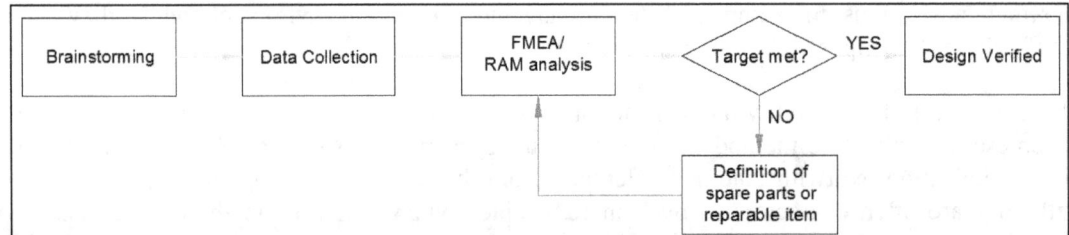

Brainstorming
The brainstorming has been conducted as a multi-disciplinary meeting aiming at the sharing of all the information regarding the analyzed system, with the following objectives:
- collection of information concerning the system under analysis;
- identification of system components and main sub components, and relevant specific functionality and roles;
- definition of the main assumption and hypothesis.

Data Collection
The second step consists of the preparation of the Data Collection Table, aiming at associating a Main Equipment Type to each item of the analyzed system, together with the relevant failure modes and related failure rates.

The Data Collection Table collects the failure modes (FMs), the Main Time To Failure (MTTF) and related Active Repair Times (ARTs, only for reparable components) to be modeled in a Reliability Block Diagram (RBD) model.
Due to the particular architecture of this subsea system, very few data are present in literature and belongs to different data bases. The data collection has been mainly based on OREDA Handbook [2], OGP Database [3], Exida Handbook [4], OTRC report [8] and SINTEF report [9].

FRs have been assumed to be constant on time and mainly characterized by their mean value, and the probabilities of failures have been characterized by the use of a negative exponential distribution over time.
For some components failure rates have been characterized by their upper value (95th percentile) due to their particular operative conditions. As for several publications from scientific literature ([5]), this option represents a reasonable and appropriate solution for technological complex systems, especially when availability modelling is conducted at high level and elements represent complete items rather than individual failure modes. No additional consideration about wear-in and wear-out impact is done, considering the equipment under study in the reliability optimal range of its lifetime when failures can be considered characterized only on a random basis.
Active Repair Times are values directly extracted from reference databases. The related Repair Rates assumed to be constant and related probabilities of repair are well characterized by a log-normal distribution, officially recognized by the scientific literature as appropriated to characterize human interventions([2], [6], [5]).

FMEA/RAM Analysis
The RAM analysis have been performed in three steps:
1) all the components have been considered "not repairable", in order to evaluate the reliability and to establish the most critical components;
2) for the most critical components a sensitivity analysis has been performed, setting the hypothesis of "repairable items" as a recommendation and the availability has been evaluated;
3) if the availability target is not met, the implementation of the possible mitigation (i.e. additional redundancies, further components set to "repairable") is evaluated.

It is to be noted that due to installation constraint the use of redundant components shall be minimized, therefore, if the availability target is not met, preference is done to the recommendation of "repairable" components.

2.3. Results

Brainstorming
Table 1 shows the identified main components and sub-components of the analyzed systems.

Table 1: Main components and sub-components of the analyzed system.

Main Component	Sub-component type	Sub-component Tag
Subsea Separator	Separator	VA-001
	Gate Valve	V-S1, V-S2
	Choke Valve	V-S4
	Gate valve	V-S3
	Lines	Line V-S3, Line V-S4
	Differential Pressure Transducer	DP-1, DP-2
	Density meter	DM-1
	Temperature and Pressure Transducer	APT-1

Main Component	Sub-component type	Sub-component Tag
	Radar Level Meter	RD-1
	Linear Displacement Transducer	LD-1
	Temperature Transducer	TT-1/2/3/4/5/6
	Level Gauge	LG-1
Subsea Riser Base	Gate Valve	V-Cl4, V-Cl5, V-Cl6
	Riser	Riser
	ESP	PS-001
	ESP Motor	PS-001 Motor
	Gate Valve	V-PI, V-BF, V-PB, V-II, V-AI, V-EX, V-Cl1, V-Cl2, V-Cl3
Subsea Submergible Pump	Choke Valve	V-PR
	Linear Displacement Transducer	LD-2
	Differential Pressure Transducer	DP-3
	Absolute Pressure transducer	AP-3, AP-2
	Check Valve	V-CHECK1/2/3
Umbilical system reel	Umbilical	U-003 (Power Supply) U-004 (Chemical/hydraulic) U-006 (Chemical/optical)

Data Collection

The following Table 2 reports a portion of the data collection used for the case-study 1. It includes only the Critical FMs and the relevant FRs as per section 2.2.

Table 2: Extract of the Data Collection ([2], [3], [4])
(NOTE: If not stated the FRs refer to the mean value)

Main Equipment Type	Modelled Items	Failure Mode	Failure Rate (h^{-1})	MTTF (h)	ART (h)	Dispersion
Valve check	Check Valves	Critical Failure	7.0E-08	14285714	-	1.4
Risers	Riser	External leakage	1.7E-07	5882353	-	1.4
		No immediate effect	1.7E-07	5882353	-	1.4
		Structural deficiency	5.0E-07	2000000	-	1.4
V-S3 Valve (FRs refer to the upper value, 95th percentile)	V-S3	Delayed operation	1.2E-06	847458	-	1.4
		External Leakage	1.9E-06	512821	-	1.4
		Fail to close on demand	1.0E-05	98522	-	1.4
		Fail to open on demand	1.8E-06	540541	-	1.4
		Spurious operation	1.6E-06	641026	-	1.4
		Structural deficiency	1.5E-06	675676	-	1.4
		Valve leakage in closed position	1.2E-06	819672	-	1.4
		Other	1.5E-06	671141	-	1.4

RAM Analysis Results

The analysis shows a Reliability of the system equal to **77.35%**. This result means that the system has a probability of 77.35% to fully perform its function, at the 100% of its capacity, continuously and without interruption over the defined mission time (6 operational months).

Table 3 reports the main items that contribute to the system un-reliability.

Table 3: Reliability Results

Item	Un-Reliability
ESP	7.9%
V-S3	4.3%
Subsea Separator (VA-001)	4.1%
V-S4	2.2%
ESP Motor	2%
V-Cl6	1.2%

The Table 3 shows that the major contribution to system un-reliability is given by the ESP.
Considering the results of Table 3, in order to achieve the availability target, it has been decided to allow the ESP repair, in particular it has been foreseen to install the ESP on the riser base and the module "riser base - ESP" has been made retrievable. Replace times are assumed to be equal to 24 hours.

The availability analysis has been carried out considering the system model described above and the module "riser base - ESP" reparable. The expected Availability of the system is equal to **87.422%**. This means that the system is expected to run on average 3766 hours over six months at full capacity.
As an additional contribution to the final results, the main critical components referred to the expected Availability are summarized in the Table 4. For each item the related criticality index is shown. This index represents for each component its direct quantitative contribution to the overall Unavailability of the system.

Table 4: Availability Results

Element	Description	Criticality Index (%)
V-S3 (all FMs)	Fast Gas Release Valve	4.0
880VA001 (all FMs)	Separator	3.7
V-S4 (all FMs)	Proportional Gas Release Valve	1.8
V-CI 6	Oil Soluble Chemical Injection Valve	1.0
V-CI 4	Methanol Injection Valve	1.0
Overall residual criticality due to all other items with CI less than 1		1.1
NOTE: (all FMs) = all FMs considered critical on FMECA.		

As it can be seen, main contributors on unavailability of the system are the not reparable items placed on Subsea Separator. In particular:
- **VS-3 Valve**. This component has a MTTF=47893 hours. It is obtained summing all failure rates of all failure modes of a generic "gate valve" taken from OREDA 2009. As this valve is characterized by an high number of operation per day, the "upper" value of failure rate (95th percentile of statistic distribution) has been considered to be more realistic knowing the high number of operation per day of this item. Moreover, this item is not reparable.
- **Subsea Separator** (MTTF = 54025 hours). This value is obtained summing all failure rates of generic Vessel Separator (10-1000 m^3) taken from OREDA 2009
- **VS-4 Valve**: This item is modelled as Choke Valve, and related failure rate (referred to "external leakage") is taken from OREDA 2009. It has a MTTF of 102881 hours.
- **V-CI 6** and **V-CI 4**, that are valves not reparable associated to chemical injection system. Their MTTF is equal to 194175 hours because they are modelled as "gate valve" (mean value of statistic distribution of failure rate).

On the basis of previous results, the methodological approach allowed to define a set of possible improvement of system availability:
- insertion of a redundancy of V-S3/V-S4 valves;

- requirements to the vendors for the reliability parameters of V-S3/V-S4, that must be characterized by a MTTF equal or better than 30 years (with related failure rate corresponding to 3.8E-06 failures per hour).

The main issue, related to case-study 1, is the lack of reliability data, due to the new technology employed and the not common use of some widely diffused items (such as ESP, valves, etc…). Anyway, following the methodology described in section 2.2, a suitable data collection have been developed. Despite this issue, the obtained results have highlighted the system criticalities, suggesting some design improvement and contributing to achieve the availability target.

3. CASE-STUDY 2

3.1. Description

Case-study 2 concerns two riser systems, a steel lazy wave riser (SLWR) and a Free Standing Hybrid Riser (FSHR). Figure 3 shows the schemes of the two riser systems.

Due to the system complexity the risers have been designed as "Maintenance Free". In particular the preventive maintenance will not be performed on the riser systems. The only type of maintenance possible on the systems is the corrective maintenance, as a result of anomalies highlighted during an inspection.

The application of a reliability analysis on these system aims at validating the design, in particular the "Maintenance Free" concept, and at defining a criticality list of the items composing the system, to establish an order of priority for the Inspection activities to be performed during the operational life of the riser systems.
It is to be noted that the definition of an inspection interval aims also at detecting the failure mechanism before item failure.

Figure 3: FSHR and SLWR scheme
(B.L. = Battery Limit, MCV = Vertical Connection Module, URTA = Upper Riser Termination Assembly, LRTA = Lower Riser Termination Assembly).

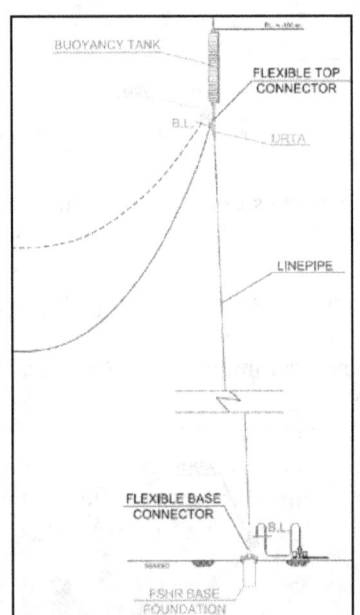
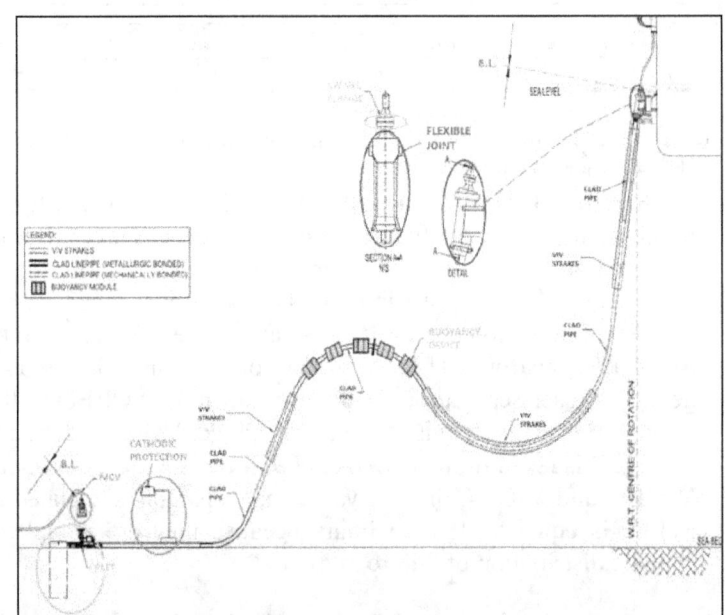

3.2. Methodology

The methodology followed to perform this study consists of the main steps listed in the following Figure 4.

Figure 4: Scheme of the methodology adopted for the Case-study 2.

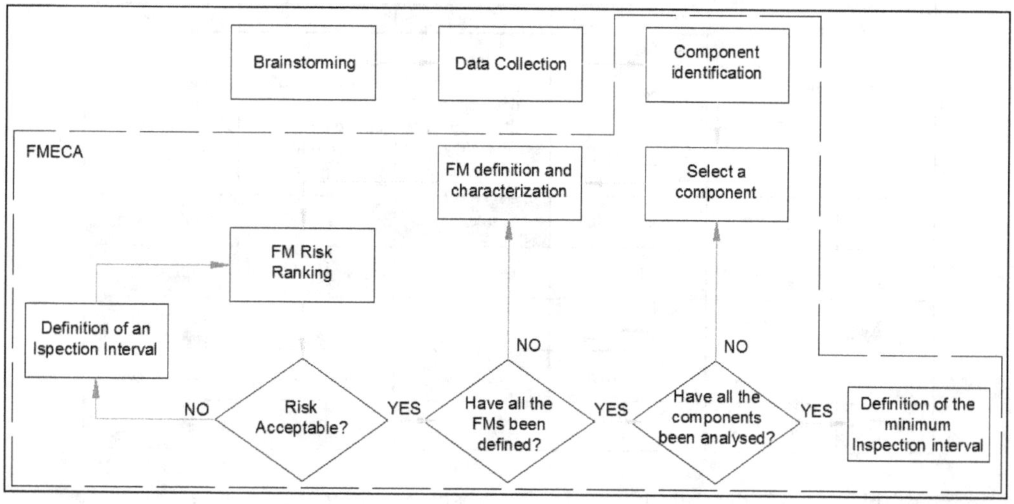

Brainstorming and Data Collection
The same approach used in the case-study 1 has been applied.

FMECA execution
The FMECA (Failure Mode, Effect and Criticality Analysis) execution consists of the assessment of the effects of each failure mode in terms of loss of system functionality, the identification of the relevant detection method and of possible compensating provisions, the assessment of the risk for each failure mode and finally the evaluation of the Minimum Inspection Interval.

The FMECA has been developed following the template shown in the Figure 5.

Figure 5: FMECA Template.

ID	Component	Function	Failure Mode	Failure Rate (h^{-1})	P_i	Failure Mechanism or cause	Failure Local Expected Consequence	Failure Detection/ Mitigation	Failure System Expected Consequence	Risk Ranking			Recommended Inspection interval (y)	P_i*	Residual Risk Ranking considering Inspection		
										C	P	R			C	P	R

Identification of the components of the riser system
The analysis has been developed at Main Equipment level: this means that the system has been subdivided in sub-items. Each sub-item has been then associated to Main Equipment set in the Data Collection Table.
In this step the column ID, Component and Function have been filled.

Identification and characterization of the FMs for each component
The failure modes for each component have been identified on the basis of the Data Collection Table. For each FM, the Failure mechanism or cause, the Failure Expected Consequence (Local and on the system) and the already foreseen Failure Detection/Mitigation have been identified.

Risk Ranking

The risk ranking have been performed by means of a risk matrix obtained from IEC standard 60812 [1] and is shown in Figure 6.

Figure 6: Risk Matrix.

Likelihood Class		Severity Class			
		1: Insignificant	2: Marginal	3: Critical	4: Catastrophic
		System performance degradation, NO damage to the system and NO threat to life or injury	System performance degradation, limited damage to system, NO threat to life or injury	Loss of system's primary functions, damage to environment, NO threat to life or injury	Loss of system's primary functions, damage to environment and/or personal injury.
5: Frequent	$Pi \geq 0.2$	Undesirable	Intolerable	Intolerable	Intolerable
4: Probable	$0.1 \leq Pi < 0.2$	Tolerable	Undesirable	Intolerable	Intolerable
3: Occasional	$0.01 \leq Pi < 0.1$	Tolerable	Undesirable	Undesirable	Intolerable
2: Remote	$0.001 \leq Pi < 0.01$	Negligible	Tolerable	Undesirable	Undesirable
1: Improbable	$0 \leq Pi < 0.001$	Negligible	Negligible	Tolerable	Tolerable

For the scope of this analysis the following rules have been considered:
- If the risk level is Negligible or Tolerable, the level of risk is broadly acceptable;
- If the risk level is Undesirable, the level of risk is tolerated only if risk reduction is impracticable or is kept as low as reasonably practicable by adopting reduction measures unless their cost is grossly disproportionate to the improvement gained. An Inspection interval shall be defined;
- If the risk level is Intolerable, the level of risk is not acceptable and risk reduction measures are required. An Inspection interval shall be defined.

P_i has been calculated by means of the following equation [1].

$$P_i = 1 - e^{-\lambda_i \cdot t_j} \qquad (1)$$

where
λ_i = failure mode failure rate.
t_j = time of active component operation.

The likelihood classes of each FM for the Risk Ranking have been determined first of all considering t_j equal to the design life (30 years).
The likelihood class, obtained in function of P_i, has been decreased by one unit if Detection measures are put in place and are able to detect/mitigate at least one of the identified Failure mechanisms or causes. This approach has been followed in order to take into account the presence of Detection/Mitigation measures, as suggested by IEC 60812 [1].

Evaluation of the Inspection interval and of the Residual Risk

For the Failure Modes, whose risk level is Undesirable or Intolerable, the Risk has been reduced considering the positive impact of Inspection activities: in particular the hypothesis of restoring the items to "as good as new" has been applied. This assumption can be considered in this case reasonable and realistic as during an accurate and systematic Inspection the status of the item is verified and, in case of any anomalies, the item can be maintained and/or substituted. Moreover the Inspection allows to detect the cause of all the failure modes identified in the FMECA, before the failure of an item.

The calculation of the Inspection interval has been performed in the following way:
- P_i^* has been calculated considering the Equation 1 and t_j equal to the Inspection interval;

- as per the first Risk Ranking, the likelihood class, obtained in function of P_i^*, has been decreased by one unit if Detection/Mitigation measures are put in place and are able to detect/mitigate at least one of the identified Failure mechanisms or causes;
- the Inspection interval has been obtained iteratively, with the purpose to reduce the risk to an acceptable level. The minimum Inspection interval has been considered corresponding to 1 year, to be in line with the interval suggested by international standards (e.g. DNV-RP-F206, [7])
- the residual risk have been evaluated considering the risk matrix reported in Figure 6, suggesting additional safety measure in case of not acceptable results, on the basis of a criticality ranking of the items in the battery limits of the study.

3.3. Results

Brainstorming
Table 5 shows the main components and sub-components of the two riser systems.

Table 5: FSHR and SLWR components and subcomponents.

FSHR			SLWR		
Main Component	Subcomponent	Spare part	Main Component	Subcomponent	Spare part
Buoyancy Tank	Tank composed by 21 compartments	YES, one compartment	Flexible Joint	Flexible Joint Support	NO
	Pipework - ballasting system	YES		Flexible Joint System	NO
	Valves - ballasting system	YES			
	Central core	NO	Swivel Flange	Swivel Flange	NO
	Flexible Top Connector (Male connection)	NO			
URTA (Upper Riser Termination Assembly)	Flexible Top Connector (Female connection)	NO	Line pipe	CLAD Pipe (Metallurgic Bonded)	NO
	Piping	NO		CLAD Pipe (Mechanically Bonded)	NO
	Structural parts	NO			
	Isolation valve	NO		VIV Strakes	YES
	Diverless connector (Male connection)	NO	Buoyancy Device	Buoyancy Module	YES
MCV (Vertical Connection Module)	Diverless connector (Female connection)	NO		Clamp	YES
	Gooseneck	NO			
	Structural reinforcements	NO	PLET (Pipeline End Terminator)	Pipe Structural parts and anchoring system	NO
	Swivel Flange	NO		Pipe	NO
	Isolation valve	NO		PLET/MCV connector (Female Connection)	NO
Linepipe	Linepipe	NO			NO
Buoyancy Foam Module	Buoyancy Module	YES		PLET/MCV connector (Male Connection)	NO
	Clamp	YES	MCV (Vertical Connection Module)		NO
LRTA (Lower Riser Termination Assembly)	Piping	NO		Pipe	NO
	Structural parts	NO		Structural part	NO
	Diverless connector Hub	NO		Swivel Flange	NO
	Isolation valve	NO	Cathodic Protection	Anode Sled	NO
	Flexible Bottom Connector (Male connection)	NO		Continuity Cable	YES

	FSHR			SLWR		
Main Component	Subcomponent	Spare part	Main Component	Subcomponent	Spare part
Foundation	Flexible Bottom Connector (Male connection)	NO		Mechanical Connection	YES
	Structural parts	NO		Anode	NO
Cathodic Protection	Anode	YES			

The following detection measures have been identified for the FSHR:
- The tension on the Flexible Top connector can be monitored by the Top Tension Monitoring system.
- The tension on the following components can be extrapolated by the measurement of the Top Tension Monitoring system: Structural components of URTA, Riser line pipe, Flexible Bottom connector.
- A Local Camera can monitor the following items: Flexible Top connector, the components of the MCV, the components of the URTA.

The following detection measure has been identified for the SLWR:
- The tension on the following components can be extrapolated by the measurement of the Top Tension Monitoring system: Flexible joint, Swivel flange, Riser line pipe, PLET.

Data Collection

The following Table 6 reports a portion of the data collection used for the case-study 2. Due to the particular features of the items in the battery limits of the system few data have been extracted from commercial databases. The most part of the failure rates have been statistically calculated starting from all input data from operational experience and from vendors. They can be considered suitable and conservative for all items belonging to systems object of analysis.

Table 6: Extract of Data Collection

Equipment	Failure Modes	Failure Rate [h^{-1}]	MTTF [y]	ART [h]
Riser Base	Structural Deficiency	1.10E-06	104	5.1
Valve - manifold	External Leakage - process medium	2.20E-07	519	15.2
	Leakage in closed position	1.10E-07	1038	26.6
	Other Failure Mode(s)	6.00E-08	1903	5
VIV	General damage	8.54E-07	1170732	240
…	…	…	…	…

FMECA

The systematic performance of the FMECA allowed the identification of a detailed list of critical items for FSHR and SLWR, (see examples in the following Table 7 and Table 8). Results are shown in terms of risk associated to the failure of each piece of equipment without inspection and related residual risk following the recommendation of a defined inspection interval.

Table 7: FSHR criticality list and recommended Inspection Interval.

Item	Risk without Inspection	Recommended Inspection interval (y)	Residual Risk
Structural Part of foundation (Riser base)	Intolerable	1	Undesirable
Structural parts LRTA	Undesirable	1	Undesirable
Flexible Bottom Connector (rotolatch)	Intolerable	1	Tolerable
Line pipe, Gooseneck, URTA piping	Intolerable	2	Tolerable

Table 8: FSHR criticality list and recommended Inspection Interval.

Item	Risk without Inspection	Recommended Inspection interval (y)	Residual Risk
LRTA piping	Undesirable	2	Tolerable
Structural parts URTA	Undesirable	2	Tolerable
Structural part MCV	Intolerable	2	Tolerable
Buoyancy tank central core	Undesirable	3	Tolerable
Isolation valves (MCV, URTA, LRTA)	Undesirable	5	Tolerable
Flexible Top Connector and Swivel Flange	Undesirable	16	Tolerable
Diverless connectors (MCV, URTA, LRTA)	Undesirable	28	Tolerable
Buoyancy Module, Buoyancy Clamp, Anode and Buoyancy Tank	Tolerable	(a)	(a)
Pipework and valves of the ballasting system	Negligible	(a)	(a)

NOTE:
[a]: The Risk without Inspection is acceptable, therefore a minimum Inspection interval is not suggested.

Item	Risk without Inspection	Recommended Inspection interval (y)	Residual Risk
PLET Structural parts and anchoring system	Intolerable	1	Undesirable
Anodes	Intolerable	1[b]	Undesirable
Flexible Joint Support and Flexible Joint System	Intolerable	1	Tolerable
CLAD Pipe (Metallurgic and Mechanically Bonded)	Intolerable	2	Tolerable
Pipe PLET and Pipe MCV	Undesirable	2	Tolerable
MCV structural reinforcements	Undesirable	2	Tolerable
Anode Sled	Undesirable	5	Tolerable
Swivel Flange and MCV Swivel Flange	Undesirable	16	Tolerable
PLET/MCV connector, VIV strakes, Mechanical connection for Anode Sled, Buoyancy Module and clamp. Continuity Cable	Tolerable	(a)	(a)

NOTE:
[a]: The Risk without Inspection is acceptable, therefore a minimum Inspection interval is not suggested.
[b]: If it is necessary to improve the inspection interval, a spare anode should be installed. In this case the risk results tolerable also without inspection.

The results reported in Table 7 and Table 8 show that the most stringent requirements in term of Inspection Intervals are connected with the components subjected to high stress (e.g. riser base, Flexible Bottom Connector, PLET etc.) and for which detection measures and/or spare parts are not foreseen. For these item the Inspections are essential and permit to reduce the risk to acceptable level.

4. CONCLUSION

In the common practice the application of the RAM techniques to subsea systems is not widely diffused, differently from the standard Oil & Gas industry (such as onshore plant or offshore platform). In this frame the reliability methods do not require the development of new approaches, but the adaptation of the existing RAM techniques to these new subsea systems.

The two case-studies reported in this paper demonstrate the effectiveness and the benefits of the application of the existing RAM tools to subsea systems. The main design improvements have been reached by means of the identification of the critical items, moreover it also possible to give recommendations and define a guideline for the maintenance and inspection activities.
Finally the RAM methods are essential to validate the design, verifying the availability target.

The main criticality connected with this tools and highlighted in the two case-studies is the lack of reliability data. The development of "ad hoc" database, based on the Oil & Gas Company experience on subsea systems, should be an important future development to allow a more and more consolidate application of the RAM techniques to increasingly wide field of application.

References

[1] IEC 60812, *"Analysis techniques for system reliability - Procedure for failure mode and effects analysis (FMEA)"*, IEC, 2006, Geneva.
[2] OREDA Participants, *"Offshore Reliability Data Handbook, Volume 2 - Subsea Equipment"*, OREDA Participants, 2009.
[3] OGP, *"Risk Assessment Data Directory, Report No. 434-1"*, OGP, 2010.
[4] EXIDA, *"Electrical & Mechanical Component Reliability Handbook, 2nd Edition"*, exida.com L.L.C., 2008, Sellersville.
[5] RIAC, *"Reliability Modeling - The RIAC Guide to Reliability Prediction, Assessment and Estimation"*, Reliability Information Analysis Center, 2010, Utica.
[6] Swain, Guttmann, *"Handbook of Human Reliability Analysis with Emphasis on Nuclear Power Plant Applications Final Report"*, Sandia National Laboratories 1983, Albuquerque.
[7] DNV-RP-F206, *"Riser Integrity Management"*, Det Norske Veritas, 2008, Norway.
[8] J. Melendez, J. Schubert, M. Amani, *"Risk Assessment of Surface vs. Subsurface BOP's on Mobile Offshore Drilling Units"*, OTRC, 2006, Texas
[9] P. Holland, *"Reliability of Subsea BOP Systems for Deepwater Application, Phase II DW"*, SINTEF, 1999, Trodheim

Statistical Analysis of Common Cause Failure Events using ICDE Data

S. Yu[a], M.D. Pandey[a], S. Yalaoui[b] and Y. Akl[b]
[a] University of Waterloo, Waterloo, Canada
[b] Canadian Nuclear Safety Commission, Ottawa, Canada

Abstract: Analysis of Common Cause Failures (CCF) is an important element of the Probabilistic Safety Assessment (PSA) of systems important to safety in a nuclear power plant. Based on the conceptualization of the CCF event, many probabilistic models have been developed in the literature. This paper utilizes a modern method, called "General Multiple Failure Rate Model", for the probabilistic modeling of CCF events. To estimate the parameters of the GMFR model, the Empirical Bayes (EB) method is adopted. A detailed case study is presented using CCF data for Motor Operated Valves (MOVs).

Keywords:
PSA, Common cause failure, Poisson Process, Alpha factor, ICDE Data base, Motor Operated valves

1. INTRODUCTION

In the Probabilistic Safety Assessments (PSA), Common Cause Failure (CCF) events arc a subset of dependent events in which two or more components fail within a short interval of time as a result of a shared (or common) cause. Common cause events are highly relevant to PSA due to their potential adverse impact on the safety and availability of critical safety systems in the nuclear plant. An accurate estimation of CCF rates is therefore important for a realistic PSA of plant safety systems.

Due to lack of data, CCF rates were estimated by expert judgment in early years of PSA. Over the years as the data were collected by the utilities and regulators world-wide, more formal statistical analysis methods for data analysis emerged to derive improved estimates of CCF rates. The estimates of CCF rates in line with the operating experience should be used in PSA in place of generic or expert judgment estimates.

International Common Cause Failure Data Exchange (ICDE) is a concerted effort by undertaken by many countries to compile the CCF event data in a consistent manner [1, 2]. This paper describes the General Failure Rate model of CCF events, and presents a detailed case study to evaluate the model parameters using the Empirical Bayes (EB) method along with the data mapping techniques. The case study is based on CCF data for motor operated valves (MOVs).

2. BASIC MODELING OF CCF EVENTS

The probabilistic basis for CCF modelling is that the occurrences of failures in a single component follows the homogeneous Poisson process (HPP). It means: (1) failures are purely random without any trend due to ageing, (2) occurrences of failure events are independent of each other, and (3) after a failure component is renewed to its original "as new" condition. In a time interval (0, t), the number of failures are given by the Poisson distribution as

$$P[N(t) = k] = \frac{(\lambda t)^k}{k!} e^{-\lambda t} \qquad (1)$$

where λ denotes the failure rate, defined as the average number of failures per unit time.

Historically, several conceptual probabilistic models of CCF events have been presented in the literature. The General Multiple Failure Rate (GMFR) model has been a new and widely accepted model of CCF events, which is also advocated by the working group of the Nordic countries [3].

The basic idea is that a failure observed at the system level involving either a single component failure or a failure of k-components is caused by an external shock generated by independent HPPs. In short,

n independent HPPs are generating external shocks that cause CFF events of various multiplicities. These HPPs are mutually exclusive, i.e., one HPP is in action at any given time. In an n-component system, failure event data are described using the following parameters ($k = 1, 2, ..., n$):

$N_{k/n}$ = Number of failures involving any k components
T = Operation time of the system in which $N_{k/n}$ failures occurred
$\Lambda_{k/n}$ = Failure rate of the HPP causing k out of n component failures
Λ_n = Sum of failure rates of all n HPP = $\sum_{k=1}^{n} \Lambda_{k/n}$

The failure events that are observed at the system level are caused by component failures. So it is assumed that component failures are caused by shocks modelled as HPPs. Since component failures produce system failure events, the failure rates at the system and component levels are related. Define $\lambda_{k/n}$ = Failure rate of an HPP causing a CCF involving k specific components, then failure rates at the system and component levels are then related as

$$\Lambda_{k/n} = \binom{n}{k} \lambda_{k/n} = \frac{N_{k/n}}{T} \tag{2}$$

$$\Lambda_n = \sum_{k=1}^{n} \binom{n}{k} \lambda_{k/n} = \sum_{k=1}^{n} \Lambda_{k/n} \tag{3}$$

Based on the GMFR model, the alpha factors are defined as the following ratios of system CCF rates

$$\alpha_{k/n} = \frac{\Lambda_{k/n}}{\Lambda_n} \tag{4}$$

3. ESTIMATION OF CCF RATES

The maximum likelihood is the simplest method for the estimation of the failure rate of a HPP model. If N_i failures are observed in the duration T_i, the failure rate and the associated standard error are estimated as [4]:

$$\hat{\lambda}_i = \frac{N_i}{T_i} \text{ and } \sigma(\hat{\lambda}_i) = \sqrt{\frac{N_i}{T_i^2}}$$

Since CCF dataset are sparse, ML estimates are not considered robust. Also the confidence interval associated with the rates tends to be fairly wide due to lack of data. Therefore, the development of the Bayesian estimation method has been actively pursued by PSA experts.

In the Bayesian Method, a prior distribution, $\pi(\lambda)$, is assigned to the failure rate λ, which is usually estimated from past experience and expert judgment. The conjugate prior, the gamma distribution, is a common choice due to mathematical simplicity and its flexibility to fit various types of data [5].

$$\pi(\lambda_i) = \frac{\beta^\alpha \lambda_i^{\alpha-1} e^{-\beta \lambda_i}}{\Gamma(\alpha)}, \quad \lambda_i, \alpha, \beta > 0 \tag{6}$$

The mean and variance of the prior are
$$M = \alpha/\beta \text{ and } V = \alpha/\beta^2$$

Poisson likelihood for the failure events is given as

$$L[N_i | \lambda_i, T_i] = \frac{e^{-\lambda_i T_i} (\lambda_i T_i)^{N_i}}{N_i!}$$

The posterior of the failure rate is also a gamma distribution:

$$p(\lambda_i | N_i, T_i) = \frac{(\beta + T_i)^{(\alpha + N_i)} \lambda_i^{\alpha + N_i - 1} e^{-(\beta + T_i)\lambda_i}}{\Gamma(\alpha + N_i)} \tag{7}$$

The mean and variance of the gamma posterior are given as
$$\hat{M}_i = (\alpha + N_i)/(\beta + T_i) \text{ and } \hat{V}_i = (\alpha + N_i)/(\beta + T_i)^2 \tag{8}$$

Typically the mean of the posterior is reported as the estimate of the failure rate.

Empirical Bayes (EB) is a method for estimating parameters of a prior distribution used in the Bayesian analysis. In this paper, the EB method proposed by Vaurio [6] has been adopted, since it is also followed by the Nordic PSA group. The basic idea is that the failure data of n components are generated by a Poisson process and the failure rate for each component is a realization from a single Gamma prior with hyper-parameters α and β. The EB method is applied to estimate the parameters α and β using the past event data. Then the distribution for a particular plant is obtained from the updated posterior of the distribution, as described in the previous Section. A novel feature is that in the pooling of the data collected from different systems, proper weights are assigned.

In PSA, methods have been developed to assimilate CCF data available from systems of various sizes to analyse a particular system, also called the "target" system. This process is in general called data mapping, i.e., mapping source data from systems of all different group sizes to the target k-oo-n system. The mapping down means the mapping of the source data from systems with CCG greater than n to a target k-oo-n system. In this paper, a mapping down method proposed by [7] has been adopted. The mapping up means the mapping of the source data from systems with CCG less than n to a target k-oo-n system. In this paper, a mapping up method proposed by Mosleh et al. [8] has been adopted.

2. CASE STUDY: MOV DATA

This Section presents a comprehensive case study of analyzing the CCF data for the motor operated valves (MOV). The purpose is to illustrate the application of data mapping and statistical estimation methods (MLE and EB) in a practical setting.

The data set consists of CCF data from MOV CCG of 2, 4, 8 and 16, as shown in Table 1. These data were collected over an 18 year period. The MOV failure mode is "failure to open" (FO).
In the ICDE database component states are defined as complete failure (C), degraded (D), incipient (I) and working (W). In a formal analysis, impact vectors are assigned depending on the state of the system. For sake of simplicity (and lack of data), no distinction is made among the states C, D and I, and they are treated as the failure.
The objective of the case study is to estimates CCF rates using this data for a target system consisting of a parallel system of 4 MOVs. The statistical analysis is based on MLE and EB methods.

Table 1: CCF Data for MOVs

System size	No. of systems	Total Time	No. of failures					
n	m	$m \times 18 \times 12$ (months)	$1/n$	$2/n$	$3/n$	$4/n$	$5/n$	$6/n$
2	49	10584	36	1	0	0	0	0
4	17	3672	18	2	10	1	0	0
8	8	1728	6	1	0	0	0	0
16	5	1080	13	1	0	0	0	1

2.1. MLE without Data Mapping

In this case only the data for CCG=4 collected from a population of 17 systems over an 18 year period is analyzed using the MLE method. There is no data mapping considered here. The mean and SD of failure rates $\Lambda_{(k/4)}$, and corresponding α-factors are given in Table 2.

Table 2: MLE results for CCF rates

Multiplicity	k	1	2	3	4
Number of failures	$N_{k/4}$	18	2	10	1
Mean of failure rate (/month)	$\Lambda_{k/4}$	4.90×10^{-3}	5.45×10^{-4}	2.72×10^{-3}	2.72×10^{-4}
Standard deviation (/month)	$\sigma(\Lambda_{k/4})$	1.16×10^{-3}	3.85×10^{-4}	8.61×10^{-4}	2.72×10^{-4}
α factor	$\alpha_{k/4}$	5.81×10^{-1}	6.45×10^{-2}	3.23×10^{-1}	3.23×10^{-2}

2.2. MLE with Data Mapping

The CCF data of CCG = 8 and 16 are mapped down to CCG 4 and data from CCG 2 is mapped to CCG 4. In the mapping up the data, parameter ρ is assumed as 0.2 according to NUREG-4780. The results of data mapping are given in Table 3.

Table 2: CCF data mapped to CCG of 4

System size		Number of failures $N_{k/n}$					
n		$1/n$	$2/n$	$3/n$	$4/n$	$5/n$	$6/n$
2	original	36	1	0	0	0	0
2	mapped	36	0.6400	0.3200	0.0400	0	0
4	original	18	2	10	1	0	0
4	mapped	18	2	10	1	0	0
8	original	6	1	0	0	0	0
8	mapped	3.5714	0.2143	0	0	0	0
16	original	13	1	0	0	0	1
16	mapped	4.0456	0.4209	0.1099	0.0082	0	0
sum	mapped	61.6170	3.2752	10.4299	1.0482	For HPP	

In MLE analysis, all the mapped number of failures of a particular multiplicity k-oo-n are summed, as well as the corresponding exposure times. The total operation time is 17064 month. The MLE analysis of mapped data leads to the results shown in Table 4. The comparison of results obtained with and without mapping is given in the next Section of the paper.

Table 3: MLE results with data mapping

Multiplicity	k	1	2	3	4
Number of failures	$N_{k/4}$	61.6170	3.2752	10.4299	1.0482
Mean of failure rate (/month)	$\Lambda_{k/4}$	3.61×10^{-3}	1.92×10^{-4}	6.11×10^{-4}	6.14×10^{-5}
Standard deviation (/month)	$\sigma(\Lambda_{k/4})$	4.60×10^{-4}	1.06×10^{-4}	1.89×10^{-4}	6.00×10^{-5}
α factor	$\alpha_{k/4}$	8.07×10^{-1}	4.29×10^{-2}	1.37×10^{-1}	1.37×10^{-2}

2.3. Empirical Bayes (EB) without Data Mapping

EB method was applied to CCF dart for CCG of 4. The parameters of the gamma prior were estimated as $\alpha = 0.9367$ and $\beta = 434.4733$. The posterior mean and SD of failure rates are given in Table 5.

Table 4: Posterior Mean and SD of failure rate without data mapping

Multiplicity	k	1	2	3	4
No. of failures	$N_{k/4}$	18	2	10	1
Mean of failure rate (/month)	$\Lambda_{k/4}$	4.61×10^{-3}	7.15×10^{-4}	2.66×10^{-3}	4.72×10^{-4}
Standard deviation (/month)	$\sigma(\Lambda_{k/4})$	1.06×10^{-3}	4.17×10^{-4}	8.05×10^{-4}	3.39×10^{-4}
α factor	$\alpha_{k/4}$	5.45×10^{-1}	8.45×10^{-2}	3.15×10^{-1}	5.57×10^{-2}

2.4. EB with Data Mapping: Multiple Priors

In this case mapped data given in Table 3 is used in EB analysis. A prior distribution is assigned to each failure rate $\Lambda_{k/4}, k = 1,..,4$. This way, there are 4 gamma priors and 4 sets of distribution parameters are estimated.

Table 5: Number and exposure times of 1oo4 failures after data mapping

System size n	$N_{1/4}$	T_n (month)
2	36	10584
4	18	3672
8	3.5714	1728
16	4.0456	1080

As an example, 1oo4 failure mapped data given in Table 3 is analyzed. The number of failures after data mapping and the corresponding exposure time are given in Table 6. The EB analysis leads to the posterior mean of $\Lambda_{1/4}$ as 4.49×10^{-3} failures per month.

Table 6: Posterior Mean and SD of failure rates with data mapping

Multiplicity	k	1	2	3	4
Mean of failure rate (/month)	$\Lambda_{k/4}$	4.49×10^{-3}	4.58×10^{-4}	2.55×10^{-3}	2.26×10^{-4}
Standard deviation (/month)	$\sigma(\Lambda_{k/4})$	8.97×10^{-4}	2.72×10^{-4}	7.93×10^{-4}	1.93×10^{-4}
α factor	$\alpha_{k/4}$	5.81×10^{-1}	5.94×10^{-2}	3.30×10^{-1}	2.93×10^{-2}

Repeating the above procedure, results were obtained remaining multiplicities as shown in Table 6. Parameters of the gamma prior are given in Table 8.

Table 7: Parameters of Priors (mapped data)

Failure rate	$\Lambda_{1/4}$	$\Lambda_{2/4}$	$\Lambda_{3/4}$	$\Lambda_{4/4}$
α	7.0049	0.8436	0.3118	0.3799
β	1902.2505	2534.1352	377.2051	2429.8139

2.5. Empirical Bayes (EB) with Data Mapping: Single Priors

In this case, it is assumed that a single gamma prior is applicable to entire mapped data given in Table 2, which contains 16 different values of the number of failures and corresponding exposure times. EB method lead to the following estimates $\alpha = 0.4909$ and $\beta = 418.5876$. Results are tabulated in Table 9 and distributions are plotted in Figure 1.

Table 8: Posterior mean and of failure rates with data mapping

Multiplicity	k	1	2	3	4
No. of failures	$N_{k/4}$	18	2	10	1
Mean of failure rate (/month)	$\Lambda_{k/4}$	4.52×10^{-3}	6.09×10^{-4}	2.56×10^{-3}	3.64×10^{-4}
Standard deviation (/month)	$\sigma(\Lambda_{k/4})$	1.05×10^{-3}	3.86×10^{-4}	7.92×10^{-4}	2.98×10^{-4}
α factor	$\alpha_{k/4}$	5.61×10^{-1}	7.56×10^{-2}	3.18×10^{-1}	4.52×10^{-2}

3. COMPARISON OF RESULTS OF THE CASE STUDY

3.1. Comparison of mean failure rate

In order to understand the effect assumptions associated with 5 estimation methods, the mean of the failure rates are compared in Table 10 and graphically shown in Figure 2.

Table 9: Mean failure rates (per month) by different methods used in the case study

No.	Method	$\Lambda_{1/4}$	$\Lambda_{2/4}$	$\Lambda_{3/4}$	$\Lambda_{4/4}$
M1	MLE without data mapping	4.90×10^{-3}	5.45×10^{-4}	2.72×10^{-3}	2.72×10^{-4}
M2	MLE with data mapping	3.61×10^{-3}	1.92×10^{-4}	6.11×10^{-4}	6.14×10^{-5}
M3	EB without data mapping	4.61×10^{-3}	7.15×10^{-4}	2.66×10^{-3}	4.72×10^{-4}
M4	EB with data mapping (multiple priors)	4.49×10^{-3}	4.58×10^{-4}	2.55×10^{-3}	2.26×10^{-4}
M5	EB with data mapping (single prior)	4.52×10^{-3}	6.09×10^{-4}	2.56×10^{-3}	3.64×10^{-4}

Figure 1: Failure rate distributions with data mapping (EB with single prior)

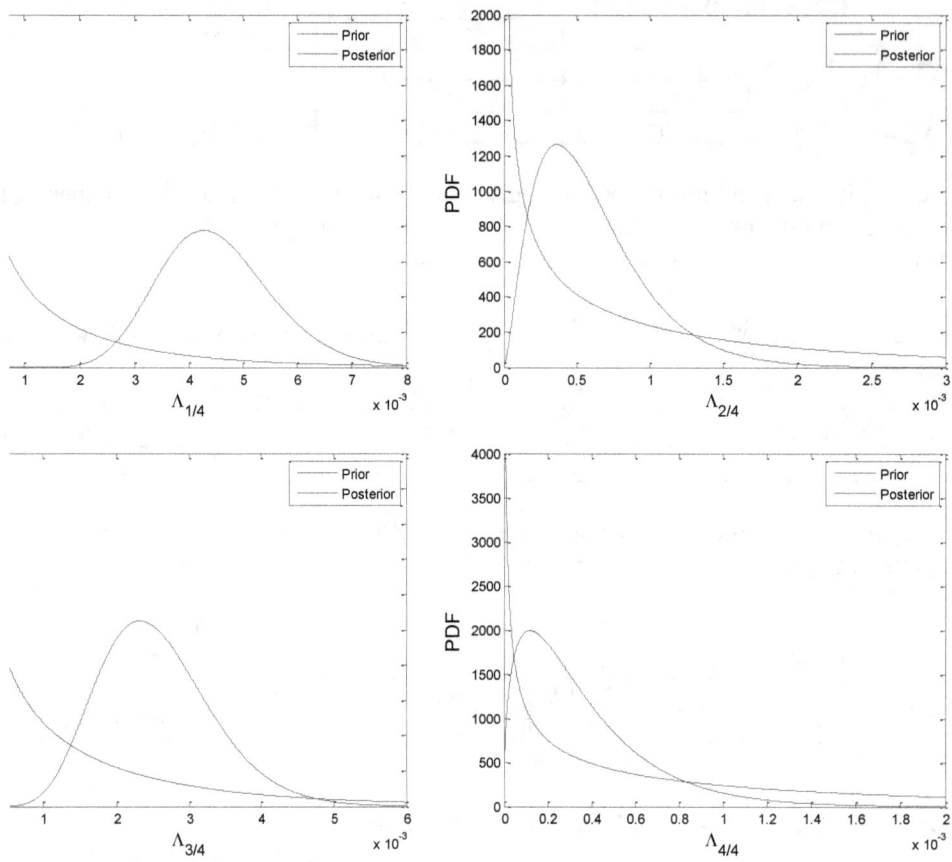

Figure 2: Comparison of mean failure rates estimated in the case study

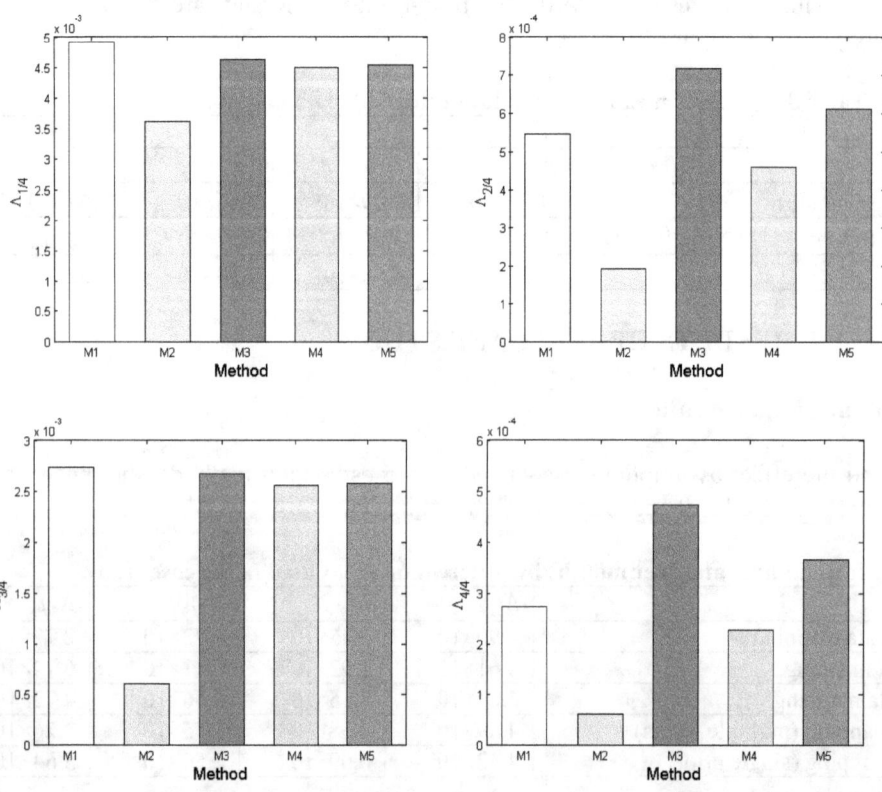

The following observations are notable:
- MLE with data mapping (M2) leads to lower values of mean failure rate as compared to that without data mapping (M1).
- EB with data mapping and single prior (M5) leads to higher mean rate than MLE with data mapping (m2).

3.2. Comparison of standard deviation of failure rate

Table 10: Standard deviation of failure rates by different methods

No.	Method	SD($\Lambda_{1/4}$)	SD($\Lambda_{2/4}$)	SD($\Lambda_{3/4}$)	SD($\Lambda_{4/4}$)
M1	MLE without data mapping	1.16×10^{-3}	3.85×10^{-4}	8.61×10^{-4}	2.72×10^{-4}
M2	MLE with data mapping	4.60×10^{-4}	1.06×10^{-4}	1.89×10^{-4}	6.00×10^{-5}
M3	EB without data mapping	1.06×10^{-3}	4.17×10^{-4}	8.05×10^{-4}	3.39×10^{-4}
M4	EB with data mapping (multiple priors)	8.97×10^{-4}	2.72×10^{-4}	7.93×10^{-4}	1.93×10^{-4}
M5	EB with data mapping (single prior)	1.05×10^{-3}	3.86×10^{-4}	7.92×10^{-4}	2.98×10^{-4}

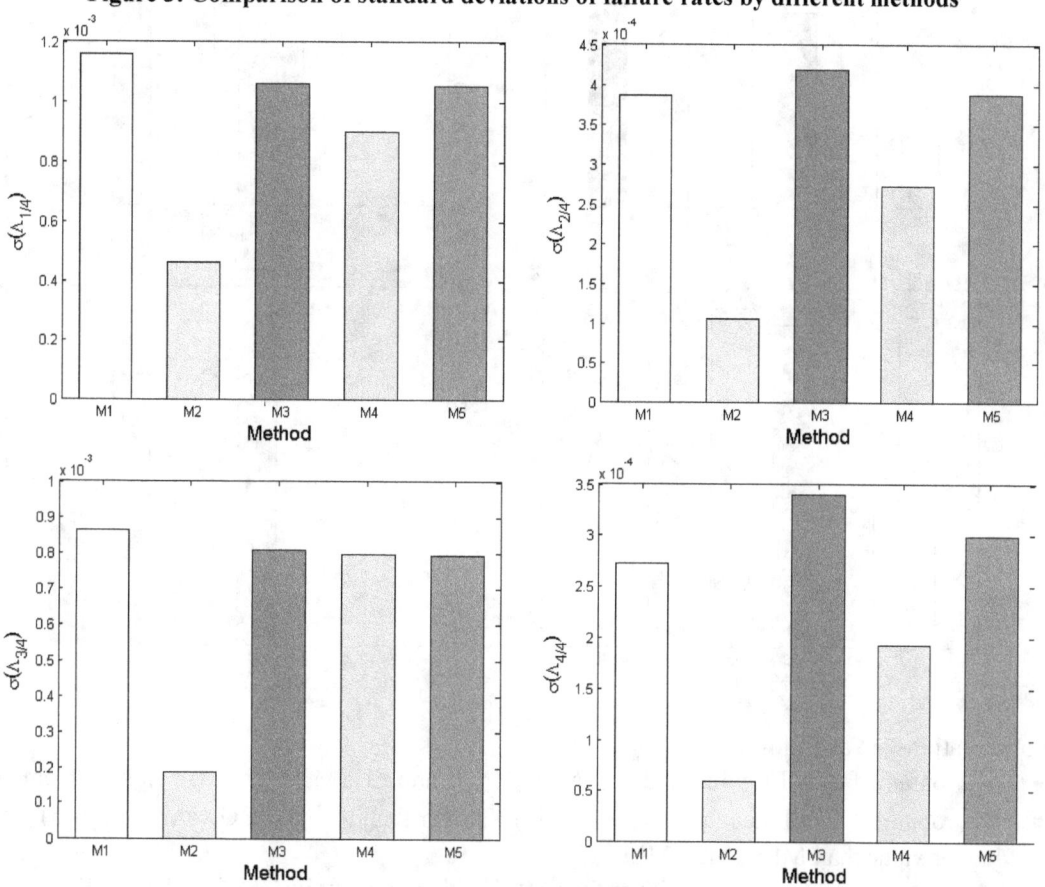

Figure 3: Comparison of standard deviations of failure rates by different methods

Key observations are
- SD of failure rate obtained by EB methods (M3 – M5) is fairly close.
- SD of EB method is slightly smaller than that of MLE without data mapping (M1).

3.3. Comparison of Alpha factors

In Table 12, alpha factors are compared along with the values given in NUREG/CR-5497 for 4-HPCI/RCIC injection MOV. Graphical comparison is shown in Figure 4.

Table 11: Alpha factors by different method

No.	Method	$\alpha_{1/4}$	$\alpha_{2/4}$	$\alpha_{3/4}$	$\alpha_{4/4}$
M1	MLE without data mapping	5.81×10^{-1}	6.45×10^{-2}	3.23×10^{-1}	3.23×10^{-2}
M2	MLE with data mapping	8.07×10^{-1}	4.29×10^{-2}	1.37×10^{-1}	1.37×10^{-2}
M3	EB without data mapping	5.45×10^{-1}	8.45×10^{-2}	3.15×10^{-1}	5.57×10^{-2}
M4	EB with data mapping (multiple priors)	5.81×10^{-1}	5.94×10^{-2}	3.30×10^{-1}	2.93×10^{-2}
M5	EB with data mapping (single prior)	5.61×10^{-1}	7.56×10^{-2}	3.18×10^{-1}	4.52×10^{-2}
Empirical	NUREG/CR-5497	9.69×10^{-1}	6.50×10^{-3}	2.50×10^{-3}	2.22×10^{-2}

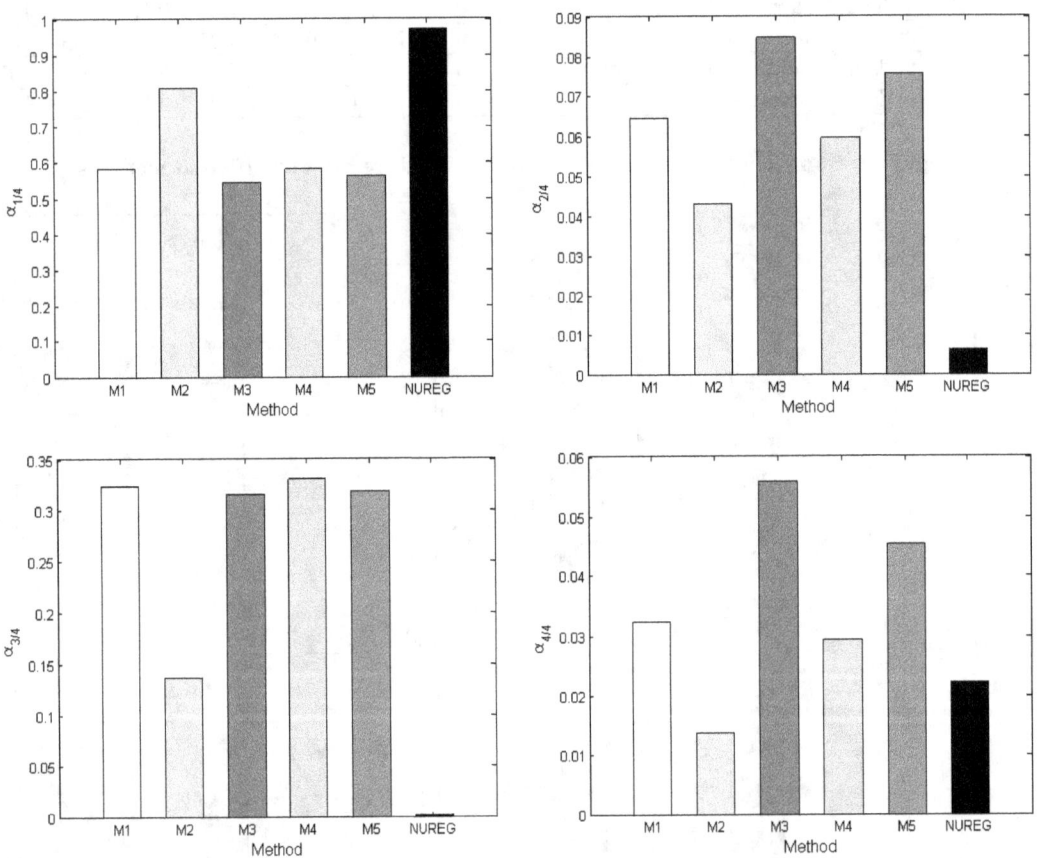

Figure 4: Comparison of alpha factors by different methods

Key observations are as follows:
- $\alpha_{1/4}$ obtained by MLE without data mapping (M1) and EB (M3-M5) are in close agreement.
- $\alpha_{2/4}$ obtained by all the methods (M1 – M5) is larger than the NUREG value. EB (M5) gives a higher value than MLE (M2). MLE
- $\alpha_{3/4}$ obtained by EB (M5) is higher than MLE with data mapping (M2). All M1-M5 estimates are higher than NUREG value.
- $\alpha_{4/4}$ obtained by EB (M5) is higher than MLE with data mapping (M2). All EB estimates (M3-M5) estimates are higher than NUREG value.
- A possible reason for CCF rate by M1-M5 being higher than NUREG is that the impact vectors are not considered in the present analysis. Because of which, estimates M1-M5 are more pessimistic (or more conservative).

4. CONCLUSION

This paper describes the basic background of probabilistic modelling of CCF events and the estimation of parameters using MLE and EB methods. A case study is presented in which CCF data related to MOVs are analyzed in detail. The CCF data can be utilized in 5 different ways depending on whether or not data mapping is done and how the Bayesian priors are selected.

It is interesting to note that when a data set has a relatively high number of failure events, EB estimates of CCF rates turn out to be quite close to those obtained by simple MLE method without data mapping. Thus, EB's utility is apparent only in cases of fairly sparse failure data.

References

[1] Becker, G., Jänkälä, K., Johanson, G., etc. (2007) SKI Report 2007:41, Dependency Analysis Guidance, Nordic/German Working group on Common Cause Failure analysis - Phase 1 project report: Comparisons and application to test cases.

[2] Becker, G., Johanson, G., Lindberg, S. and Vaurio J. K. (2009) SKI Report 2009:07, Dependency Analysis Guidance, Nordic/German Working group on Common Cause Failure analysis - Phase 2, Development of harmonized approach and applications for common cause failure quantification.

[3] Vaurio, J. K. (1995) The Probabilistic modeling of external common cause failure shocks in redundant systems. Reliability Engineering and System Safety 50, pp. 97-107.

[4] Crowder, M. J., Kimber, A. C., Smith, R. L. and Sweeting, T. J. (1991) Statistical Analysis of Reliability Data. Chapman & Hall, London.

[5] Carlin, B. P. and Louis T. A. (2000) Bayes and Empirical Bayes Methods for Data Analysis. 2nd edition. London: Chapman & Hall/CRC.

[6] Vaurio, J. K. (1987) On analytic empirical Bayes estimation of failure rates. Risk Analysis. Vol 17, No. 3, pp. 329-338.

[7] Vaurio, J. K. (2007) Consistent mapping of common cause failure rates and alpha factors. Reliability Engineering and System Safety 92, pp. 628-645.

[8] Mosleh, A., Fleming, K. N., Parry, G. W., Paula, H. M., Worledge, D. H., Rasmuson, D. M. (1989) *Procedures for Treating Common Cause Failures in Safety and Reliability Studies*. NUREG/CR-4780. Washington, DC: U.S. Nuclear Regulatory Commission.

Extending the Alpha Factor Model for Cause Based Treatment of Common Cause Failure Events in PRA and Event Assessment

Andrew O'Connor[a*], Ali Mosleh[b]

[a1] Center for Risk and Reliability, University of Maryland, College Park, United States
[b2] Center for Risk and Reliability, University of Maryland, College Park, United States

Abstract: Common Cause Failure modeling for Probability Safety Assessments has become standard practice in many industries. Of the numerous models proposed to include common cause, one of the most widely adopted has been the Alpha Factor Model, which is supported by the US Nuclear Regulatory Commission CCF database and software tools.

The Alpha Factor Model (AFM) uses an empirical ratio between the independent failures and CCF failures to quantify the model parameters. While this has been advantageous in allowing the prediction of system reliability with little or no data, it has been limiting in other applications such as modeling the characteristics of a system design or including the characteristics of failure when assessing the risk significance of a failure or degraded performance event (known as an event assessment).

This paper proposes a new CCF model called the Partial Alpha Factor Model (PAFM), which extends the AFM to allow the explicit modeling of coupling factors between components such as shared maintenance, or shared location. Using this more explicit modeling allows the model to be tailored depending on how far the system design defends against such dependencies. By using the principles of the AFM as the basis for this new model, its implementation may be feasible without modification to existing PRA software or significant changes in data collection requirements.

Keywords: Common Cause Failure, Alpha Factor Model, Partial Alpha Factor Model, Dependency Modelling.

1. INTRODUCTION

Common Cause Failures (CCFs) are 'simultaneous' failures of a two or more components due to a shared event. These types of failures have the ability to cut through multiple layers of redundancy and cause unforeseen coincidental events that will put safety critical systems in jeopardy. During the 1980s the PRA community first developed quantitative CCF models, however with CCF event data being so scarce, a consolidated effort by the nuclear industry established data collection methods to support the quantitative models. These data collection activities aligned themselves to quantitative method which used impact vectors such as the Alpha Factor Model (AFM). With software and data to support the use of the AFM it quickly became one of the most common CCF models to be used within the nuclear industry.

Since the AFM was first proposed, PRAs have gradually expanded as a management and decision tool beyond the simple quantification of system failure probabilities. In the nuclear industry, PRAs are increasingly used to support the following activities [1]:

- decisions on safety and performance improvement,
- evaluation of proposed modifications,
- assessment of new designs, and
- event assessment and significance determination.

Whilst the AFM allow for the quantification of soft dependencies at a system level, they are inadequate for providing further insight into the causes of CCF which are necessary to inform mitigation strategy

---

[1] Current affiliation, Relken Engineering Pty Ltd, Australia
[2] Current affiliation, University of California, Los Angeles, USA

development. Furthermore, without any cause information contained within the model, it cannot provide an accurate assessment of the system response to failure (also known as event assessments). It has become evident that the commonly accepted CCF modeling methodology [2] and corresponding tools need to be enhanced to meet these emerging PRA activities.

2. RATIO AND ALPHA FACTOR CCF MODELS

Ratio models are based on the hypothesis that system specific estimates for CCF can be made by combining generic average ratio parameters with system specific single/total failure rates [3]. This provides the advantage that ratio models can be estimated from specific data collection activities such as the Common Cause Failure Database [4] and applied to areas where CCF data may not exist. Popular ratio models include the Beta Factor Model, Alpha Factor Model and the Multiple Greek Letter Model.

Limitations of ratio models include:
- Assumes a transferable empirical ratio between failure rates and Common Cause Failure rate.
- Assumes component symmetry.
- Assumes all failure causes have the ability to propagate within the Common Cause Component Group (CCCG)
- No inference can be made given knowledge of the failure cause.
- The model cannot account for unique system architectures which may contribute or defend against CCF.
- Ambiguity in the interpretation of single failures being modeled as independent failures, particularly when applying impact mapping rules.

The alpha factor model (AFM) is a failure event ratio model that was first proposed by Mosleh and Siu in 1987 [3]. Each α_k factor is the probability that given a failure it will fail k components out of m components within the Common Cause Component Group (CCCG). The AFM parameters are defined and calculated as (Mosleh et al. 1998):

$$\alpha_k = \frac{n_k}{\sum_{k=1}^{m} n_k} \qquad (1)$$

$m =$ the number of redundant components
$n_k =$ the number of failure events/frequency which resulted in k components failing within a common cause component group of size m, ($1 \leq k \leq m$).
$\alpha_k =$ the fraction of total failure events/frequency that occur in the system resulting in k out of m failures.

The main difference between the AFM and some other ratio models (such as the Beta Factor Model) is the ability for the AFM to model multiplicities of failure.

CCF data collection activities such as the ICDE database make quantitative CCF modeling accessible to analysts. However these databases were designed around the concept of an impact vector which does not explicitly use information about the failure causes. Causal information has only been used to obtain insights into the CCF phenomena through qualitative analysis [5–9]. Since 2004 the failure causes for single failure data have also been recorded within the NRC CCFDB [10]. This has provided an opportunity to further conditionalize the Alpha Factor Model (AFM) parameters based on failure causes.

3. OBJECTIVE OF PARTIAL ALPHA FACTOR MODEL

The development of the Partial Alpha Factor Model (PAFM) is motivated by a need to conduct event assessments using knowledge of the failure cause, whilst minimizing changes to the popular AFM methodology. The PAFM will add consideration of failure cause as shown in Figure 1.

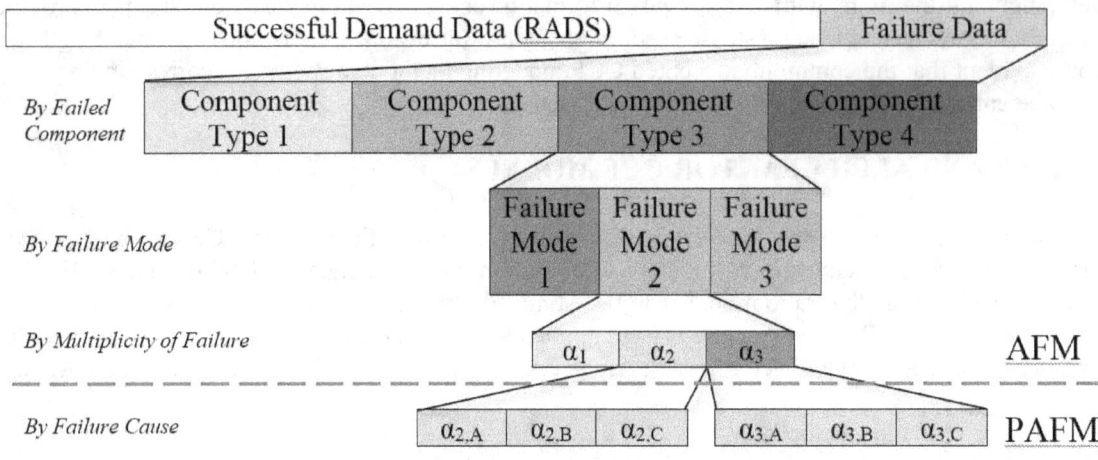

Figure 1: Failure Event Conditionalization

This conditionalization allows the use of the failure cause during event assessments and allows system specific coupling factor dependencies to be added or removed in the model. By dividing the data into a further category the estimates become more difficult when limited data is available. Despite this limitation, the total system failure probabilities will be no worse than using the AFM.

3. PARTIAL ALPHA FACTOR MODEL OVERVIEW

The PAFM follows the same basic principles as the Alpha Factor Model, that the quantification of CCFs is a ratio between the component failure rate and the quantity of CCF events. However the PAFM uses this concept for each possible failure cause. This introduces two parameter types.
- *Partial Alpha Factors* (PAF) which represent the propensity of a failure to propagate to other components through a particular coupling factor.
- *Gamma Factors* (GF) represent the portion of system failure which have the potential to propagate through the coupling factor.

An *Assessed Alpha Factor* is calculated using a combination of gamma factors and partial alpha factors. Assessed alpha factors can be used in exactly the same way as the AFM parameters. The AFM is a special case of the PAFM when the components within a Common Cause Component Group share all coupling factors.

With these revised definitions, the AFM CCF analysis procedure detailed in NUREG/CR-5485 [2] is modified to implement the PAFM:

1. *Qualitative Screening.* A component may be a member of multiple CCCGs based on its coupling factor with other components (noting that the component can only be in one CCCG for each coupling factor). For example, if pump 1 shares its maintenance team and location with pump 2 and shares its installation procedure with pump 3, pump 1 will be part of two common cause component groups: $CCCG_X = \{P_1, P_2\}$ and $CCCG_Y = \{P_1, P_3\}$.

2. *Identification of Common Cause Basic Events.* The CCCBEs will be constructed with consideration for all CCCGs which the component is a part of. E.g. $P_1 = P_{1,i} + X_{P_1 P_2} + Y_{P_1 P_3}$.

3. *Parameter Representation of CCBEs.* The CCBEs are quantified using the Basic Parameter which accounts for multiple CCCGs. For example $P(P_{1,i}) = Q_1^{(2)}$, $P(X_{P_1 P_2}) = Q_2^{(2)[X]}$ and $P(Y_{AC}) = Q_2^{(2)[Y]}$

4. *Alpha Factor Model Parameterization.* The new AFs are quantified using a combination of PAFs. E.g. $Q_2^{(2)[X]} = \binom{m_X - 1}{k_X - 1}^{-1} \cdot \alpha_k^{[X]} \cdot Q_t$.

5. *Parameter Estimation – Impact Vectors.* The system total impact vectors are calculated for each failure cause within the database.

6. *Parameter Estimation – Partial Alpha Factor Model.* The PAFS for each failure cause are calculated. The GF representing the frequency of each cause is calculated. The alpha factor for each CCCG (i.e $\alpha_2^{[Y]}$) is calculated.

7. *System Quantification and Results Interpretation.* The remainder of the CCF analysis process is identical to using the AFM.

4. PARAMETER ESTIMATION

NUREG/CR-6823 [4] discusses a number of methods for conducting data analysis and parameter assessments, however this section will use two formulations of parameter estimation, a classical (frequentist) interpretation using Maximum Likelihood Estimates and Bayesian methodology with conjugate priors. The following estimates have been formulated with the assumption that only one cause could propagate through one coupling factor. Estimates have been developed when this is not the case and will be presented in future papers.

4.1. Partial Alpha Factor Estimation

The maximum likelihood estimate for the PAF is:

$$\alpha_{k,i} = E\left[\frac{n_{k,i}}{\sum_{k=1}^{m} n_{k,i}}\right] = E\left[\frac{n_{k,i}}{n_{t,i}}\right] \quad (2)$$

$\alpha_{k,i}$ = a partial alpha factor which represents the portion of system failure events which resulted in k components failing within a common cause component group of size m, $(1 \leq k \leq m)$ when there was a potential for failure propagation through coupling factor i where $i \in \{1,2,3,...,w\}$

$n_{k,i}$ = the number of failure events which resulted in k components failing within a common cause component group of size m, $(1 \leq k \leq m)$ of coupling factor i where $i \in \{1,2,3,...,w\}$

$n_{t,i}$ = the total number of common cause failure events for coupling factor/cause i where $i \in \{1,2,3,...,w\}$.

The Bayesian estimates for PAF are:

$$\boldsymbol{\alpha}_i \sim \text{Dirichlet}_m\{\boldsymbol{\psi}_i\} \quad (3)$$

$\boldsymbol{\alpha}_i$ = the portion of failure events for each multiplicity of failure $[\alpha_{1,i}, \alpha_{2,i}, ..., \alpha_{m,i}]$ for failure cause i.

$\boldsymbol{\psi}_i$ = the equivalent count of failure events for each multiplicity of failure with cause I $[\psi_{1,i}, \psi_{2,i}, ..., \psi_{m,i}]$

The parameter $\boldsymbol{\alpha}_i$, which is the unknown of interest (UOI), can be estimated using Bayes' rule:

$$f(\boldsymbol{\alpha}_i|\boldsymbol{n}_i) = \frac{f(\boldsymbol{\alpha}_i)L(\boldsymbol{n}_i|\boldsymbol{\alpha}_i)}{\sum_{j=1}^{m} L(\boldsymbol{n}_i|\alpha_{j,i})f(\alpha_{j,i})} \quad (4)$$

$f(\boldsymbol{\alpha}_i)$ = is the prior distribution of the parameter $\boldsymbol{\alpha}_i$
$L(\boldsymbol{n}_i|\boldsymbol{\alpha}_i)$ = is the likelihood equation for observing the evidence $\boldsymbol{n}_i$ given the parameters $\boldsymbol{\alpha}_i$.

$f(\boldsymbol{\alpha_i}|\boldsymbol{n_i})$ = the posterior distribution of $\boldsymbol{\alpha_i}$ given the evidence $\boldsymbol{n_i}$
$\boldsymbol{n_i}$ = the number of failure events for each multiplicity of failure $[n_{1,i}, n_{2,i}, ..., n_{m,i}]$ for failure cause i.
$\boldsymbol{\alpha_i}$ = the portion of failure events for each multiplicity of failure $[\alpha_{1,i}, \alpha_{2,i}, ..., \alpha_{m,i}]$ for failure cause i.

The likelihood equation of observing the number of failures in each failure cause category, $\boldsymbol{n_i} = [n_{1,i}, n_{2,i}, ..., n_{m,i}]$ is distributed as a multinomial distribution with parameters $\boldsymbol{\alpha_i} = [\alpha_{1,i}, \alpha_{2,i}, ..., \alpha_{m,i}]$.

$$\boldsymbol{n_i} \sim \text{Multinomial}_m\{n_{t,i}, \boldsymbol{\alpha_i}\} \tag{5}$$
$$\tag{6}$$

Therefore the hyper parameters, $\boldsymbol{\psi_i}$, for the posterior $\boldsymbol{\alpha_i}$ given evidence $\boldsymbol{n_i}$ using a Dirichlet prior with parameters, $\boldsymbol{\psi_{i,0}}$, is:

$$\boldsymbol{\psi_i} = \boldsymbol{\psi_{i,0}} + \boldsymbol{n_i} \tag{7}$$

The choice of a prior distribution parameters, $\boldsymbol{\psi_{i,0}}$, depends on the availability of data as discussed in detail by Siu and Kelly [11].

4.1. Gamma Factor Estimation

The maximum likelihood estimate for the GF is:

$$\gamma_i = E\left[\frac{n_{t,i}}{\sum_{i=1}^{w} n_{t,i}}\right] = E\left[\frac{n_{t,i}}{n_t}\right] \tag{8}$$

γ_i = the portion of failure events which had the potential to propagate through coupling factor i where $i \in \{1,2,3,...,w\}$
n_t = the total number of failure events/frequency.

The Bayesian estimates for GF is:

$$\boldsymbol{\gamma} \sim \text{Dirichlet}_w\{\boldsymbol{\varphi}\} \tag{9}$$

$\boldsymbol{\gamma}$ = the portion of failure events for each cause $[\gamma_1, \gamma_2, ..., \gamma_w]$
$\boldsymbol{\varphi}$ = the equivalent count of failure events for each cause $[\varphi_1, \varphi_2, ..., \varphi_w]$

The point estimates for each gamma factor can be obtained using:

$$\hat{\gamma}_i = \frac{\varphi_i}{\sum_{i=1}^{w} \varphi_i} \tag{10}$$

The parameter $\boldsymbol{\gamma}$, which is the unknown of interest (UOI), can be estimated using Bayes' rule:

$$f(\boldsymbol{\gamma}|\boldsymbol{n_t}) = \frac{f(\boldsymbol{\gamma})L(\boldsymbol{n_t}|\boldsymbol{\gamma})}{\sum_{j=1}^{w} L(\boldsymbol{n_t}|\boldsymbol{\gamma_j})f(\boldsymbol{\gamma_j})} \tag{11}$$

$f(\boldsymbol{\gamma})$ = is the prior distribution of the parameter $\boldsymbol{\gamma}$
$L(\boldsymbol{n_t}|\boldsymbol{\gamma})$ = is the likelihood equation for observing the evidence
$f(\boldsymbol{\gamma}|\boldsymbol{n_t})$ = the posterior distribution of $\boldsymbol{\gamma}$ given the evidence $\boldsymbol{n_t}$
$\boldsymbol{n_t}$ = the number of failure events for each cause $[n_{t,1}, n_{t,2}, ..., n_{t,w}]$
$\boldsymbol{\gamma}$ = the portion of failure events for each cause $[\gamma_1, \gamma_2, ..., \gamma_w]$

The likelihood equation of observing the number of failures in each failure cause category, $\boldsymbol{n_t} = [n_{t,1}, n_{t,2}, ..., n_{t,w}]$ is distributed as a multinomial distribution with parameters $\boldsymbol{\gamma} = [\gamma_1, \gamma_2, ..., \gamma_w]$.

$$\boldsymbol{n_t} \sim \text{Multinomial}_w\{n_t, \boldsymbol{\gamma}\} \tag{12}$$

Therefore the hyper parameters, $\boldsymbol{\varphi}$, for the posterior $\boldsymbol{\gamma}$ given evidence $\boldsymbol{n_t}$ using a Dirichlet prior with parameters, $\boldsymbol{\varphi_0}$, is:

$$\boldsymbol{\varphi} = \boldsymbol{\varphi_0} + \boldsymbol{n_t} \tag{13}$$

The choice of a prior distribution parameters, φ_0, depends on the availability of data as discussed in detail by Siu and Kelly [11].

5. EXAMPLE

A cooling systems objective is to provide water to a cooling system using three pumps and two generators. Only one pump needs to be running in order to provide sufficient water. A pump requires power from only one generator to operate. The failure probability for each generator is also assumed to be $Q_t^{[E]} = 0.006$ and the failure probability for a pump is assumed to be $Q_t^{[P]} = 0.002$. The fault tree with the system failure rate is shown in Figure 2.

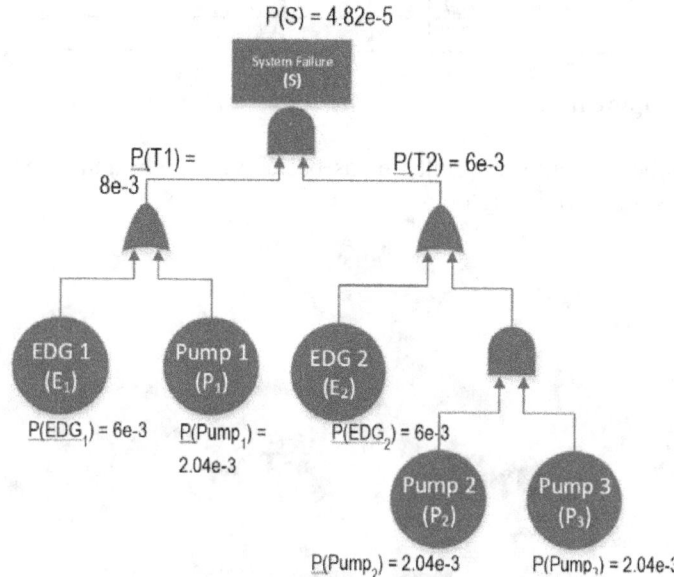

Figure 2: Fault tree for example 2 – Two train Emergency Diesel Generator and pump system

The minimal cut sets for the second example system are:

$$\{E_1, E_2\}; \ \{P_1, E_2\}; \{P_1, P_2, P_3\}; \ \{E_1, P_2, P_3\}$$

5.1. Qualitative Analysis

Qualitative analysis involves finding CFs that can help define CCCGs. This is achieved by finding qualitative dependencies between components as illustrated in Table 1.

Table 1: Qualitative dependency assessment

Component	Install Procedure	Maintenance Staff	Location
EDG 1 (E1)	EDG	Team X	Room Y
EDG 2 (E2)	EDG	Team X	Room Y
Pump 1 (P1)	Pump V1.1	Team X	Room Y
Pump 2 (P2)	Pump V2.8	Team X	Room Y
Pump 3 (P3)	Pump V1.1	Team Y	Room X

While the EDGs are symmetrical (in that they share all CFs), pumps 1 and 2 only share the same maintenance team and location (two coupling factors). Pumps 1 and 3 share an installation procedure only (one coupling factor). Note that as per the AFM, CCCGs can only formed with like components. Therefore the following Common Cause Component Groups Exist:

$$CCCG_X = \{E_1, E_2\}, CCCG_Z = \{P_1, P_2\}, CCCG_Y = \{P_1, P_3\}.$$

5.2. Identification of Common Cause Basic Events

The CCBEs will be constructed with consideration for all CCCGs which the component is a part of. If a component belongs in multiple CCCGs, then the CCBEs for both CCCGs would be added. The CCBE events for the example system are shown in Table **2**.

Table 2: CCBE for example system

Component	Common Cause Basic Events
EDG 1 (E_1)	$E_{1,i}, X_{E1,E2}$
EDG 2 (E_2)	$E_{2,i}, X_{E1,E2}$
Pump 1 (P_1)	$P_{1,i}, Z_{P1,P2}, Y_{P1,P3}$
Pump 2 (P_2)	$P_{2,i}, Z_{P1,P2}$
Pump 3 (P_3)	$P_{3,i}, Y_{P1,P3}$

5.3. Fault Tree Development

The CCBEs are incorporated into the fault tree as basic events. The fault tree for the example after substitution of CCBEs is shown in Figure **3**.

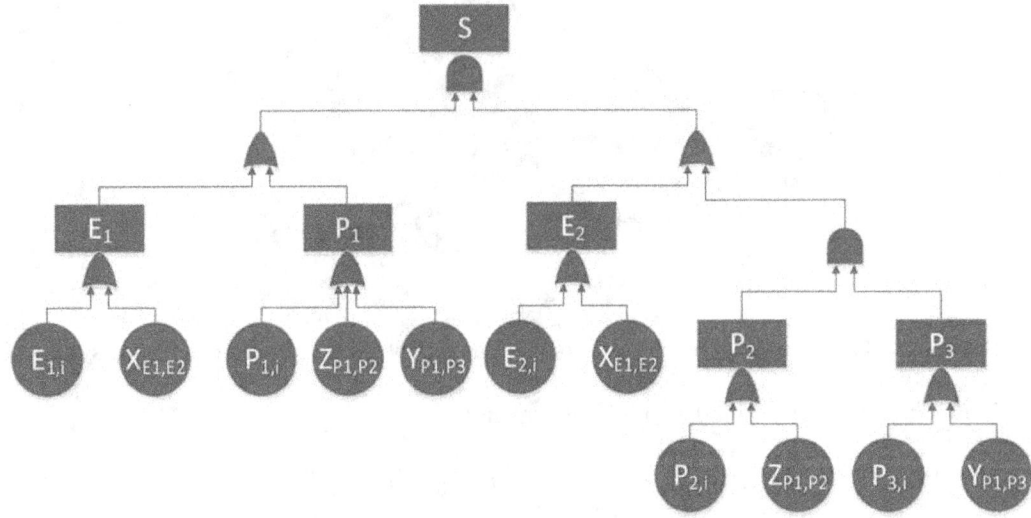

Figure 3: Fault tree for example with CCBEs

The cut sets with CCBEs are:

$$\{E_{1,i}, E_{2,i}\}; \{P_{1,i}, E_{2,i}\}; \{P_{1,i}, P_{2,i}, P_{3,i}\}; \{E_{1,i}, P_{2,i}, P_{3,i}\};$$
$$\{P_{3,i}, Z_{P1,P2}\}; \{E_{2,i}, Z_{P1,P2}\}; \{P_{2,i}, Y_{P1,P3}\}; \{E_{2,i}, Y_{P1,P3}\};$$
$$\{Z_{P1,P2}, Y_{P1,P3}\}; \{X_{E1,E2}\}$$

Note that the cut set $\{Z_{P1,P2}, Y_{P1,P3}\}$ treats the failure of a single component due to different causes in separate CCCGs as independent, not mutually exclusive. This is the same assumption used to separate individual and common cause basic events. Generally, as CCBEs are treated as independent events (not mutually exclusive), this may produce a lower estimate of the component failure rates.

5.4. Parameter Representation of CCBEs

For the example system, the CCBEs are equal to:

$$P(E_{1,i}) = P(E_{2,i}) = Q_1^{(2)[E]}$$
$$P(X_{E1,E2}) = Q_2^{(2)[X]} \qquad (14)$$
$$P(P_{1,i}) = Q_1^{(2)[P_1]}, \ P(P_{2,i}) = Q_1^{(2)[P_2]}, \ P(P_{3,i}) = Q_1^{(2)[P_3]}$$

$$P(Z_{P1,P2}) = Q_2^{(2)[Z]}, \quad P(Y_{P1,P3}) = Q_2^{(2)[Y]}$$

In comparison to the AFM, the pumps are not symmetrical and so:
$$Q_1^{(2)[P_1]} \neq Q_1^{(2)[P_2]} \neq Q_1^{(2)[P_3]} \tag{15}$$

Using rare event approximation, the system equation is now:
$$\begin{aligned} P(S) &= \left(Q_1^{(2)[E]}\right)^2 + Q_1^{(2)[P_1]}Q_1^{(2)[E]} + Q_1^{(2)[P_1]}Q_1^{(2)[P_2]}Q_1^{(2)[P_3]} \\ &+ Q_1^{(2)[E]}Q_1^{(2)[P_2]}Q_1^{(2)[P_3]} + Q_1^{(2)[P_3]}Q_2^{(2)[Z]} + Q_1^{(2)[E]}Q_2^{(2)[Z]} + Q_1^{(2)[P_2]}Q_2^{(2)[Y]} \\ &+ Q_1^{(2)[E]}Q_2^{(2)[Y]} + Q_2^{(2)[Z]}Q_2^{(2)[Y]} + Q_2^{(2)[X]} \end{aligned} \tag{16}$$

5.5. Partial Alpha Factor Model Parameterization

The Basic Parameter values are quantified using assessed AFs which add the contribution from the relevant dependencies. The CCBEs are quantified using assessed alpha factors.
$$Q_k^{(m)} = \alpha_k Q_t \binom{m-1}{k-1}^{-1} \tag{17}$$

For the example, the basic events for each component are:
$$Q_1^{(2)[P]} = \alpha'^{[P]}_1 Q_t^{[P]}, \quad Q_2^{(2)[P]} = \alpha'^{[P]}_2 \cdot Q_t^{[P]} \tag{18}$$

The EDGs are symmetrical and therefore the basic events are:
$$Q_1^{(2)[E]} = \alpha'^{[E]}_1 Q_t^{[E]}, \quad Q_2^{(2)[X]} = \alpha'^{[X]}_2 \cdot Q_t^{[E]} \tag{19}$$

$$\alpha'^{[X]}_2 = \sum_{i=\{IP,MH,EE\}} \gamma_i^{[E]} \alpha_{2,i}^{[E]} \tag{20}$$

$$\alpha'^{[E]}_1 = 1 - \alpha'^{[X]}_2 \tag{21}$$

The pumps in $CCCG_Z$ only share the maintenance team and external environment. The assessed alpha factors will not include the contribution from installation procedures.
$$Q_2^{(2)[Z]} = \alpha'^{[Z]}_2 \cdot Q_t^{[P]} \tag{22}$$

$$\alpha'^{[Z]}_2 = \sum_{i=\{MH,EE\}} \gamma_i^{[P]} \alpha_{2,i}^{[P]} \tag{23}$$

The pumps in $CCCG_Y$ only share installation procedure and the assessed alpha factor will not include the contribution from maintenance team and external environment.
$$Q_2^{(2)[Y]} = \alpha'^{[Y]}_2 \cdot Q_t^{[P]} \tag{24}$$

$$\alpha'^{[Y]}_2 = \sum_{i=\{IP\}} \gamma_i^{[P]} \alpha_{2,i}^{[P]} \tag{25}$$

The assessed alpha factors for each pump may now be calculated as the remaining failure events.
$$\begin{aligned} Q_1^{(2)[P_1]} &= \alpha'^{[P_1]}_1 \cdot Q_t^{[P]}, & \alpha'^{[P_1]}_1 &= 1 - \alpha'^{[Z]}_2 - \alpha'^{[Y]}_2 \\ Q_1^{(2)[P_2]} &= \alpha'^{[P_2]}_1 \cdot Q_t^{[P]}, & \alpha'^{[P_2]}_1 &= 1 - \alpha'^{[Z]}_2 \\ Q_1^{(2)[P_3]} &= \alpha'^{[P_3]}_1 \cdot Q_t^{[P]}, & \alpha'^{[P_3]}_1 &= 1 - \alpha'^{[Y]}_2 \end{aligned} \tag{26}$$

5.6. Parameter Estimation – Impact Vectors

Impact vectors are calculated for each CCF in accordance with NUREG/CR-5485 [2]. The primary difference is that the average impact vectors are calculated for each cause. Assume for this example that the sum of the EDG average impact vectors, for each cause is:

$$n_{\Omega,IP}^{[E]} = [0, \quad 172.2, \quad 2.8]$$
$$n_{\Omega,MH}^{[E]} = [0, \quad 154.35, \quad 3.15]$$
$$n_{\Omega,EE}^{[E]} = [0, \quad 16.45, \quad 1.05] \tag{27}$$
$$n_0^{[E]} = 29400$$

The sum of the pump average impact vectors for each cause is:

$$n_{\Omega,IP}^{[P]} = [0, \quad 26.0663, \quad 0.1838]$$
$$n_{\Omega,MH}^{[P]} = [0, \quad 59.4125, \quad 1.838]$$
$$n_{\Omega,EE}^{[P]} = [0, \quad 82.5213, \quad 4.9788] \tag{28}$$
$$n_0^{[P]} = 44433$$

5.7. Parameter Estimation – Partial Alpha Factor Model

In order to calculate the assessed alpha factors, the partial alpha factors and gamma factors must be calculated. For the example, the PAFs and GFs can be calculated as:

$$\alpha_{2,IP}^{[P]} = \frac{0.18375}{26.06625 + 0.18375} = 0.007 \quad \gamma_{IP}^{[P]} = \frac{26.25}{26.25 + 61.25 + 87.5} = 0.15$$
$$\alpha_{2,MH}^{[P]} = \frac{1.8375}{59.4125 + 1.8375} = 0.03 \quad \gamma_{MH}^{[P]} = \frac{61.25}{26.25 + 61.25 + 87.5} = 0.35 \tag{29}$$
$$\alpha_{2,EE}^{[P]} = \frac{4.97875}{82.52125 + 4.97875} = 0.0569 \quad \gamma_{EE}^{[P]} = \frac{87.5}{26.25 + 61.25 + 87.5} = 0.50$$

The assessed alpha factors can now be calculated as:

$$\alpha'^{[X]}_2 = \sum_{i=\{IP,MH,EE\}} \gamma_i^{[E]} \alpha_{2,i}^{[E]} = 0.02$$
$$\alpha'^{[Z]}_2 = \sum_{i=\{MH,EE\}} \gamma_i^{[P]} \alpha_{2,i}^{[P]} = 0.03895$$
$$\alpha'^{[Y]}_2 = \sum_{i=\{IP\}} \gamma_i^{[P]} \alpha_{2,i}^{[P]} = 0.00105 \tag{30}$$
$$\alpha'^{[P_1]}_1 = 1 - \alpha'^{[Z]}_2 - \alpha'^{[Y]}_2 = 0.96$$
$$\alpha'^{[P_2]}_1 = 1 - \alpha'^{[Z]}_2 = 0.96105$$
$$\alpha'^{[P_3]}_1 = 1 - \alpha'^{[Y]}_2 = 0.99895$$

5.8. System Quantification and Results Interpretation

The parameter estimates may now be substituted back into the system equations. Instead of calculating the system probability of failure using the system equation, the following quantities can be placed into the fault tree for calculation:

$$P(E_{1,i}) = P(E_{2,i}) = Q_1^{(2)[E]} = \alpha'^{[E]}_1 Q_t^{[E]} = 5.88e\text{-}3$$
$$P(X_{E1,E2}) = Q_2^{(2)[X]} = \alpha'^{[X]}_2 \cdot Q_t^{[E]} = 1.2e\text{-}4$$
$$P(P_{1,i}) = Q_1^{(2)[P_1]} = \alpha'^{[P_1]}_1 \cdot Q_t^{[P]} = 1.9584\text{-}3 \tag{31}$$
$$P(P_{2,i}) = Q_1^{(2)[P_2]} = \alpha'^{[P_2]}_1 \cdot Q_t^{[P]} = 1.9605e\text{-}3$$
$$P(P_{3,i}) = Q_1^{(2)[P_3]} = \alpha'^{[P_3]}_1 \cdot Q_t^{[P]} = 2.0379e\text{-}3$$

$$P(Z_{P1,P2}) = Q_2^{(2)[Z]} = \alpha'^{[Z]}_2 \cdot Q_t^{[P]} = 7.95e\text{-}5$$
$$P(Y_{P1,P3}) = Q_2^{(2)[Y]} = \alpha'^{[Y]}_2 \cdot Q_t^{[P]} = 2.1e\text{-}6$$

Substitution back into the system equation gives:

$$\begin{aligned}P(S) = &\left(Q_1^{(2)[E]}\right)^2 + Q_1^{(2)[P_1]}Q_1^{(2)[E]} + Q_1^{(2)[P_1]}Q_1^{(2)[P_2]}Q_1^{(2)[P_3]} \\ &+ Q_1^{(2)[E]}Q_1^{(2)[P_2]}Q_1^{(2)[P_3]} + Q_1^{(2)[P_3]}Q_2^{(2)[Z]} + Q_1^{(2)[E]}Q_2^{(2)[Z]} + Q_1^{(2)[P_2]}Q_2^{(2)[Y]} \\ &+ Q_1^{(2)[E]}Q_2^{(2)[Y]} + Q_2^{(2)[Z]}Q_2^{(2)[Y]} + Q_2^{(2)[X]} \\ = &\ 1.668e\text{-}4 \end{aligned} \quad (32)$$

The system failure probability is calculated as 1.668e-4.

6. CONCLUSION

The PAFM aims to provide a cause-based CCF model which uses the same analysis methodology and data sources as the popular AFM. While minimized increases in complexity, the PAFM has the ability to create quantitative models using impact vectors, recognize efforts to create coupling factor defenses, and allow more sophisticated event assessments based on the observed failure cause. The PAFM does however use the same empirical ratio methods of the AFM and therefore most of the issues which exist with the current methodology remain.

The PAFM is quantified using two parameters, the partial alpha factors ($\alpha_{k,i}$) and the gamma factors (γ_i). The partial alpha factors represent the multiplicity of failures in a common cause component group, given a failure has occurred from cause i. The gamma factors represent the portion of failure events which are due to cause i. These two parameters can be combined to provide an 'assessed' alpha factor which may be implemented in an identical way to the AFM for system modelling.

8. ACKNOWLEDGEMENTS

This work was funded through Collaborative Research Grant NRC-04-10-164 by the U.S. Nuclear Regulatory Commission (USNRC). The authors thank the NRC Project Technical Monitor, Dr. Jeffrey Wood for his interest and support throughout the project. The authors also appreciate valuable technical guidance and feedback offered by Dr. Song Hua Shen, Dr. Kevin Coyne, Mr. Gary Demoss, Dr. Nathan Siu, and Mr. Don Marksberry, from the NRC Office of Research. While this report was prepared under a collaborative research grant and benefitted from extensive discussions with the NRC staff, USNRC has neither approved nor disapproved its technical content.

9. REFERENCES

[1] A. Bates, Use of Probabilistic Risk Assessment Methods in Nuclear Regulatory Activities; Final Policy Statament, US Nuclear Regulatory Commission, 1995.
[2] A. Mosleh, D.. Rasmuson, F.M. Marshall, Guidelines on Modeling Common Cause Failures in Probabilistic Risk Assessments, U.S. Nuclear Regulatory Commission, Washington DC, 1998.
[3] J.K. Vaurio, Common Cause Failure Modeling, Encycl. Quant. Risk Anal. Assess. (2008).
[4] T.. Wierman, D.M. Rasmuson, A. Mosleh, Common-Cause Failure Database and Analysis System: Event Data Collection, Classification, and Coding, U.S. Nuclear Regulatory Commission, 2007.
[5] S. Lindberg, Common Cause Failure Analysis: Methodology evaluation using Nordic experience data, Uppsala Universitet, 2007.
[6] T.. Wierman, D.. Rasmuson, N.. Stockton, Common-Cause Failure Event Insights: Circuit Breakers, U.S. Nuclear Regulatory Commission, Washington DC, 2003.
[7] T.. Wierman, D.. Rasmuson, N.. Stockton, Common-Cause Failure Event Insights: Emergency Diesel Generators, U.S. Nuclear Regulatory Commission, Washington DC, 2003.
[8] T.. Wierman, D.. Rasmuson, N.. Stockton, Common-Cause Failure Event Insights: Pumps, U.S.

Nuclear Regulatory Commission, Washington DC, 2003.
[9]   T.. Wierman, D.. Rasmuson, N.. Stockton, Common-Cause Failure Event Insights: Motor-Operated Valves, U.S. Nuclear Regulatory Commission, Washington DC, 2003.
[10] T.. Wierman, Causes for Single Failures, (2013).
[11] N.O. Siu, D.L. Kelly, Bayesian parameter estimation in probabilistic risk assessment, Reliab. Eng. Syst. Saf. 62 (1998) 89–116. doi:10.1016/S0951-8320(97)00159-2.

Estimating Common Cause Failure Probabilities for a PRA Taking into account Different Detection Methods

Kalle E. Jänkälä
Fortum Power and Heat Oy, Espoo, Finland

Abstract: The methodology to estimate residual parametric common cause failure (CCF) probabilities consists of the selection of the data source, source plants, source systems and component type, failure mode, assessment of the impact vectors, determination of equivalent observations, calculation of CCF rates of different multiplicities with uncertainties using an empirical Bayes estimation method and finally determining explicit CCF basic events and their probabilities to be used in the probabilistic safety assessment model. The CCF probabilities are obtained as the result of unavailability estimation accounting for different detection methods and corresponding outage times. Typically CCF events of safety system components are detected by tests during plant operation or during annual overhaul. In CCF quantification this is often regarded as the only way of detection. This leads into CCF unavailability quantification in which the CCF rate is based on all kinds of CCF events and the corresponding outage time is always determined by the test interval and testing scheme. This approach might be overly conservative or sometimes optimistic. This paper improves CCF unavailability estimation by taking into account monitoring and different kinds of tests and outage times and considering failure modes in the failure rate estimation.

Keywords: PRA, CCF, unavailability, probability, failure rate.

1. INTRODUCTION

Typically CCF events of safety system components are detected by testing during plant operation or during annual overhaul. In CCF quantification this is often regarded as the only way of detection. This leads into CCF unavailability quantification in which the CCF rate is based on all kinds of CCF events having outage times from minutes to years. This approach might yield overly conservative CCF unavailability estimates or sometimes optimistic.

For example diesel generators of Loviisa Nuclear Power Plant (NPP) have once started spuriously due to a false signal (this event is recorded in the ICDE database [1]). The diesel generators stopped simultaneously when the large leakage signal disappeared, because at that time there was no delay circuit of the start-up signal. All the diesel generators were simultaneously completely unavailable a couple of minutes. CCF coupling factor and time factor measuring the simultaneity were high. Thus, this event is clearly a complete CCF and it has a high effect when estimating the CCF rate of diesel generators, if it is taken us such in the estimation. However, such an event is always detected immediately which means that the unavailability due to such complete CCF failures is negligible. Therefore it is important to take into account detection methods and corresponding outage times. It is clear that immediately detected failures have a minor effect on the overall unavailability when the repair times are short compared to the test intervals.

Sometimes CCF events are not found out in the normal tests during power operation, but in the annual or more rarely made more profound tests. These kinds of events can be important contributors to the unavailability. They should be properly taken into account in the estimation, or the result could be optimistic.

Detection methods are recorded in the ICDE CCF event data collection. If there are many events it might be possible to estimate rates of different types of events. However, the data is usually sparse so that we do not get sufficient amounts of data to estimate dedicated rates of those different types. If we only have events with immediate detection does not mean that other kinds of events would have zero failure rate. Another problem in the data handling is that other plants might have different testing methods.

Plant-specific testing methods and schemes have to be taken into account. Their effectiveness can be studied by analyzing single failure data. In case of Loviisa NPP we have studied how the events have been detected in the failure histories of the safety systems. Scheduled tests were found to be the most important way of detection, except in case of measurements. The results of this study indicate that there are differences in the detection methods not only between component types but especially between different systems. This study of single failure histories was utilized in estimating the shares of different kinds of events with different detection methods.

The methodology to estimate residual parametric common cause failure (CCF) rates consists of the selection of the data source (OECD ICDE, EPRI, other), source plants, source systems and component type, failure mode and detection methods, assessment of the impact vectors, calculation of CCF rates with uncertainties using a parametric empirical Bayes estimation method and finally determining CCF basic event probabilities to be used in the probabilistic safety assessment model. The methodology was first introduced in 2001 [2], later to a wider audience [3] and a more comprehensive description and further developed methodology was presented in 2005 [4]. This paper takes into account failure modes and improves the unavailability estimation by taking into account different detection methods and corresponding outage times.

Plant-specific test intervals and schemes and detection methods were used to calculate the common-cause unavailabilities. All different CCF multiplicities were modeled as basic events in the system fault trees, connected by OR-gates to the components affected. This approach facilitates many PRA applications like the evaluation of test interval modifications and risk-informed optimization of allowed outage times.

This paper gives at first an overview of the CCF methodology that has been used for Loviisa PRA. CCF rate estimation is briefly presented pointing out some principles in the data handling. More emphasis is put on the unavailability estimation which yields the probability values that are used in the PRA.

2. CCF METHODOLOGY

2.1 Overview

The analysis of dependencies has consisted of the following subtasks:
A. Designed dependencies
    A.1 Initiator dependencies
    A.2 System dependencies and interactions
B. Statistical dependencies
    B.1 Dependent/repeated human errors
    B.2 Plant-specific hardware dependencies
    B.3 Residual parametric common cause failures.

The designed dependencies are to be taken into account in the normal process of logical modelling of the plant. This includes identification of specific initiating events that degrade one or more safety functions, development of event trees that account for mutual dependencies of safety systems, and linking fault trees for systems that depend on each other or common components for functional performance.

Statistical dependencies are not necessarily recognized or quantitatively accounted for in the design stage. Yet they can significantly increase the probability of multiple failures, which is the main motive for dependency analysis.

The human reliability assessment includes special consideration of factors affecting the likelihood of repeated errors in maintenance activities [5].

The analysis of plant-specific hardware dependencies covers mainly dependencies caused by physical phenomena, like risen temperature, humidity, vibration etc. which increase failure probability in other systems or components [6]. Cascade and propagating failures are also covered as well as such human errors, which are not covered by a systematic study of scheduled periodic tests or maintenance measures [5]. Cascade failures occur when one equipment failure causes changes in operating conditions, environments or requirements to cause another item or items of equipment to fail or to increase the failure probability, which might fail a third item and so on. Examples of such cases are missiles from a failing turbine, steam jets or a leakage in a room where a flood may cover and fail a pump.

Residual parametric common cause failures are dependent simultaneous multiple failures that have not been specifically identified and quantified as vulnerabilities through plant-specific walk-throughs in a normal fault tree modelling process. Their causes are typically a priori unforeseen events, conditions or phenomena. They affect normally identical redundant components within each system. The focus of this paper is in the estimation of these probabilities.

2.2 Factors of unavailability

One basic observation even with single failures is that they occur at random times, mostly due to time-related stresses (corrosion, temperature, humidity, sticking, loosening, wear) rather than stresses associated with true demands (initiating events) or test demands. Consequently, they are modelled by failure rates λ_j, probability of failure of component j per unit time, rather than by probability per demand. When the test interval is T, the basic event probability $z_j = Pr(Z_j) = \frac{1}{2} \lambda_j T$ is the average probability for single-failed state of component j. When used in a fault tree model it approximately yields the system average unavailability. In a more exact quantification repair time and logistic delay contributions have to be taken into account, approximately

$$u = q + \tfrac{1}{2}\lambda T + \lambda\tau + p\tau/T + f\tau + h(T), \qquad (1)$$

where q = probability of a demand (startup and operation stress) to cause a critical failure
  λ = rate of critical failures caused by stresses during standby and revealed by tests; probability per unit time
  T = test interval
  τ = restoration time including repair time and logical delays
  p = probability of a demand to cause a non-critical failure
  f = rate of non-critical failures caused by stresses during standby
  $h(T)$ = human error probability, some may depend on T.

For simplicity the terms like $\lambda\tau/(1+\lambda\tau)$ are replaced by the usual approximation. Some earlier studies [7, 8, 9] have indicated that the standby failure rate term $\frac{1}{2}\lambda T$ typically dominates for motor-operated valves and diesel generators, with virtually zero q, and f about 3 to 6 times larger than λ, and similarly p larger than q.

In addition to the unavailability according to Eq. 1 immediately detected failures cause an unavailability $\lambda_o(\tau+T_o)$ where T_o is the mission time. These are typically modeled as "fail to run" or "fail to remain open" or corresponding running time failures.

2.3 Equivalent observations

With similar arguments the common cause failures occur at random times and are modelled by general multiple-failure rates λ_i, λ_{ij}, λ_{ijk} etc. so that $\lambda_{ij}\,dt$ is the failure probability of exactly components $i, j, ..$ in dt due to a common cause. Such rates can easily be estimated directly from the observed total number of k/n-events, $N_{k/n}$, over the system observation time T_n: $\Lambda_{k/n} = \binom{n}{k}\lambda_{k/n} \sim X^2(2Nk/n + 1)/(2T_n)$. Here we assume symmetry, i.e. $\lambda_{ij} = \lambda_{2/n}$, $\lambda_{ijk} = \lambda_{3/n}$ etc. for all i, j, k etc. The mean value and variance of this gamma distribution are

$$E(\Lambda_{k/n}) = [N_{k/n} + \alpha]/T_n, \quad \sigma^2(\Lambda_{k/n}) = [N_{k/n} + \alpha]/T_n^2, \quad (2)$$

where $0 < \alpha \leq \frac{1}{2}$, usually $\alpha = \frac{1}{2}$. Unfortunately, due to uncertainties in observations and interpretations each event is not with certainty known to be a certain k/n-event. One has to assess for each plant v for each observation i the impact vector weights $w_{k/n}(i, v)$ = probability (conditional on symptoms) that in observation i at plant v exactly k components out of n identical components failed due to a common cause (CCF–group size n).

Assessing and quantification of these weights are based on component degradations, shared causes and timing to measure the simultaneity of the failures. Vaurio [10] has shown that the moments for plant v in terms of the weights are

$$E(\Lambda_{k/n}) = [\sum_{i=1}^{N_n} w_{k/n}(i,v) + \alpha]/T_n, \quad \sigma^2(\Lambda_{k/n}) = \{\sum_{i=1}^{N_n} w_{k/n}(i,v)[2 - w_{k/n}(i,v)] + \alpha\}/T_n^2, \quad (3)$$

where N_n is the total number of events in time T_n for plant v. The variance can be written in the form

$$\sigma^2(\Lambda_{k/n}) = \left[\sum_{i=1}^{N_n} w_{k/n}(i,v) + \alpha\right]/T_n^2 + \{\sum_{i=1}^{N_n} w_{k/n}(i,v)[1 - w_{k/n}(i,v)]\}/T_n^2$$

where the first term corresponds to the statistical variance and the second term can be viewed as the subjective variance. Making Eqs. 2 and 3 equal yields the "equivalent" data pairs $N_{k/n}$ and T_n that without uncertainties would give the same moments

$$\tilde{N}_{k/n}(v) = \frac{(\sum_{i=1}^{N_n} w_i)^2 + \alpha \sum_{i=1}^{N_n} w_i^2}{\sum_{i=1}^{N_n} w_i(2 - w_i) + \alpha}, \quad \tilde{T}_n(v) = \frac{\sum_{i=1}^{N_n} w_i + \alpha}{\sum_{i=1}^{N_n} w_i(2 - w_i) + \alpha} T_n, \quad (4)$$

where N_n is the total number of events at plant v in time T_n and $w_i = w_{k/n}(i, v)$. Due to the variance of $N_{k/n}$ the virtual observation time is shorter than the real time T_n.

This CCF estimation methodology applying the idea of equivalent observations has been extended to cover cases in which multiple events of different multiplicities k/n are allowed [11, 4], because there is a need to accept impacts of even two events of the same multiplicity k in a single observation. How this affects the equations can be found from [4].

2.4 Empirical Bayes estimation

The equivalent data pairs for selected plants v are input to a robust parametric moment matching method that yields the population distribution of the rate $\Lambda_{k/n}$ of k/n–events for the whole plant population. This is the empirical prior distribution used in the Robust Parametric Empirical Bayes (PREB) estimation process [12, 13] to obtain the posterior distribution of $\Lambda_{k/n}(v)$ for the target plant. The computerised procedure calculates the posterior distributions for all plants included in the prior calculation, not only for the target plant. The distribution of the rate of *specific* k failures out of n, $\lambda_{k/n}(v) = \Lambda_{k/n}(v)/\binom{n}{k}$ is obtained easily by dividing the mean value and the standard deviation by the Binomial factor. The key idea in PREB is to use a biased positive variance estimate to guarantee realistic solutions with all kinds of data, even for small samples with small empirical variance. PREB gives efficiently realistic results with all kinds of data.

The PREB method has been developed also for probabilities per demand, $Q_{k/n}$, in which case $T_n(v)$ is the total number of system–demands (opportunities) [14, 13]. Then the parameters $q_{k/n} = Q_{k/n}/\binom{n}{k}$ are directly the basic event probabilities, but this approach ignores the dependencies on detection methods, test interval and test staggering, which may be important for optimisation.

2.5 Quantification tasks

The quantification tasks that are needed for estimating CCF probabilities for Loviisa PRA are presented in Fig. 1. This procedure is slightly modified from the one presented by Vaurio [4]. At first the database is selected. Then we select a specific system and component type for the analysis. The target plant has a certain number n of redundant trains or components in that system. This is the size of the Common Cause Component Group (CCCG).

Figure 1: CCF quantification procedure

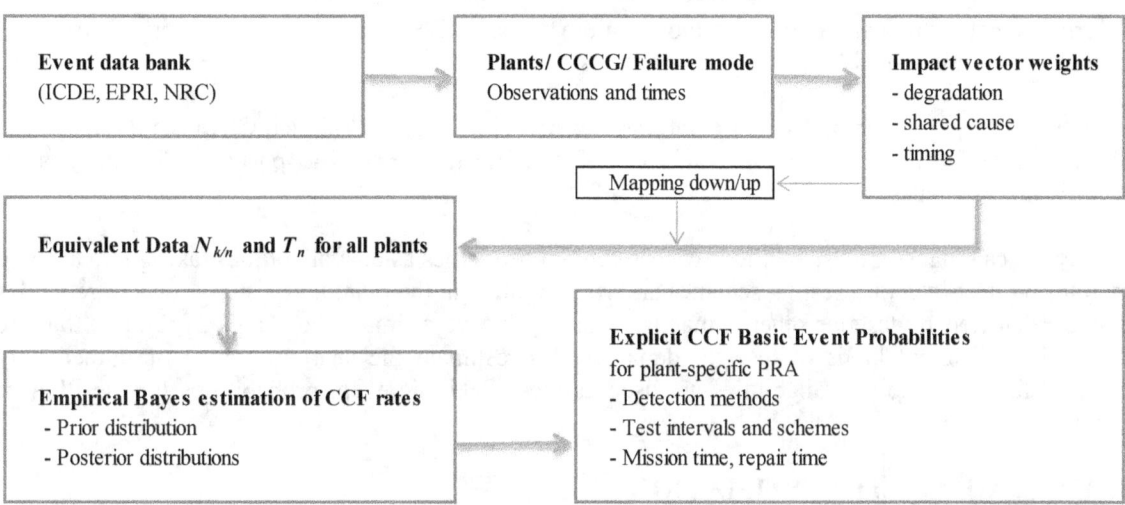

The next task is to select a set of relevant plants and systems, i.e. CCCG that has components similar enough to those of the target plant. We define as detailed CCCG as reasonable, e.g. centrifugal pumps are divided into several groups like high pressure safety injection pumps, low pressure safety injection pumps, containment spray pumps, auxiliary feed water pumps, component cooling water pumps and service water pumps. Especially service water pumps and component cooling water pumps have very different failure histories from the other pumps. For each of those pump groups we estimate specific CCF probabilities.

The next step is to select the failure mode that we are interested in. Fail to start and fail to run as well as fail to open and fail to close can have completely different failure cause, entry and detection mechanisms. Therefore these different failure modes can have different failure rates and different unavailability factors leading to different unavailability times.

In order to avoid mapping up and mapping down we take into account only such CCCG's that have the same redundancy level n as the plant under study. If data is too scarce in this subset, an option is to accept data from all plants and use "mapping up" (when $n' < n$) and "mapping down" (when $n' > n$) rules to obtain weights supposed to be valid for plants with the same system size n as the target plant [15]. This option, especially mapping up is avoided for Loviisa PRA, because it is based on some severe assumptions about the external nature of causes, and assume the same frequencies and consequences of cause events, independent of the system size n.

Then for each observation we determine the impact vector weights $w_{k/n}(i, v)$ for $k = 2, 3,.., n$ by studying the event descriptions and other relevant information like degradation values, shared cause factors and time factors. We determine them for the plant where the event took place, not for the plant for which the CCF probability is to be estimated. In estimating the weight probabilities the weighting procedure by Vaurio [4] is applied. The method is slightly modified from [16]: More than one cause-event can be involved in one observation, and component degradations are not assumed mutually independent but coupled (e.g. the smallest degradation rather than the product of degradations determines the probability of a complete n/n-failure).

After defining the impact vector weights we estimate effective data pairs (K_i, T_i) for each CCCG (plant), failure mode and failure multiplicity.

In the next step PREB calculations are performed separately for each CCCG and failure multiplicity to obtain posterior CCF rates with uncertainties for the plant under study and for all other plants as well. We obtain also the sampling (prior) distribution that describes the variability between all the plants. The rates of specific k failures out of n are obtained by dividing with the binomial factor: $\lambda_{k/n}(v) = \Lambda_{k/n}(v)/\binom{n}{k}$. The PREB procedure yields individual failure rates as posterior distributions, which describe the statistical variabilities of those failure rates. PREB gives also the sampling distribution from which all individual failure rates are samples.

Thus, the data pairs, numbers of observations and observation times, of each CCCG is the only data that is needed in this estimation to obtain CCCG-specific failure rates as posterior distributions for specific failure modes.

Finally we calculate CCF probabilities of different multiplicities as unavailabilities taking into account monitoring, test intervals, testing schemes and rules applied in the plant under study for failure modes that are detected by tests or other demands. That is why we estimate the shares of failures that are detected in different kinds of tests or demands. We estimate the shares or rates of immediately detected failures. The unavailabilities are used as the CCF basic event probabilities for the PRA of Loviisa NPP.

3. UNAVAILABILITY ESTIMATION

3.1 CCF unavailability

Like single failures common cause failures have also different kinds of restoration times. Some of the failures are immediately detected during standby or mission, some in demands at scheduled tests, maintenance tests or other start-ups. Scheduled tests at plant shutdown annually or more rarely can be more profound and reveal other kinds of failures than the more frequently performed tests during power operation. Tests can be staggered or consecutive. In case of a staggered testing scheme a complete common cause failure can only have a latent unavailability time that is the test interval of one train divided by the number of redundant trains.

When we consider only failures that are detected by tests with test interval T we define the probabilities $z_{ij..}$ of the basic CCF-events $Z_{ij..}$ (failing exactly specific k components $i, j,..$ out of n similar components) needed in the system fault tree. For standby safety components tested with test interval T these values are

$$P(Z_{ij..}) = c_{k/n}\lambda_{k/n}T, \tag{5}$$

where $0 < c_{k/n} < 1$. The coefficients $c_{k/n}$ depend on k, n, test staggering, repair policy and the system success criterion [17, 4]. They can be determined so that correct time-average risk is obtained by a single fault tree calculation. In case of sequential or simultaneous testing the average residence time of the failures would be approximately one half of the test interval, which gives us a general approximation similar to the single failure practice, $c_{k/n} = ½$, for $n = 1, 2, 3,...$ and $1 \le k \le n$.

When we consider also other detection methods we obtain a more general expression for the probabilities $z_{ij..}$

$$P(Z_{ij..}) = a \cdot \lambda_{k/n} \cdot (c_{k/n} \cdot T + \tau) + b \cdot \lambda_{k/n} \cdot (T_L/2 + \tau) + \lambda_{dk/n} \cdot (T_o + \tau) + \lambda_{ok/n} \cdot (T_o + \tau), \quad (6)$$

where
- $\lambda_{k/n}$ = the rate of multiplicity k/n failures (detected in periodic tests),
- a = share of failures that are detected by scheduled tests during power operation,
- b = share of failures that are detected in broader tests that are performed during plant shutdown,
- $c_{k/n}$ = coefficient that depends on k, n, test staggering, repair policy and the system success criterion,
- T = test interval during power operation (typically 4 weeks or 12 weeks),
- T_L = test interval of broader, consecutive tests, typically one year or longer,
- $\lambda_{dk/n}$ = the rate of multiplicity k/n failures that are immediately detected,
- $\lambda_{ok/n}$ = the rate of running time multiplicity k/n failures,
- T_o = mission time.

The share of failures that are detected in broader tests that are performed during plant shutdown maybe split into two parts: those that are only detected in the annual tests and those that are detected in a different broad test after maintenance or due to some other reason. These latter ones are important but difficult to estimate, because the latent time can be long, even several years, and it can vary a lot. One possibility is to estimate an average latent time based on the CCF experiences of a component type.

For some components $\lambda_{dk/n}\tau$ -term can be dominating like for measurements and automation modules but typically the contribution of $\lambda_{dk/n}\tau$ is small. Those failures tend to be more often complete common cause failures and that is a good reason to estimate specific rates for them also in case of other components. Their observation time is the same as the calendar time from the beginning to the end of the observation period. An example of such a case is the diesel generator failure due to a spurious signal described in the introduction. It was a complete failure but it had a very short unavailability time.

If there is not enough data to estimate $\lambda_{dk/n}$ it is possible to estimate the overall CCF rate $\lambda_{k/n}$ that covers failures detected in periodic tests or demands and immediately detected. Then these different contributions are estimated by using shares that are based on CCF histories and/or on plant specific data on single failures.

The contribution of $\lambda_{ok/n}$ cases can be large because their observation time is only the running time of the component, which is usually very short for standby components. This is modeled as separate basic events representing running failure modes. Because of the short running times it is not always possible to estimate specific rates for them and they are included e.g. in the fail to start estimation.

It is also possible that a failure is detected in a scheduled test and it disappears in a short time. For example two out of four Loviisa diesel generators had a relay jamming failure that disappeared by itself after 14 seconds in the first case and after 150 seconds in the second case 2 weeks later. One diesel generator has a test interval of four weeks and the testing scheme is staggered. Thus, in this case the CCF was effective only 14 seconds. In case of Loviisa so short unavailability does not prevent an emergency operation, except maybe in case of a large LOCA, and therefore this case must not be taken into account as a CCF. The replacement times of the relays are then taken into account as single failure unavailabilities. If such a short time was critical this case must be considered as a CCF and the test interval contribution must be taken into account in the unavailability estimation. Similar single event cases were found from other diesel generators and those relays were found to be from the same batch. Therefore eventually we consider this case of two failures to be an incipient CCF having an impairment vector IIWW, where I refers to Incipient and W refers to Working.

The non-critical failure terms $p\tau/T$ and $f\tau$ of Eq. 1 can be neglected in the CCF unavailability estimation. Incipient failures are such defects that a component is still able to perform its intended function (if needed), but the repair itself takes the component out of service, and this is how we quantify such unavailabilities for single failures: $f\tau/(1+f\tau)$. Correspondingly we could consider the same approach for CCF events. However, redundant components are not repaired at the same time during power operation. Only single incipient failures can cause unavailability when they are repaired. CCF events are considered to be incipient when at least two incipient failures due to a same reason have been found. If the other redundant components have also been repaired or replaced it is a rule in the ICDE database to classify them as incipiently failed. This affects the impact vector as a degradation value of 0.1. Thus, the incipient failures are taken into account in the CCF rate estimation.

The average values according to Eq. 6 are applied in the power operation PRA. In the shutdown PRA we take into account the latest testing time of each component prior to each plant operating state. So instead of the term $c_{k/n} T$ we use time from the previous test. This is more realistic than assuming the same constant unavailabilities as in full power operation. However, 10 % of the average unavailability is assumed to represent a constant unavailability term q (see Eq. 1) that is otherwise neglected. Thus, the failure probabilities vary from state to state and can differ considerably from those of power operation. Components that have been tested already a year ago have two times higher unavailabilities than in the average in power operation. Components that have been tested in the previous plant operating state might have an order of magnitude smaller unavailability value because the time-related unavailability is minimal.

Restoration times τ are estimated based on the single failure experiences of the target plant. In this way they represent the maintenance practices in the target plant. The last term of Eq. 1, $h(T)$, is not presented in Eq. 6 because it is not estimated in the process of CCF unavailability quantification based on histories, but as a part of human reliability assessment (see Ch. 2.1).

3.2. Detection methods

Failures of standby safety systems are usually considered to be detected by scheduled tests. This is not always true. In case of Loviisa NPP we have studied how the events have been detected in the failure histories of the safety systems. The first unit of Loviisa NPP has been in operation since 1977 and the second unit since 1980. Scheduled tests were found to be the most important way of detection, except in case of measurements.

The following detection methods have been recorded in the maintenance history: 1) monitoring in the control room from symptoms or alarms, 2) direct observation immediately after failure, 3) monitoring on shift walk down, 4) scheduled test, 5) scheduled maintenance, 6) demand event, 7) QA/C inspection, condition monitoring, 8) repair or trial run after repair, 9) random check. Altogether in the history scheduled tests revealed 68 % of all the failures, demand events revealed 7 %, monitoring in control room found out 14 % and other immediate detections 5 %. The results [18] are presented in Table 1 classified according to component types and in Table 2 according to systems.

Table 1: Detection methods of component types

Detection method	1 Imm.	2 Imm.	3 Shift	4 Test	5 Maint.	6 Dem.	7 Insp.	8 Repair	9 Rand.
Alternating pumps	4.5 %	4.5 %	0.0 %	68.2 %	0.0 %	22.7 %	0.0 %	0.0 %	0.0 %
Standby pumps	5.4 %	1.8 %	1.8 %	80.4 %	1.8 %	1.8 %	1.8 %	3.6 %	1.8 %
Diesel generators	12.6 %	5.3 %	1.1 %	72.6 %	2.1 %	0.0 %	1.1 %	5.3 %	0.0 %
Check valves	0.0 %	0.0 %	0.0 %	100 %	0.0 %	0.0 %	0.0 %	0.0 %	0.0 %
Isolation valves	0.0 %	7.1 %	0.0 %	92.9 %	0.0 %	0.0 %	0.0 %	0.0 %	0.0 %
Relief valves	14.3 %	14.3 %	14.3 %	57.1 %	0.0 %	0.0 %	0.0 %	0.0 %	0.0 %
Motor operated valves	15.8 %	5.5 %	0.7 %	66.3 %	0.7 %	10.0 %	0.3 %	0.3 %	0.3 %
Control valves	36.7 %	10.0 %	3.3 %	40.0 %	0.0 %	6.7 %	0.0 %	3.3 %	0.0 %
Pressure and level m.	78.7 %	1.9 %	5.2 %	5.8 %	3.2 %	1.3 %	1.3 %	0.6 %	1.9 %
Temperature m.	88.9 %	0.0 %	5.6 %	0.0 %	0.0 %	2.8 %	2.8 %	0.0 %	0.0 %

Imm. = immediate, Maint.= maintenance, Dem.=demand, Insp.=Inspection, Rand.=random

Table 2: Detection methods of systems

Detection method	1 Imm.	2 Imm.	3 Shift	4 Test	5 Maint.	6 Dem.	7 Insp.	8 Repair	9 Rand.
Main steam	4.9 %	4.9 %	4.9 %	80.5 %	0.0 %	0.0 %	0.0 %	2.4 %	2.4 %
Feed water	40.2 %	2.4 %	1.2 %	37.8 %	0.0 %	17.1 %	0.0 %	1.2 %	0.0 %
Residual heat removal	20.0 %	20.0 %	0.0 %	40.0 %	0.0 %	20.0 %	0.0 %	0.0 %	0.0 %
Secondary makeup	28.6 %	28.6 %	4.8 %	28.6 %	0.0 %	9.5 %	0.0 %	0.0 %	0.0 %
Primary coolant purification	0.0 %	0.0 %	0.0 %	100 %	0.0 %	0.0 %	0.0 %	0.0 %	0.0 %
Component cooling water	2.3 %	2.3 %	0.0 %	84.1 %	0.0 %	9.1 %	0.0 %	2.3 %	0.0 %
LPSI[a]	3.8 %	0.0 %	0.0 %	90.6 %	3.8 %	1.9 %	0.0 %	0.0 %	0.0 %
HPSI[b]	8.3 %	0.0 %	0.0 %	87.5 %	0.0 %	0.0 %	4.2 %	0.0 %	0.0 %
Primary make-up	27.8 %	13.9 %	2.8 %	36.1 %	0.0 %	13.9 %	0.0 %	5.6 %	0.0 %
Containment spray	2.7 %	0.0 %	0.0 %	95.9 %	0.0 %	1.4 %	0.0 %	0.0 %	0.0 %
Service water	8.6 %	14.3 %	2.9 %	48.6 %	0.0 %	25.7 %	0.0 %	0.0 %	0.0 %

Imm. = immediate, Maint.= maintenance, Dem.=demand, Insp.=Inspection, Rand.=random
a) Low Pressure Safety Injection   b) High Pressure Safety Injection

The differences between such component types as pumps, diesel generators and valves are rather small. The role of scheduled tests is only 6 % in case of measurement failures. Monitoring in control room found out 78 % in case of measurement failures. On the other hand the differences between systems are statistically significant (Table 2). An essential factor is the usage of the system. Scheduled tests are important for standby systems, but not so important for operating systems.

The detection methods of single failures can be different from those of CCFs. The role of more profound or exceptional tests can be more important with CCFs. For example in Loviisa NPP the testing procedure of high pressure safety injection pumps was changed in 1993 leading to increased test durations. This revealed a CCF that caused the temperatures of the motor bearings to rise unacceptably high. Therefore it is important to study the detection methods of the CCF events even if the testing procedures were different from those of the plant under study. If the CCF database has CCF events with long latent times we take those into account and estimate their shares on the basis of CCF data, unless there is a very good reason to believe that the plant under study has better tests or detection methods. We quantify the shares of annual overhaul tests and exceptional more rare tests according to the CCF data.

Concerning immediately detected failures we estimate their shares, typically for valves, or we estimate separately their failure rates, typically for some pumps. Monitoring on shift walk down is treated as an immediate detection in the unavailability estimation, because the latent time is so short. When estimating a failure rate it is important to take into account the correct observation time. For running time failures only the running time is considered, whereas for standby time failures the calendar time is considered. One should also notice that running failures are immediately detected. This must be taken into account when estimating and utilizing the shares of detection methods.

The detection methods of the common cause failures in ICDE database seem not to differ too much from those of the Loviisa failure history. Therefore we consider that we can utilize Loviisa shares of detection methods when estimating the CCF unavailabilities for Loviisa NPP, except in case of failures found out by more profound tests. Long latent times are estimated based on CCF histories.

3.3. Examples

An example of check valve CCF probabilities is given in Table 3 concerning low pressure safety injection (LPSI) system pump discharge lines. The PRA model has a basic event for each CCF event presented in the left column. Failure multiplicity and the corresponding failure rate is given in the next two columns. Then the coefficient $c_{k/n}$ and testing scheme are given. The last column gives the unavailability value that is used as the corresponding basic event probability. These components have a staggered testing with a test interval of four weeks and success criterion 1/4. The share of the immediately detected failures is 4 %, 95 % in monthly tests and the share of failures detected only in more rare tests, in this case annual tests, is 1 %. In this case the number of CCF events is so small (17) that we have pooled histories of all systems together, including events from reactor core isolation cooling, auxiliary and emergency feed water, residual heat removal, chemical and volume control, emergency core cooling, component cooling water and essential service water. Concerning the emergency core cooling system this approach is conservative. The rate estimates are

- $1.76 \cdot 10^{-8}$ ($2.45 \cdot 10^{-7}$) for 2/4 failures,
- $1.18 \cdot 10^{-8}$ ($2.07 \cdot 10^{-7}$) for 3/4 failures and
- $9.74 \cdot 10^{-9}$ ($1.62 \cdot 10^{-7}$) for 4/4 failures

per hour with standard deviations shown in the parentheses. The shares of the detection methods were estimated based on all the 88 check valve CCF events, not only fail to open cases. The $u_{m/n}$ values would be 15…27 % lower if we assumed that all failures are detected by scheduled tests.

Table 3: CCF Unavailabilities of LPSI Pump Discharge Line Check Valves

CCF event	m/n	$\lambda_{k/n}(v)$ /h	$c_{k/n}$	Test	$u_{m/n}$
TH11/12S02 fail to open	2/4	$2.94 \cdot 10^{-9}$	0.25	Staggered	$6.68 \cdot 10^{-7}$
TH11/51S02 fail to open	2/4	$2.94 \cdot 10^{-9}$	0.25	Staggered	$6.68 \cdot 10^{-7}$
TH11/52S02 fail to open	2/4	$2.94 \cdot 10^{-9}$	0.25	Staggered	$6.68 \cdot 10^{-7}$
TH12/51S02 fail to open	2/4	$2.94 \cdot 10^{-9}$	0.25	Staggered	$6.68 \cdot 10^{-7}$
TH12/52S02 fail to open	2/4	$2.94 \cdot 10^{-9}$	0.25	Staggered	$6.68 \cdot 10^{-7}$
TH51/52S02 fail to open	2/4	$2.94 \cdot 10^{-9}$	0.25	Staggered	$6.68 \cdot 10^{-7}$
TH11/12/51S02 fail to open	3/4	$2.95 \cdot 10^{-9}$	0.178	Staggered	$5.34 \cdot 10^{-7}$
TH11/12/52S02 fail to open	3/4	$2.95 \cdot 10^{-9}$	0.178	Staggered	$5.34 \cdot 10^{-7}$
TH11/51/52S02 fail to open	3/4	$2.95 \cdot 10^{-9}$	0.178	Staggered	$5.34 \cdot 10^{-7}$
TH12/51/52S02 fail to open	3/4	$2.95 \cdot 10^{-9}$	0.178	Staggered	$5.34 \cdot 10^{-7}$
TH11/12/51/52S02 fail to open	4/4	$9.74 \cdot 10^{-9}$	0.125	Staggered	$1.44 \cdot 10^{-6}$

Another example of check valves of LPSI system is given in Table 4 concerning accumulator line check valves. In this case the test interval is two years so that two of the four valves are tested in the same outage, so that they have sequential testing. For the other valve combinations the testing scheme is staggered, as shown in Table 4. For those two valve CCF combinations with sequential testing the unavailability estimate is more than two times higher. The shares of the detection methods are the same as above assuming that the more rare tests are two times longer. This estimate is not considered to be optimistic even if in the CCF history of check valves the long latent time was in the average three times longer than the test interval. The $u_{m/n}$ values would be 1…4 % higher if we assumed that all failures are detected by scheduled tests.

Table 4: CCF unavailabilities of LPSI accumulator line check valves

CCF event	m/n	$\lambda_{k/n}(v)$ /h	$c_{k/n}$	Test	$u_{m/n}$
TH41/42S01 fail to open	2/4	$2.94 \cdot 10^{-9}$	0.25	Staggered	$1.28 \cdot 10^{-5}$
TH41/81S01 fail to open	2/4	$2.94 \cdot 10^{-9}$	0.577	Sequential	$2.88 \cdot 10^{-5}$
TH41/82S01 fail to open	2/4	$2.94 \cdot 10^{-9}$	0.25	Staggered	$1.28 \cdot 10^{-5}$
TH42/81S01 fail to open	2/4	$2.94 \cdot 10^{-9}$	0.25	Staggered	$1.28 \cdot 10^{-5}$
TH42/82S01 fail to open	2/4	$2.94 \cdot 10^{-9}$	0.577	Sequential	$2.88 \cdot 10^{-5}$
TH81/82S01 fail to open	2/4	$2.94 \cdot 10^{-9}$	0.25	Staggered	$1.28 \cdot 10^{-5}$
TH41/42/81S01 fail to open	3/4	$2.95 \cdot 10^{-9}$	0.178	Staggered	$9.32 \cdot 10^{-6}$
TH41/42/82S01 fail to open	3/4	$2.95 \cdot 10^{-9}$	0.178	Staggered	$9.32 \cdot 10^{-6}$
TH41/42/82S01 fail to open	3/4	$2.95 \cdot 10^{-9}$	0.178	Staggered	$9.32 \cdot 10^{-6}$
TH42/81/82S01 fail to open	3/4	$2.95 \cdot 10^{-9}$	0.178	Staggered	$9.32 \cdot 10^{-6}$
TH41/42/81/82S01 fail to open	4/4	$9.74 \cdot 10^{-9}$	0.25	Staggered	$4.25 \cdot 10^{-5}$

Some safety related components have considerable running times. For example Lo NPP has four essential service water and four component cooling water pumps of which two are running and two are standby. The running time is two weeks and then the pump is stopped and the redundant pump is taken into use. For them we have modelled separately fail to run and fail to start failures: six 2/4 failures, four 3/4 failures and a 4/4 failure for both failure modes. In case of fail to run of the service water pumps the rate estimates are

- $1.43 \cdot 10^{-7}$ ($1.03 \cdot 10^{-6}$) for 2/4 failures,
- $1.50 \cdot 10^{-7}$ ($6.43 \cdot 10^{-7}$) for 3/4 failures and
- $1.31 \cdot 10^{-7}$ ($8.73 \cdot 10^{-7}$) for 4/4 failures

per hour with standard deviations shown in the parentheses. The corresponding unavailability is obtained by multiplying by the mission time of 24 hours. In case of fail to start we take into account the test intervals of Loviisa NPP and the essential detection methods for quantification are considered to be: 24 % immediately detected, 75 % detected in scheduled tests and demands, 1 % detected only annually. These long latent times, e.g. due to design errors, require special conditions, because pumps are started several times per year in normal conditions and failures are revealed. Special conditions are such that the pumps should operate in them according to the technical specifications. The CCF histories include such long latent times in two and three line systems, less than 10 % of all CCF cases. Although Loviisa NPP has four pumps we take this possibility into account with 1 % value. In this way we obtain for the two standby pumps 2/4 fail to start event a probability $1.98 \cdot 10^{-5}$. This probability would be 20 % lower if we assume 0 % for long latent times and 76 % for scheduled tests according to Loviisa history and it would be five times higher if we assume 4 % for long latent times, 76 % for scheduled tests and demands and 20 % for immediate detection according to all CCF history of such pumps. If we assumed that all failures are detected by scheduled tests the estimate would be $1.97 \cdot 10^{-5}$, which is almost the same as the basic case.

This approach is useful for different kinds of PRA applications. For example the knowledge of the effect of the test intervals on the basic event probabilities has been utilized in the risk-informed evaluation of test intervals and limiting conditions of operation.

4. CONCLUSIONS

CCF events of safety system components are mostly detected by testing during plant operation or during annual overhaul. In CCF quantification this is often regarded as the only way of detection. This leads into CCF unavailability quantification in which the CCF rate is based on all kinds of CCF events having outage times from minutes to years. This approach might yield overly conservative CCF unavailability estimates or sometimes optimistic.

Plant-specific testing methods and schemes have to be taken into account. Their effectiveness can be studied by analyzing single failure data of the target plant in addition to the common cause failure histories of a world-wide database. Scheduled tests were found to be the most important way of detection, except in case of measurements and automation. The results of this study indicate that there are differences in the detection methods not only between component types but especially between different systems.

Measurement failures are usually found out immediately via alarms or symptoms in the control room. Scheduled tests play an important role in finding out failures of conventional components like valves, pumps and diesel generators. However, the systems in which these components are used affect significantly to the detection methods. Detection methods are different in systems that are normally in operation compared to the systems that are normally standby.

If we only have events with immediate detection does not mean that other kinds of events would have zero failure rate. Failures that are detected in the annual overhaul or more rare and profound tests can affect the CCF probability more even if their rates were thousand times smaller.

Failure modes and detection methods have to be taken into account already in the estimation of CCF rates, not only when estimating the CCF unavailabilities.

This approach of estimating CCF probabilities by estimating CCF rates and accounting for failure entering and detection methods when estimating unavailabilities facilitates many PRA applications. This approach does not assume anything about single failure probabilities and CCF dependency on them, what is not shown in the CCF histories. We can think that common cause failure probabilities are not dependent on single failure probabilities.

Changes in the test intervals and in the testing schemes affect the CCF probability estimates in a consistent way. CCF probabilities can be estimated more correctly for different shutdown plant operating states taking into account that the components have been tested some hours ago.

References

[1] *"International common cause failure data exchange (ICDE)"*. NEA/CSNI/R(2011)12, OECD Nuclear Energy Agency, Committee on the Safety of Nuclear Installations, February 2012.
[2] J.K. Vaurio. *"From failure data to CCF-rates and basic event probabilities"*. Proceedings ICDE seminar and workshop on qualitative and quantitative use of ICDE data, 12-13 June 2001, Stockholm. Report NEA/CSNI/R(2001)8, OECD Nuclear Energy Agency, CSNI 2012.
[3] J.K. Vaurio & K.E. Jänkälä. *"Quantification of common cause failure rates and probabilities for standby-system fault trees using international event data sources"*, Proceedings of PSAM 6 Conference, San Juan, Puerto Rico, 23-28 June, 2002. Amsterdam, Elsevier.
[4] J.K. Vaurio. *"Uncertainties and quantification of common cause failure rates and probabilities for system analyses"*, Reliability Engineering and System Safety, 90, pp. 186-195, (2005).
[5] J. K. Vaurio & U. M. Vuorio. *"Human Reliability Assessment in Loviisa 1 PSA"*, Probabilistic Safety Assessment and Management, Editor G. Apostolakis, Elsevier, 1991, New York. Pp. 841-846.
[6] K. E. Jänkälä and J. K. Vaurio, *"Dependent failure analysis in a PSA"*, Probabilistic Safety Assessment and Management, Editor G. Apostolakis, Elsevier, New York. Pp.607-612
[7] E. V. Lofgren and M. Thaggard, *"Analysis of standby and demand stress failure modes,"* NUREG/CR-5823, 1992.
[8] U. Pulkkinen et al., *"Reliability of diesel generators in the Finnish and Swedish nuclear power plants"*, Research notes 1070, Technical Research Centre of Finland, 1989, Espoo.
[9] W. E. Vesely et al., *"Evaluation of diesel unavailability and risk effective surveillance test intervals"*, NUREG/CR-4810, USNRC, 1987.
[10] J. K. Vaurio, *"Estimation of Common Cause Failure Rates Based on Uncertain Event Data"*, Risk Analysis, 14, pp. 383-387, (1994).
[11] J.K. Vaurio, (2002). *"Extensions of the uncertainty quantification of common cause failure rates"*, Reliability Engineering and System Safety, 78, pp. 63-69, (2002).
[12] J.K. Vaurio, *"On analytic empirical Bayes estimation of failure rates"*, Risk Analysis, 7, pp. 329-338, (1987).
[13] J.K. Vaurio & K.E. Jänkälä. *"Evaluation and comparison of estimation methods for failure rates and probabilities"*, Reliability Engineering and System Safety, 91, pp. 209–221, (2006).
[14] K.E. Jänkälä and J.K. Vaurio, *"Empirical Bayes data analysis for plant specific safety assessment"*, Probabilistic Safety Assessment and Risk Management PSA'87, Volume I, 281-286, Verlag TÜV Rheinland GmbH, Köln. Proc. of PSA'87, Zurich, Switzerland, Aug. 30 to Sep. 4, 1987.
[15] J.K. Vaurio, *"Consistent mapping of common cause failure rates and alpha factors,"* Reliability Engineering and System Safety, 92, pp. 628-645, (2007).
[16] T.E. Wierman, D.M. Rasmuson and A. Mosleh, *"Common-cause failure database and analysis system: event data collection, classification and coding,"*. NUREG/CR-6268, Rev. 1. US NRC, 2007. Original version published in June 1998.
[17] J.K. Vaurio, *"Implicit and explicit modelling and quantification of dependencies in system reliability and availability analysis"*, EN A - 48, Lappeenranta University of Technology, 2000.
[18] R. Kleinberg, *"Turvallisuudelle tärkeiden laitteiden koestusten merkitys vikojen havaitsemisessa,"* Bachelor's thesis (in Finnish), Aalto University, Espoo 2012, Finland.

Time Dependent Analysis with Common Cause Failure Events in RiskSpectrum

Pavel Krcal[a,b] and Ola Bäckström[a]
[a] Lloyd's Register Consulting, Stockholm, Sweden
[b] Uppsala University, Uppsala, Sweden

Abstract: Testing of components with common cause failures presents a challenge to a realistic analysis of failure probabilities. In reality, the most commonly used testing scheme is staggered testing. Common Cause Failure (CCF) models in Probabilistic Safety Assessment (PSA) studies often assume a sequential testing scheme. This might be overly conservative if the actual testing scheme is staggered. Some software tools, e.g., RiskSpectrum, offer time dependent analysis where one can model testing of components in time explicitly. This paper deals with effects of different testing schemes on the quantification of CCF events in time dependent analysis.

Determining which formulae shall be used by software tools in time dependent analysis requires an in-depth understanding of how to model effects of tests on the common cause parts of failures. We analyze assumptions which lie behind different ways of modeling tests of common cause failure events.

Keywords: PSA, Time Dependent Analysis, Common Cause Failures.

1. INTRODUCTION

Testing of components with common cause failures presents a challenge to a realistic analysis of failure probabilities. In reality, the most commonly used testing scheme is staggered testing. Common Cause Failure (CCF) models in Probabilistic Safety Assessment (PSA) studies often assume a sequential testing scheme. This might be overly conservative if the actual testing scheme is staggered. Some software tools, e.g., RiskSpectrum, offer a mean probability calculation and also time dependent analysis where one can model testing of components in time explicitly.

Quantification of common cause failure events depends on the chosen testing scheme, both in mean probability calculations and in a time dependent analysis. We focus mainly on the alpha model for common cause failure events in this paper. NUREG/CR-5485 [2] presents formulae for mean value quantification of common cause failure events with the alpha model under assumption of both sequential and staggered testing schemes. We discuss the following issue: How does an explicit modeling of a testing scheme in a time dependent analysis relate to the mean value estimates obtained by formulae from NUREG/CR-5485? This requires an in-depth understanding of how to model effects of tests on the common cause parts of failures.

This leads to more detailed questions about the analysis algorithm. Which formulae shall be used by software tools in a time dependent analysis? Under which assumptions are they correct? It is of great importance to understand the underlying assumptions when interpreting the numerical results in order to avoid taking an unjustified credit for top frequency decrease with staggered testing [3].

The main topic of this paper can be generalized to the following issue. The simplified models for staggering only consider staggering within one CCF group, but this is still a significant simplification since testing of important systems is often also staggered. Time dependent analyses provide a better solution, since these can also take staggering between different systems into account.

The paper is organized as follows. First, we present the background for common cause failure modeling and tested basic events. Then we formulate the main problem in Section 3. Section 4 shows the relation between time dependent analysis and mean value calculations. Section 5 discusses assumptions on which both methods are based. Finally, we conclude the paper.

2. BACKGROUND

The classical approach in PSAs assumes that basic failures modeled by basic events are independent of each other. Dependent failures are then either modeled explicitly by adding the root cause into the model as a new basic event or by defining so called common cause failures. Since its introduction in the sixties, common cause failure analysis became an integral part of PSA studies, with a mature methodology described in many standards and procedures [1,2,5].

2.1 Parametric models

Common cause failures are defined by parametric models. This means that we identify groups of basic events that might fail together as a result of a common cause. Then we choose a model and assign values to model parameters. The model together with the parameter values already define how to complete the fault tree model by so-called common cause failure events, and also defines how to quantify them. Fault tree analysis results will then reflect the contribution of common cause failures to the top failure. In fact, common cause failures often significantly contribute to core damage frequency.

Basic events included in the same common cause failure group have to be identical. In this paper, we assume that they have the same reliability model with the same reliability parameters.

Parametric models for common cause failure modeling divide into shock and non-shock models. Shock models assume that common cause failures result from the impact of an external shock which occurs with a given frequency. Non-shock models do not have any such assumption and directly determine probabilities of common cause failure events.

The most commonly used non-shock parametric models include the Beta Factor, the Alpha Factor, and the Multiple Greek Letters (MGL) model. The Beta Factor model is the simplest one, distinguishing only between independent failures and a common failure of all basic events from the common cause failure group. This model requires only one parameter value. The Alpha Factor and MGL models offer an extension of this model to all combinations of common cause failures. An advantage of the Alpha Factor model over the MGL model is that its parameters can be directly estimated from the failure data obtained from tests. This paper focuses on the Alpha Factor model, since it is recommended in, e.g., [2,5], and since other models can be transformed to it.

2.2 Alpha Factor model

The Alpha Factor model [7] for a group of m basic events is determined by m parameters denoted by $\alpha_1, \alpha_2, ..., \alpha_m$. Each parameter α_k is the fraction of failures that occur together in groups of k components in total failures. In other words, it is a "probability that when a common cause basic event occurs in a common cause group of size m, it involves failure of k components" [2]. Let us by Q_k^m denote probability of a common cause failure event representing that k components from a common cause failure group with m components fail because of a common cause involving these k components. Since components in a common cause failure group are assumed to be equivalent, this probability is well-defined. Now we can define α_k formally by

$$\alpha_k = \frac{\binom{m}{k} Q_k^m}{\sum_{k=1}^{m} \binom{m}{k} Q_k^m} \qquad (1)$$

We assume that the common cause failure events are mutually exclusive to each other (i.e., a common cause failure of components A and B models a situation where A and B fail together and components

C and D from the same common cause failure group function correctly) and that they add up together with independent failures to the original failure probabilities of the basic events included in the common cause failure group. Formally,

$$Q_t = \sum_{k=1}^{m} \binom{m-1}{k-1} Q_k^m \qquad (2)$$

Quantification of common cause failure events based on the alpha parameters and the independent basic event probability is derived from the two basic equations above. The probability of Q_k^m is determined by

$$Q_k^m = \frac{k}{\binom{m-1}{k-1}} \frac{\alpha_k}{\alpha_t} Q_t \qquad (3)$$

where

$$\alpha_t = \sum_{k=1}^{m} k\alpha_k \qquad (4)$$

Complete derivation is described in, e.g., [2], Appendix A.

So far, we did not consider any time dependent behavior of basic events. Equation 3 above is valid for mean value calculations. Now, let us consider basic events representing stand-by components which are periodically tested. Such basic events are defined in [4] by the following reliability model. This model assumes an exponential distribution for the failure process (constant failure rate λ), a constant fixed test interval (*TI*) with optional different time to first test (*TF*). To simplify the understanding of this model, we will first present it with only the failure rate and test interval parameters. The unavailability in this case is given by:

$$Q(t) = 1 - e^{-\lambda(t-T_i)} \qquad T_i = 0, TI, 2TI, \ldots \qquad (5)$$

This model results in the classical saw-tooth curve for the unavailability (Figure 1). If a *TF* (time to first test) parameter is given, the model is identical except that the time points for the tests are "offset" by the value *TF*, i.e. the test time points are $T_i = 0, TF, TF + TI, TF + 2TI, \ldots$

Figure 1. Graph of the unavailability over time of a tested basic event.

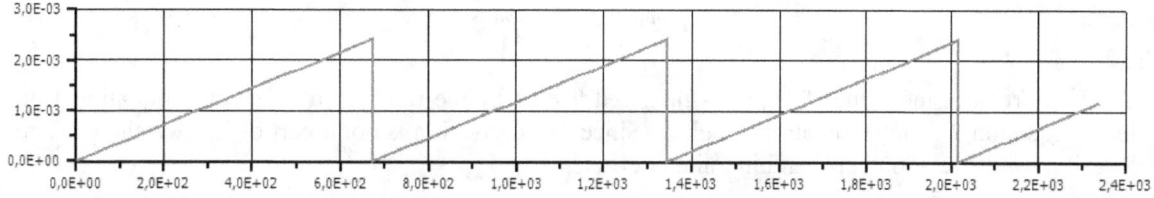

The mean unavailability Q_{mean} is obtained by integrating the unavailability $Q(t)$ over a complete test cycle:

$$Q_{mean} = \frac{1}{TI} \int_0^{TI} Q(t)dt = 1 - \frac{1}{\lambda TI}(1 - e^{-\lambda TI}) \qquad (6)$$

We assume that a repair occurs directly after a test if the component is failed at the test. The unavailability of the component immediately before a test, $Q(TI)$, is, in other words, the probability that a repair is needed after the test.

Clearly, testing influences the mean failure probability of such basic events. If these basic events form a common cause failure group then they are implicitly divided into several other basic events modeling common cause failures. Testing of such events is determined by testing of all components which are included in this common cause failure. As long as all components are tested synchronously at the same time points, this does not alter assumptions for the mean value formula above. When we test components at different time points then, from the logic of common cause failure modeling, each successful test also demonstrates that there is no common cause failure involving the tested component. This means that testing one component affects other components.

Testing components from one common cause failure group at different time points evenly spread over the testing calendar is called *staggered testing*. The test procedure works as follows. If a tested component functions normally then we wait until the next test time point where another component is tested. If a failure is observed during a test of one component then all other components from the same common cause failure group are tested as well. All malfunctioning components are repaired.

Our aim is to describe different ways of taking staggered testing into account both in mean value and time dependent analysis.

3. PROBLEM STATEMENT

We present a formula from [2] for the Alpha Factor model that takes staggered testing into account. Then we introduce a way to perform time dependent analysis. Finally, we state the problem as establishing a relation between the two approaches.

3.1. Mean Value Calculation with Staggered Testing

Equation 3 does not explicitly depend on a testing scheme, because we did not need to make any assumptions about it in Equations 1 and 2. However, values of alpha parameters and the total basic event probability might depend on the actual testing scheme. [2] presents a formula for the Alpha Factor model with staggered testing:

$$Q_k^m = \frac{1}{\binom{m-1}{k-1}} \alpha_k Q_t \qquad (7)$$

To obtain this, we have to use the relation between estimators for Q_k^m in both schemes. We have that

$$Q_k^{m\,[NS]} = k \cdot Q_k^{m\,[S]} \qquad (8)$$

and if we replace the values from non-staggered testing by the right hand side in the Equation 1 and leave Equation 2 then we obtain Formula 7. Since this derivation is not a part of [2], we show it here. Recall a simple relation between binomial coefficients.

$$\binom{m}{k} = \frac{m}{k}\binom{m-1}{k-1} \qquad (9)$$

Applying Equation 8 on Equation 1 together with Equation 9 gives us

$$\alpha_k = \frac{\binom{m}{k} k Q_k^m}{\sum_{k=1}^{m} \binom{m}{k} k Q_k^m} = \frac{\binom{m-1}{k-1} m Q_k^m}{\sum_{k=1}^{m} \binom{m-1}{k-1} m Q_k^m} = \frac{\binom{m-1}{k-1} Q_k^m}{\sum_{k=1}^{m} \binom{m-1}{k-1} Q_k^m} \qquad (10)$$

Now we can substitute the denominator according to Equation 2.

$$\alpha_k = \frac{\binom{m-1}{k-1} Q_k^m}{Q_t} \qquad (11)$$

From here we obtain Formula 7.

3.2. Time dependent Analysis

Time dependent analysis in RiskSpectrum is based on an assumption that alpha factors are the same over time. Component failure probabilities are established according to Formula 5 and it is always the latest tested component that determines probabilities of common cause failure events containing this component.

Common cause failure modeling works in the following way. We split the original event(s) into several new ones. Each new event:
- Represents failure of one or more events from the common cause failure group.
- Contains reliability parameters from the original event.
- Has its failure probability (unavailability) at a given time multiplied by a factor obtained from alpha parameters depending on its multiplicity according to Formula 3.

After this, each new event represents a clearly defined physical event (e.g., components A and B fail simultaneously because of a common cause failure) and behaves as an independent event. Its failure probability at a given time is independent of the probability of other common cause failure events at this time point. It is determined only by its parameters, alpha parameters, and the time point, which in turn determines the time point of the last test. An event representing a simultaneous failure of several components is tested each time one of these components is tested. If this tested component works then it means that the simultaneous failure of these components has not happened and therefore its probability is equal to zero.

The mean value of top event unavailability is obtained by calculating unavailability at different time points (Figure 2), integrating it over time numerically and dividing it by the interval length.

Figure 2. A graph of unavailability over time for an example emergency feed water system.

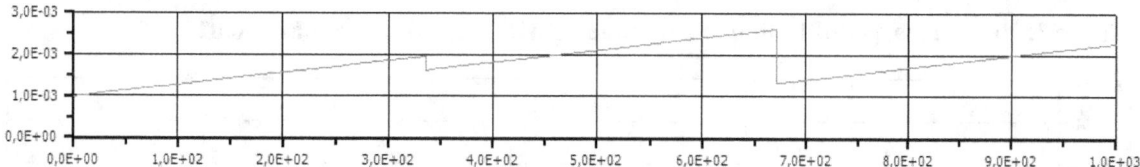

3.3. Final Problem Statement

We shall deal with the following questions in the rest of the paper:
- Which assumptions lie behind the time dependent analysis in RiskSpectrum?
- How does this time dependent analysis relate to the staggered mean value formula from NUREG/CR-5485?
- What are the advantages and disadvantages of either approach?

4. RELATING TIME DEPENDENT ANALYSIS AND MEAN VALUE FORMULAE

Recall that failure probability of a common cause failure event at a given time is determined only by its parameters, alpha parameters, and the last time point when a component included in this event was tested. We call this a *perfect testing assumption*. Each test excludes the possibility of all common cause failure events including the tested component, leaving other common cause failure event probabilities unaffected.

One consequence of this is that probabilities of common cause failure events at a random time point might not satisfy Equation 2 for Q_t. First, values Q_k^m are not well-defined anymore. They can be different for different common cause failure events with the same multiplicity. Secondly, if we, at a specific time point, sum up values of all common cause events including failure of the original component then we might get a value smaller than the original value Q_t. For example, in a group with two components A and B, if we test B at a time point t then the total failure probability of A decreases, because the probability of a common cause failure of A and B is set to zero.

The mean probability value of a common cause failure event in time dependent analysis with staggered testing might be up to k times, where k is the event's multiplicity, smaller than its mean value with non-staggered testing. There are two reasons why it is not always k times smaller:
- Tests of this common cause failure might not be evenly distributed over the test interval.
- For larger lambda values, the probability function over time deviates significantly from a linear function. See Figure 3 for an example of both phenomena.

Since the mean value formula does not take exact test time points into account, and the effects of the two items above are small if we have small lambda values, we can say that this decrease of mean probability for common cause failure events corresponds to dividing the mean value by k. This decreases the mean total probability of the component failure (modeled by the original basic event) by a factor of α_t. If we denote the mean failure probability of the component in question under assumption of staggered testing by $Q_t^{[S]}$ and under assumption of non-staggered testing by $Q_t^{[NS]}$ then we have that

$$Q_t^{[S]} = Q_t^{[NS]}/\alpha_t \qquad (12)$$

All these considerations assume that we use Formula 3 (non-staggered testing) to calculate probabilities in the time dependent analysis. On the other hand, using $Q_t^{[S]}$ or Formula 7 (staggered testing) for the time dependent analysis, i.e., to obtain event probabilities at a certain time point, would be applying staggered testing on data obtained from staggered testing and it would result in an unjustified decrease of failure probability.

Figure 3. Uneven staggered testing of components with failure rate equal to 0.02

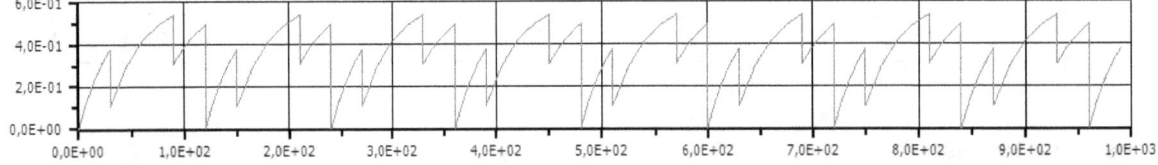

5. DISCUSSION

In this section we discuss advantages and problems with time dependent analysis of the background of Formula 7.

5.1 Perfect Testing Assumption

Time dependent analysis in RiskSpectrum works under the *perfect testing assumption*. As a result, the component failure probability obtained from the time dependent analysis is somewhat lower than the original one. For realistic alpha parameters where α_1 (the independent failure) dominates the other parameters, the difference rather small. See Table 1 for some examples of alpha parameters taken from [6].

Table 1: Decrease of Component Failure Probability in Time dependent Analysis with Staggered Testing

α_1	α_2	α_3	α_4	Decrease (%)
0.9771	1.38E-2	8.09E-3	5.59E-3	4.9
0.9766	1.49E-2	6.81E-3	1.67E-3	3.3
0.9636	1.34E-2	1.11E-2	1.13E-2	6.5
0.9612	2.42E-2	1.07E-2	3.89E-3	5.5
0.8395	8.33E-2	2.68E-2	4.75E-2	21.7

Time dependent analysis offers multiple advantages over mean value calculations. It gives complete flexibility for modeling of staggered testing. We can model different testing schemes, e.g., a two-train system with two redundant components in each train where testing of trains is staggered. We test both redundant components in a train in one test episode and the other two redundant components from the second train in the next test episode. Here, using the Formula 7 for a CCF group of size four might lead to underestimating the actual failure probabilities.

The mean probability model for staggering only considers staggering within one CCF group, but this is still a significant simplification since testing of important systems is often also staggered. Time dependent analyses provide a better solution, since these can also take staggering between different systems into account. Therefore, time dependent analysis can serve as a verification tool for designing a test scheme. Additionally, it makes time dependent analysis a natural method for on-line risk monitors.

Finally, can we use some of the equations above to obtain the same result also in mean value calculations? Formula 7 is derived from the unmodified Equation 2 which states that all common cause failure probabilities sum up to the original component failure probability. To reconcile this with the perfect testing assumption, we have to replace the original component probability by the staggered one from Equation 12. This means that we consciously distinguish between mean failure probability of a component with non-staggered and with staggered testing and we accept that this probability is lower under staggered testing. Formula 7 then gives us the same mean probability values for common cause failure events as the time dependent analysis (this follows directly from a comparison between Formula 3 and Formula 7 after application of Equation 12).

5.2 Failure Probability Preservation Assumption

In spite of the arguments above, one might argue that the perfect testing assumption is giving us an unreasonable risk decrease bonus. By a successful test of a component A, we can exclude the possibility that components A and B from the same common cause failure group fail together. If we were now in the position to partition the total failure probability for the whole component, maybe we would simply assign a bit more to other common cause failure events instead of the currently excluded event AB. By this, we would preserve the original total probability also under the staggered testing scheme. The effect of this testing scheme would be a different probability partitioning in favor of common cause failure events with lower multiplicities. This corresponds to a modification of the alpha parameters.

This is a conservative way of accounting for staggered testing. It is skeptical to our knowledge about common cause failures and does not want to take the full credit for testing of common cause failure events. Probability of a common cause failure event modeling a dependent failure of k components with staggered testing will be higher than the probability of the same event with non-staggered testing divided by k. This approach might still lead to a decrease of the top event probability/frequency, because it decreases failure probability of common cause failure events with higher multiplicities. These events are usually greater contributors to the top event probability/frequency than the independent component failures.

A disadvantage of the failure probability preservation assumption is that it is not clear how to perform the time dependent analysis leading to the same quantitative results.

6. CONCLUSIONS

We have discussed advantages and disadvantages of time dependent analysis of a system with common cause failures in comparison to mean value calculation according to formulae from NUREG/CR-5485. The main advantage of the time dependent analysis is the great flexibility which it offers for modeling of staggered testing, not only between components from one common cause failure group, but also between different common cause failure groups and whole systems. We have formulated two assumptions for staggered testing which lie behind either the time dependent analysis or the mean value calculation. We leave it for further discussion under which circumstances these assumptions are reasonable.

References

[1] A. Mosleh et al., *"Procedures for Treating Common Cause Failure in Safety and Reliability Studies"*, NUREG/CR-4780, 1988.
[2] A. Mosleh et al., *"Procedures and Guidelines in Modeling Common Cause Failures in Probabilistic Risk Assessment"*, NUREG/CR-5485, 1998.
[3] J. E. Stott et al., *"Common Cause Failure Modeling: Aerospace vs. Nuclear"*, In Proc. of PSAM10, Seattle, 2010.
[4] RiskSpectrum Analysis Tools Theory Manual, Lloyd's Register Consulting, 2012.
[5] International Atomic Energy Agency, *"Procedures for conducting common cause failure analysis in probabilistic safety assessment"*, IAEA-TECDOC-648, 1992.
[6] F. M. Marshall et al., *"Common-Cause Failure Parameter Estimations"*, NUREG/CR-5497, 1998
[7] A. Mosleh and N Siu, , *"A Multi-Parameter Common Cause Failure Model"*, In proc. of the 9th International Conference on Structural Mechanics in Reactor Technology, Lausanne, 1987.

A State of the Practice Investigation Guiding the Development of Visualizations for Minimal Cut Set Analysis

Yasmin I. Al-Zokari[a], Liliana Guzman[a], Barboros Can Conar[b], Dirk Zeckzer[c], Hans Hagen[a]

[a]TU Kaiserslautern, Kaiserslautern, Germany
[b]University of Applied Sciences, Kaiserslautern, Germany
[c]Leipzig University, Leipzig, Germany

Abstract: Minimal Cut Set (MCS) analysis is used for the qualitative and quantitative safety and reliability analysis of systems. While many studies concerning MCS computation in the safety domain are found, no study gives a complete and detailed description of the tasks performed by practitioners during MCS analysis. The goals of this study are (1) to elicit the context (including the tasks) of MCS analysis; (2) to obtain the requirements and needs of the safety analysts, and the tools used; (3) and to assess the quality of the tools from the point of view of the safety engineers regarding their (3a) representation, (3b) interaction, (3c) performance, and (3d) usability. We found that the main purpose is finding improvements to increase the hazard's safety. The main tasks are identifying critical basic events, the related system components, and single points of failure. The stakeholders are mainly decision makers and system engineers. The main requirements are finding single points of failure, determining MCS order, and finding basic events with high failure probability and related components. The results show that the usability of the tools is accepted but their information presentation can be improved by providing overviews and the missing interactions.

Keywords: Safety Visualization, user study, minimal cut set analysis, minimal cut set visualization, state of the practice.

1 INTRODUCTION

Safety is a very important property of embedded systems. Safety analyses are performed to avoid human injury or a negative impact on the environment during system usage. Fault Tree Analysis is a well-known, commonly used method for the safety analysis of embedded systems. In addition to examining the safety of embedded systems, fault tree analysis is also used in reliability and availability analysis [31, 25, 19], among others.

Minimal Cut Set (MCS) analysis is used in fault tree analysis for identifying combinations of so called basic events (BE) that cause a safety problem [6]: p. 9, 58, 188-189, [11, 13, 31]. Here, basic events (BE) are the root causes of a problem. The MCSs help to "identify failure events whose exclusion secures the system" [27]. Excluding these events will considerably reduce the probability of the top level hazard. Additionally, MCSs help in computing the failure probability (FP) of the hazard (if the failure probabilities of the basic events are provided).

Most studies about MCS analysis define MCSs, illustrate their structure, and investigate their computation, such as [42, 39, 38, 12, 20, 40, 41]. However, most studies only refer to a limited number of tasks that are involved in MCSs analysis. For example, Papadopoulos et al. [21] consider only two MCS analysis tasks regarding single points of failure (SPoF), and areas of the design that contributed to the overall failure probability (i.e., BEs), which guides improvement strategies of the system. Patterson et al. [22]: p. 18 consider the susceptibility (i.e., FP) of MCSs for ranking them. This prioritization will help the decision makers in their judgments.

Cepin [7] identified three MCS analysis tasks. Listing MCSs, finding high values of probabilities, and finding a large number of failure events (BEs) are considered important MCS analysis tasks for evaluating a fault tree. However, no studies were found giving a complete detailed task description or describing which information is necessary for MCS analysis from the point of view of the users of tools.

The IBM & Industry Studies, Costumer Interviews [16]: p. 8 reports that 30% of the users' time is spent on searching for and exploring relevant information. Textbooks like the one by Ward et al. [33] show, that interactive data visualization can be used to improve the exploration and understanding of large amounts of data. This is due to the inherently parallel process of perceiving the environment. This process is used by visualizations to convey a large amount of information to the user. It is supported by effective interactions that allow switching between overview and detail, as well as selecting, marking, sorting, filtering, and other mechanisms. Eward Tufte showed not only how to construct effective visualizations [29], but also how visualizations can help to analyze data [28]. Thus, visualizations together with matching interactions allow to deal with huge data sets and allow to analyze them effectively and efficiently.

"Safety Visualization" is a new, emerging domain in visualization meeting the challenges of providing visualizations and advanced interactions for supporting the safety analysis of (embedded) systems. In this domain, innovative visualizations and interaction mechanisms were developed to ease the safety analysts tasks by Al-Zokari et al. [2, 3, 5, 4], Khan et al. [18], and Yang et al. [37, 35, 36]. These visualizations support the practitioners in MCS analysis for assessing the safety, reliability, or availability of (embedded) systems and for suggesting improvements as well as comparing different improvement alternatives or different states of the system. Similarly, Høiset et al. [14] used virtual reality to reveal the consequence of hazards and ease the communication between analysts and stakeholders.

To guide the development and the improvement of visualizations for MCS analysis, this user study was performed. The goals of this study are: (1) to give an overview over the current state of the practice in MCS analysis, (2) to support sharing the experience of safety analysts with the visualization domain in the near future, and (3) to understand the requirements of the users of safety tools, aiming at understanding the world as experienced by the safety analysts.

2 RESEARCH PURPOSE AND RESEARCH QUESTIONS

This research aims at exploring the state of the practice regarding MCS analysis from the perspective of practitioners, i.e., safety engineers. The results of this research are intended to provide guidelines for enhancing the visualizations and the interaction capabilities of current tools supporting MCS analysis and for developing new visualizations and new tools that support MCS analysis. In particular, it focuses on the following research questions:

RQ-1 In which context do safety engineers perform MCS analysis?

RQ-2 What are the requirements of the safety engineers for performing MCS analysis?

RQ-3 How is the quality of current tools for performing MCS analysis perceived by the safety engineers?

Here, quality means to which extent the visualization and the interaction capabilities provided by the tools for performing MCS analysis support the safety engineers while performing MCS analysis. We decompose the quality of tools for performing MCS analysis as follows:

RQ-3a What is the practitioners' perception of the quality of information representation? The degree to which the representation of the information provided by a tool for performing MCS analysis is adequate, understandable, and necessary for performing MCS analysis. A good representation of the information generated by MCS analysis should help reducing the time and the effort required to perform the analysis tasks while reducing errors by, e.g., giving a good overview.

RQ-3b What is the practitioners' perception of the quality of interaction capabilities? The degree to which the interaction capabilities provided by a tool facilitate the MCS analysis, such as navigating through views, sorting data, and filtering data.

RQ-3c What is the practitioners' perception of the usability? Usefulness: The degree to which a safety engineer believes that using the tool for performing MCS analysis will help him or her to attain gains in job performance [32]: p. 447. Ease of use: The degree of ease associated with the use of the tool for performing MCS analysis [32]: p. 450. Usability: The tools' clarity, understandability, becoming skillful with, easiness, and the ability to learn their functionality. Usability comprises both usefulness and ease of use.

RQ-3d What is the practitioners' level of satisfaction with current tools for performing MCS analysis?

Comparing the capabilities of the current tools performing MCS analysis and the requirements of the safety analysts enables to identify the strengths and the drawbacks of these tools and then to suggest improvements for supporting the tasks of the analysts. Finally, the practitioners were asked whether they prefer 2D or 3D interactions when analyzing MCSs.

3 RESEARCH METHODOLOGY

3.1 Survey Design

We designed an online survey in order to reach a large sample of safety engineers. The related questionnaire was designed according the guidelines given in [9, 10, 23] and can be accessed using this link [1]. The final version included up to 33 questions (34 for the second survey). The questionnaire comprised open, dichotomous, polychotomous, and multiple choice questions. All opinion questions included the alternative answer "I do not know". Filters were used to avoid overwhelming respondents with unnecessary questions and to ensure that they answered based on firsthand experience. Furthermore, definitions and examples were added to avoid misinterpretation of the questions.

To elicit the context in which safety engineers perform MCS analysis (**RQ-1**), 10 questions were designed comprising questions regarding target purposes, tasks, stakeholders and information exchange with them, hazard sizes (time), effort, and tools for performing MCS analysis they have used.

Seven questions were prepared for gathering the requirements (**RQ-2**) regarding the identification, analysis, and comparison of basic events and minimal cut sets. Moreover, respondents were asked about the importance of the required information. We identified possible needs of information based on a literature review and in collaboration with a safety expert.

Another 13 questions were designed for collecting information about the quality of current tools for performing MCS analysis (**RQ-3**). These questions were asked only to safety engineers with experience in using at least one tool for performing MCS analysis. Four questions were designed for eliciting the quality of the information representation by focusing on its understandability, appropriateness, and support for getting relevant information in one sight (**RQ-3a**). For analyzing completeness, we compared the importance of required information with the information provided by current tools. Three questions aimed at eliciting the quality of interaction capabilities, i.e., navigating, filtering, and sorting (**RQ-3b**). Two questions measured the usability according to the TAM model [32] (**RQ-3c**). Another three questions were specified to elicit the satisfaction level of safety engineers regarding current tools for performing MCS analysis (**RQ-3d**). There was one additional question whether the participants prefer 2D or 3D interactions.

Finally, five demographic questions (marked 'DQ') were prepared for eliciting the respondents' age, position, working experience, and experience and frequency in performing MCS analysis. They allow to assess if the sample matches the target group of experienced MCS analysts.

3.2 Population and Sampling Strategy

Practitioners and safety experts were invited to participate in an online user study that elicits their purposes, their requirements, and their tasks, as well as their experience with tools regarding visual representation, interaction, and usability. The target population includes safety engineers with experience in MCS analysis and in using tools for MCS analysis. We first drew a purposive sample consisting of 29 safety experts who were identified using Google search and personal email contacts. This sample comprised researchers and practitioners. We included researchers who closely worked with practitioners in the context of MCS analysis or use tools for performing MCS analysis. As an additional sample, we invited 11 safety experts contacted during the conference PSAM11&ESREL 2012. In both cases, all participants were encouraged to invite other safety experts to the survey (snowball technique).

3.3 Implementation and Pre-Evaluation

The survey was implemented using the on-line, free, open source tool LimeSurvey [8]. The data was retrieved using an export facility of this tool. After implementing the survey in LimeSurvey, we performed a pre-evaluation to appraise its completeness, understandability, and consistency [10]. In particular, one expert in experimental design and one expert in visualization reviewed the survey. All questions were classified as necessary, but the reviewers considered the survey being too long and some questions being too complex. Thus, we simplified the questions' wording, restructured the survey into seven sections to reduce its complexity, and added filters for not overwhelming respondents with unnecessary questions. Furthermore, we added the fields "Other" and "None", and fields for comments to make the survey more flexible. The answers were randomized.

3.4 Execution

The survey was performed twice. On February 17, 2012, we sent 29 invitations to experts working in safety analysis and developing safety analysis tools. The deadline for participation was March 9, 2012. On June 25, 2012, we sent 11 additional invitations to the safety experts contacted during the conference PSAM11&ESREL 2012. The deadline for these participants was July 5, 2012. An additional participant was invited on December 18, 2012. In all cases, the invitation subsumed the research purpose and scientific value, the target sample, the confidentiality and the anonymity of responses, the expiry date, and the contact person. Anonymity was guaranteed through the use of tokens. Aiming at increasing the response rate, a reminder email was sent one week after the original invitation.

3.5 Data Analysis

The data analysis reported in the subsequent sections was performed using SPSS Statistics 17 [15]. Because several questions were asked to a specific group of respondents or were optional, we explicitly report the total number of subjects (N) who answered a question and the frequency an option is selected (n). For each research question, the question types used and the related questions are listed in Table 1. For each question type, the response scale and the reported statistics are listed in Table 2. In particular, we report descriptive statistics including the sample median (Mdn), p, Z, and frequencies.

Table 1: Research Question, Question Type, and Related Questions. (Note: The question q6-1 did not exist in the on-line survey).

Research Question	Question Type	Questions
Demographics (DQ) 4.1	Polychotomous	q1-1, q1-2, q7-1, q7-3
	Multiple choice	q7-2
RQ-1 4.2	Polychotomous	q1-4
	Selection	q2-3
	Multiple choice	q1-3, q1-5, q1-6 q2-1, q2-2
	Quote	q1-7S1, q1-7S2, q1-8S1, q1-8S2
RQ-2 4.3	Polychotomous	q3-2a, q5-8, q6-2, q6-3, q6-5, q6-6
	Dichotomous	q3-2b, q5-3
	Quote	q5-4, q6-4
RQ-3 4.4	Polychotomous	q3-1, q3-4, q4-3, q5-1, q5-2, q5-5, q5-6, q5-7
	Dichotomous	q4-1, q4-2
	Multiple choice	q3-3, q3-5

Table 2: Question Type, Response Scale, and Statistics Reported.

Question Type	Response Scale	Statistics
Dichotomous	Yes, No	N, #Yes, #No
	Complete, Incomplete	N, #Complete, #Incomplete
	2D, 3D	N, #2D, #3D
Selection	Nominal	n
Multiple choice	Nominal	n
Polychotomous	5-point Likert scale	n, Mdn, p, Z
Open	Quotes	Quotes

Taking into account the sample size (17) and that variables are not normally distributed (normal distribution was tested using the Shapiro-Wilk test and analyzing the histograms), we analyzed the difference of the tendencies using the Sign two-tailed significance test with a default value of $\alpha = 0.05$ (95% confidence level). For the five-point Likert scale data, we were interested in which items produced a meaningful opinion from the respondents, i.e., a response significantly different from a test-value, e.g., H0: $Mdn(x) =$ test-value and H1: $Mdn(x) \neq$ test-value. If not mentioned, the test-value is the midpoint (i.e., neutral = 3), according to the guidelines defined by Utts and Heckard [30]: p. 610. For these tests, we report the significance level (p) and the Sign test observed value (Z). If the p- or the Z-value are significant, then they are printed in **bold**. If no significant difference is found, the tendency of the answers is reported. This tendency is the comparison to the central tendency Mdn of the 5-point Likert scale (i.e., value 3) since the data is ordinal as recommended by Russell [24]: p. 118, Szafran [26]: p. 97, and Wohlin et al. [34]: p. 124. The final data set and the survey are stored in the repository available from [1].

4 RESULTS

Out of 40 invitations sent, 18 safety experts answered the survey. This is a response rate of 45%. The responses of one participant were eliminated because he or she did not complete the survey. Thus, we have a sample size of 17. The average time for completing the survey was 26 minutes (SD = 17). 16 participants were invited using the purposive technique and one additional participant was invited through the snowball technique.

4.1 DQ: Demographic Questions and Experience in MCS Analysis

Out of 17 participants, six were working in Industry, six at the University, three at a research institute, one was working in both industry and at a research institute, and one was working in both industry and at the university. The participants are assumed to be experienced if they have at least 3 years of working experience, and are performing MCS analysis at least for 3 years and at least sometimes. 82% of the participants ($n = 14$) have at least 3 years working experience. 64.7% ($n = 11$) of the participants have at least 3 years of experience in performing MCS analysis (i.e., 6 participants have more than 6 years and 5 have from 3 to 6 years). Finally, one participant performs MCS analysis always, three very often, eight sometimes, and five rarely. Overall, the results show that they are experienced in work and experienced in MCS analysis with statistical significant difference: work experience ($Mdn = 5$: "More than 6 years", $p < \mathbf{0.001}$, $Z = 3.47$, test-value = 2: "1-2 years"), MCS analysis experience duration ($Mdn = 3$: "3-4 years", $p = 0.057$, $Z = 2.94$, test-value = 2: "1-2 years"), and MCS analysis frequency ($Mdn = 3$: "Sometimes", $p < \mathbf{0.001}$, $Z = 3.17$, test-value = 4: "Rarely").

4.2 RQ-1: Context of Performing MCS Analysis

4.2.1 Practitioners' Main Purposes and Main Tasks While Performing MCS Analysis

Purposes of using minimal cut set analysis: Out of 17 participants, 11 use MCS analysis to "Find improvements to increase the hazard's safety", 5 use it to "determine the safety level of the hazard being analyzed", 5 to "compare alternative improvements to increase the hazards safety", and 5 to "analyze the hazards safety evolution after system changes". Seven respondents added open comments stating that they have also used MCS analysis for: fault diagnosis (frequency = 1), verifying model consistency (1), reviewing a fault tree (1), identifying single points of failures (SPoF) (1), customer request (1), determining accurate representation of the accident sequence and identifying dependencies (1), and reliability analysis (1).

MCS analysis tasks: According to the participants ($N = 17$), the tasks performed *always or very often* are: identifying the most critical BEs ($Mdn = 2$, $p < \mathbf{0.001}$, $Z = 3.474$); identifying system components related to BEs (2, **0.002**, 2.94); identifying SPoF (1, **0.007**, 2.582); and calculating the failure probability (FP) of a hazard (2, 0.804, 0.25). Tasks that are performed *sometimes* include: calculating the failure probability (FP) of all MCSs (3, 0.388, 0.866); calculating the FP of some MCSs (3, 0.549, 0.6); identifying the number of MCSs causing a hazard (3, 0.774, 0.289); classifying MCSs regarding their order (3, 0.774, 0.289); and identifying all BEs of the hazard (3, 1.000, 0). Finally, only one MCS analysis task is performed *rarely*: classifying MCSs regarding their FP (4, 0.267, 1.1).

4.2.2 Stakeholders of the Practitioners and Required Information

Stakeholders: The participants working in industry ($N = 8$) provide the information gained from their MCS analysis to decision makers ($n = 4$), customers (4), system engineers (4), safety engineers (3), and software engineers (2). Additionally, the participants working at a *university* ($N = 7$) report to safety engineers ($n = 5$), system engineers (4), decision makers (2), and software engineers (1). Finally, the participants working at a *research institute* ($N = 4$) report to safety engineers ($n = 3$), customers (3), software engineers (2), decision makers (2), and system engineers (2).

Required information: Stakeholders ask the participants ($N = 17$) for: a list of prioritized possible causes of a hazard (i.e., a list of prioritized MCSs) ($n = 12$), a list of SPoF that may cause a hazard (11), the combination of failures that may cause a hazard (i.e., MCSs) (10), a list of failures (i.e., BEs) with larger impact on hazard's safety (8), and suggestions for increasing the hazard's safety (7). Approximately one third of the respondents provide the hazards' safety evolution after system improvements (5). Less than a third provide: the safety level of the hazard being analyzed (4), the generated information from MCS analysis (4), and the properties of the system components related to the failures (3).

4.2.3 Hazard sizes and Time Spent for MCS Analysis

Hazard size: Up to six participants ($N = 9$) of the first survey defined a hazard size in terms of its number of BEs and its number of MCSs (Table 3). At the time of the second survey, the participants ($N = 8$) have analyzed hazards of size 1-30, 1-100, 1-100, 1-400, 1-500, 1-1,000, 1,000-10,000, 10,000-100,000 MCSs.

Table 3: Hazard Size. "Format": number given (number of times mentioned).

Hazard size	Number of BEs	Number of MCSs
Small	1-3, 10 (3), 20	3, 10 (3)
Medium	4-20, 50, 100 (3), 100-1,000	25, 30, 100 (2)
Large	≥20, 100 or more, 1,000 (2), ≥1,000, 10,000	50 or more, 1,000 (2), 3,000

Time spent on comparing two hazards of an average (medium) size regarding their SPoF: All participants ($N = 8$) of the second survey answered this question (Table 4).

Table 4: Time Spent for Comparing Two Hazards Regarding SPoF.

Hazard Size	Range	Average Hazard Size	Time for Performing Comparison
Small	1-30 MCSs	15 MCSs	5 minutes
	1-100 MCSs	50 MCSs	10 minutes, 2 days
Medium	1-400 MCSs	200 MCSs	0 minutes
	1-500 MCSs	250 MCSs	2 hours
	1-1,000 MCSs	500 MCSs	2-10
Large	1,000-10,000 MCSs	5,000 MCSs	2-4 hours
	10,000-100,000 MCSs	50,000 MCSs	2-10

Time spent in performing MCS analysis tasks: Up to six of the participants ($N = 9$) from the first survey answered this question. The estimation time for analyzing: *small* hazards are: 2 minutes using FaultTree+ (3 MCSs), 2 hours using C^2FT (10 MCSs), a range of minutes using self-developed tool (10 MCSs), and 1 day using ESSaRel (10 MCSs). The time spent for analyzing *medium* hazards are: 30 to 60 minutes using MagicDraw, a range from minutes to hours using self-developed tool (100 MCS), and a range from 2 to 3 days using FaultTree+ (30 MCSs); and for analyzing large hazards it takes from minutes to hours using self-developed tool (1000 MCSs), and from 2 to 3 days using FaultTree+ (3000 MCSs). The remaining answers were stated as "depends".

4.2.4 Tools Used for Performing Minimal Cut Set Analysis

Tools used: For performing MCS analysis, respondents working in *industry* ($N = 8$) have used RiskSpectrum ($n = 3$), CAFTA (1), FaultTree+ (1), ESSaRel (1), CARA (1), BlockSim (1), C^2FT (1), self-developed tools (1), and WinNUPRA (1). At *universities* ($N = 7$) respondents have used ESSaRel ($n = 4$), BlockSim (3), FaultTree+ (2), RAM Commander (1), self-developed tools (1), Item toolkit (1), RiskSpectrum (1), and WinNUPRA (1). At *research institutes* ($N = 4$) they used ESSaRel ($n = 4$), FaultTree+ (3), C^2FT (1), and MagicDraw (1). Thus, the most used tools are ESSaRel (8), FaultTree+ (5), BlockSim (4), and RiskSpectrum (3).

Best tools: Respondents working in *industry* consider the best tools for performing MCS analysis to be CAFTA (2), RiskSpectrum (2), Fault Tree+ (1), BlockSim (1), SAPHIRE (1), C^2FT (1), and WinNUPRA (1). The participants from *universities* consider that the best tools are ESSaRel (3), BlockSim (2), CAFTA (1), and RAM Commander (1). The participants from *research institutes* reported the best tools being Fault Tree+ (3), ESSaRel (1), C^2FT (1). Therefore, the best tools for performing MCS analysis of the majority of the participants are ESSaRel (4), BlockSim (3), Fault Tree+ (3), CAFTA (2), and RiskSpectrum (2).

Best tool currently used: Participants working in *industry* ($N = 6$) consider the best tool currently used for performing MCS analysis to be RiskSpectrum ($n = 2$), CARA (1), Blocksim (1), C^2FT (1), and WinNUPRA (1). At *universities* ($N = 7$) they report the best tool to be ESSaRel ($n = 3$), BlockSim (1), RAM Commander (1), RiskSpectrum (1). At *research institutes* ($N = 4$) the best tools are reported to be FaultTree+ ($n = 2$), MagicDraw (1), and C^2FT (1).

The following analysis refers to the best tool that participants were using for performing MCS analysis at the moment of the survey.

4.3 RQ-2: Requirements of the Safety Engineers for Performing MCS Analysis

Necessary information: The most *necessary* information from the point of view of the respondents for performing MCS analysis are shown in the first four rows in Table 5: SPoF, MCSs' order, BEs with high FP, and system components related to BEs. The information that tend to be necessary is shown in the fifth to the tenth row in this table.

Additional information: Out of 15 respondents, 11 do not need additional information for performing MCS analysis. However, the remaining respondents need information regarding: "*Fault Tree structure*", "*The path from hazard to a specific basic event*", "*What are the architectural components providing the largest amount of Basic failures or MCS*", "*Accident sequences*", "*Dependency matrices*", "*System Layouts*", "*Sensitivity and uncertainty analysis*", "*Test data to try to (re)create hazards*", "*Qualification data*", "*Heritage data*", and "*Standards*".

MCS analysis tasks: The first seven tasks in Table 6 are the most important tasks needed to be performed during MCS analysis: finding mistakes in the structure and incomplete information, prioritizing MCSs, showing the system parts related to BEs and their properties, knowing

Table 5: Necessary information during MCS analysis.

No.	Information option	n	Mdn, p, Z
1	SPoF	13	1, **0.003**, 2.774
2	MCSs order	13	1, **0.012**, 2.412
3	BEs with a high FP	14	1, **0.022**, 2.219
4	system components related to BEs	13	1, **0.039**, 2.021
5	FP of BEs	13	1, 0.092, 1.664
6	FP of the MCSs	12	1, 0.334, 0.949
7	BEs with high number of occurrence	14	2, 0.146, 1.443
8	FP of a hazard	12	1.5, 0.227, 1.206
9	MCSs with a high FP	14	1.5, 0.267, 1.109
10	properties of system components	11	2, 0.745, 0.316
11	total number of basic events	13	3, 1.00, 0.00
12	MCSs with identical basic events characteristics	12	3.5, 1.00, 0.00
13	total number of MCSs	14	5, 0.581, 0.555
14	MCSs with identical order	12	4.5, 0.227, 1.200
15	MCSs with identical failure probabilities	12	5, **0.021**, 2.214

Response scale: 5-point Likert scale from 1: necessary to 5: unnecessary.
Results with statistical significance (i.e., $p < 0.05$) are marked in **bold**.

BEs with high number of occurrence, exploring the systems' model and interacting with it, and comparing MCSs regarding their order. All other tasks except the last four in Table 6 tend to be needed for performing MCS analysis from the point of view of the respondents.

One participant did not use any tool at the time of the survey. Therefore, the respondent answered only the questions related to the information of MCS analysis and not related to the tools. This participant considers the following as *necessary information:* SPoF, basic events (BEs) with high number of occurrence, BEs with high FP, system components related to BEs, MCSs with high FP, properties of system components, MCSs with identical FPSs, and the total number of BEs. Finally, this participant reported the *MCSs tasks* to be: to know the system being analyzed, to know and to present the parts of the system that are related to the BEs in a hazard, and to know if the hazard is safe or not.

4.4 RQ-3: Quality of Current Tools for Performing MCS Analysis as Perceived by the Safety Engineers

4.4.1 RQ-3a: Practitioners' Perception of the Quality of Information Representation

Information representation: The results from performing MCS analysis are represented differently by the tools at the time of the survey. Nine out of 15 respondents report that this information is represented as text (using FaultTree+, ESSaRel, CARA, RiskSpectrum, or BlockSim tools). Additionally, nine report that the information is represented in a tabular form (using FaultTree+, ESSaRel, MagicDraw, RiskSpectrum, WinNUPRA, C^2FT, or RAM Commander). Only five stated that the information is represented in fault tree structures (FaultTree+ or ESSaRel). Finally, only two stated that the information is represented using block diagrams (using FaultTree+, ESSaRel, C^2FT, or WinNUPRA).

Understandability of information: The respondents find the information provided by the tools to be *understandable* ($n = 15$, $Mdn = 2$, $p = \mathbf{0.012}$, $Z = 2.41$).

Table 6: Degree of necessity of MCS analysis tasks.

No.	When performing MCS analysis I need to:	n	Mdn, p, Z
1	to find mistakes in the hazard structure	11	2, **0.001**, 3.015
2	to prioritize MCSs for detailed analysis	14	2, **0.003**, 2.774
3	to find incomplete information	11	2, **0.004**, 2.667
4	to show the system being analyzed, its physical parts related to basic events, and their properties	13	1, **0.012**, 1.66
5	not to know the BEs with high number of occurrence	8	4, **0.031**, 2.04
6	to explore the system's model and to interact with it	14	2, **0.039**, 2.021
7	to compare between MCSs regarding their order	14	2, **0.039**, 2.021
8	to compare BEs regarding the MCSs' quality (i.e., criticality: e.g., FP, order) that they affect	6	2, 0.063, 1.78
9	do not want to manually calculate the failure probabilities	10	2, 0.070, 1.76
10	to compare between MCSs regarding their failure probability (FP)	14	2, 0.092, 1.664
11	to print and show the physical parts related to basic events (BEs) to the stakeholders	11	2, 0.125, 1.512
12	to know the MCSs that have low order with high FP	8	2, 0.219, 1.22
13	to compare between BEs regarding their number of occurrence	14	2.5, 0.180, 1.333
14	to compare between BEs regarding their FP	14	2, 0.267, 1.109
15	to know the MCSs within a FP range that I specify	9	3, 1, 0
16	to know the MCSs with an order of a range that I specify	7	3, 1, 0
17	not to prioritize MCSs for detailed analysis	7	3, 1, 0
18	not to know the BEs with high FP	8	3.5, 1, 0

Response scale: 5-point Likert scale from 1: strongly agree to 5: strongly disagree.
Results with statistical significance (i.e., $p < 0.05$) are marked in **bold**.

Quality of information representation: The respondents rate the representation of all information as *fair* with respect to: the detail level ($N = 15$, $Mdn = 3$, $p = 0.18$, $Z = 1.33$), the data representation in general (15, 3, 0.125, 1.512), the sequence order in which the information is presented (14, 3, 0.754, 0.316), the arrangement of the information in a single view (15, 3, 0.774, 0.289), and getting an overview in one sight of the hazard (15, 3, 1, 0).

Immediate overview (information in one sight): Nine out of 17 participants state that they can gain an immediate overview regarding: MCSs with the highest FP (using FaultTree+, ESSaRel, CARA, C^2FT, MagicDraw, RiskSpectrum, WinNUPRA, or BlockSim), and MCSs with SPoF (using FaultTree+, ESSaRel, MagicDraw, CARA, C^2FT, or BlockSim). Six of the participants ($N = 17$) reported that they are able to get an immediate overview regarding: MCSs with low order (using FaultTree+, ESSaRel, MagicDraw, WinNUPRA, C^2FT, or BlockSim). Five of the participants ($N = 17$) reported that they get an immediate overview regarding BEs' number of occurrence (using FaultTree+, RiskSpectrum, or BlockSim), and system design (using FaultTree+, ESSaRel, C^2FT, RiskSpectrum, or BlockSim). Finally, only four respondents of ($N = 17$) get an immediate overview regarding: Safety level of the hazard being analyzed (using FaultTree+, ESSaRel, WinNUPRA, or BlockSim), and MCSs affected by a BE (using FaultTree+, C^2FT, WinNUPRA, or BlockSim).

4.4.2 RQ-3b: Practitioners' Perception of the Quality of Interaction Capabilities

Sorting the information: The participants (N = 17) report that the best tool used at the time of the survey provides the interaction capability of sorting for: MCS's FP (no = 2, yes = 8, provided by FaultTree+, BlockSim, RAM Commander, C^2FT, MagicDraw, RiskSpectrum, WinNUPRA) and MCS's order (no = 3, yes = 9, provided by FaultTree+, BlockSim, RAM

Commander, C²FT, MagicDraw, RiskSpectrum, self-developed). However, most of the tools do not provide sorting regarding: BE's FP (no = 7, yes = 4, only provided by BlockSim, RiskSpectrum, self-developed), both MCS's FP and order (no = 6, yes = 3, only provided by FaultTree+, BlockSim, RiskSpectrum), and BE's number of occurrence (no = 8, yes = 2, only provided by FaultTree+, RiskSpectrum). Furthermore, no tool can provide the sorting of both BE's FP and BE's number of occurrence (no = 9).

Filtering the information: Six out of 10 of the participants stated that the best tool used for performing MCS analysis supports filtering regarding: MCS's order (using FaultTree+, BlockSim, RAM Commander, C²FT, RiskSpectrum, WinNUPRA). The respondents were half split regarding the filtering options: MCS's FP and the BE's FP. Respondents using BlockSim, RAM Commander, RiskSpectrum, WinNUPRA reported the availability of filtering by MCSs FP. Respondents using BlockSim, RAM Commander, RiskSpectrum, and WinNUPRA report the availability of filtering by BE's FP. Only 2 of the participants state that the best tool can filter by BEs' number of occurrence (using FaultTree+, WinNUPRA).

Interaction amount: The participants report that the amount of interaction needed for getting the complete information in *navigation*, i.e., navigating through views ($n = 10$, $Mdn = 3$, $p = 1$, $Z = 0$) and navigating through the hazards' structure (10, 3, 1, 0) are considered to be *just right*. However, they rated the amount of *scrolling up/down in the same view* to be *too much* (11, 2, **0.031**, 2.04).

4.4.3 RQ-3c: Practitioners' Perception of Usability

Ease of use: The respondents find the best tool they are currently using: easy to learn ($n = 13$, $Mdn = 2$, $p = \mathbf{0.039}$, $Z = 2.0$). They tend to find it: clear and understandable (13, 2, 0.065, 1.8), easy to become skillful in using it (12, 4, 0.065, 1.8 negated), and easy to become skillful in MCS analysis when using it (12, 2.5, 0.289, 1.06). However, the participants are unsure, if the tools are easy to use (13, 3, 0.754, 0.316 negated).

Usefulness: The respondents find the best tool they are currently using: useful in their job ($n = 15$, $Mdn = 4$, $p = \mathbf{0.003}$, $Z = 2.77$ negated), enables them to accomplish tasks more quickly (15, 2, **0.004**, 3.17), enables them in understanding the system being analyzed (14, 2, **0.022**, 2.2), and increases their productivity (14, 4, **0.039**, 2.02 negated). However, they are unsure, if using the tools enables them to communicate better with the stakeholders (14, 3, 0.125, 1.5).

4.4.4 RQ-3d: Level of Satisfaction with Current Tools for Performing MCS Analysis

Satisfaction level: The respondents are confident that their results are correct because of the information generated by the tool for performing minimal cut set analysis ($n = 15$, $Mdn = 2$, $p < \mathbf{0.001}$, $Z = 3.175$), would recommend the tools to colleagues ($n = 14$, $Mdn = 2$, $p = \mathbf{0.039}$, $Z = 2.02$), and tend to be *satisfied* regarding the best tool they are currently using ($n = 15$, $Mdn = 2$, $p = 0.109$, $Z = 1.58$). The tools used by the respondents that reported their *satisfaction* are: FaultTree+, ESSaRel, CARA, C²FT, MagicDraw, and RiskSpectrum.

4.5 Do Practitioners Prefer 2D or 3D Interactions When Analyzing MCSs

Five out of seven participants prefer 2D interactions over 3D interactions. Two comments were provided by the respondents to this optional question. One respondent commented: "*In 3D there is always the possibility of hiding information in the Z-direction. This bears risks for missing information."*. Another stated: *"In my case it may be relevant to go from the cut-set list to the event tree (or sequences) to understand the failure roots. It is some how more difficult to do in the case of dependencies, but the idea is to help the user to identify the scenario realized through the cutset."* This question was only asked in the second survey.

5 DISCUSSION OF THE RESULTS

From the results of the demographic section 4.1, we conclude that most of our participants are experienced in MCS analysis, and from different working environments with at least three years of experience. Thus, the sample is representative of the target population. Moreover, most participants were frequently performing MCS analysis tasks (Section 4.1) and all but one were using MCS analysis tools (Section 4.2.4). We conclude that their insights into tools and procedures is up-to-date.

5.1 RQ-1: Context of Performing MCS Analysis

We identified eleven single *purposes* for performing MCS analysis. The participants confirmed all purposes and added additional ones (Section 4.2.1). The use of MCS analysis for several different purposes indicates its usefulness in safety and reliability analysis.

The *purposes and tasks* (Section 4.2.1) identified as frequent in this survey indicate the need of representations for critical BEs, components related to BEs, SPoF, FP of a hazard, etc., to support the analysis. Only the *task* "classifying MCSs regarding their FP" is performed rarely by the participants during MCS analysis. One plausible explanation is that sometimes the FPs of MCSs are not provided to the analysts, because, e.g., the project is still in the design phase or the FPs are not available. FPs might not be available for self-built components, or they are not provided by the component producer.

The *stakeholders* of the participants are mostly safety engineers, system engineers, decision makers, customers, and software engineers (Section 4.2.2).

Furthermore, according to the *information* required by these stakeholders (Section 4.2.2), it is important to provide representations of the prioritized possible causes of a hazard (i.e., MCSs), SPoF that may cause a hazard, the combination of failures that may cause a hazard (i.e., MCSs), failures with larger impact on the hazard's safety (i.e., BEs), suggestions for increasing the hazard's safety, the hazards' safety evolution after system improvements, the safety level of the hazard being analyzed, generated information from MCS analysis, and properties of the system components related to the failures. However, all proposed information is asked for by at least 3 stakeholders of the respondents.

The *hazard size* can be grouped according to its number of MCSs and its number of BEs: small hazards have in the order of magnitude of 10 MCSs or 10 BEs, medium having 100 MCSs or [100-1000[BEs, and large having 1000 MCSs or [1,000-10,000[BEs.

Our respondents have worked with different *sizes of hazards* in the range of [1-30 to 10,000-100,000] MCSs. This broad range is explained by the workplace of the respondents. Participants who worked with smaller hazards are working at a university, therefore, dealing with smaller systems, while the others are working in industry and at research institutes, thus, working with larger systems.

The *time needed to compare two hazards* of an average size (medium) regarding their single points of failure (SPoF) ranges from 5 minutes to 2 days. This means the respondents are using different tools with different facilities. However, because the time spent for comparing only two hazards reaches up to 2 days, there is a need for better visualizations to support this task and reduce the time and the effort spent by the analysts.

The most used *tools for MCS analysis* by the respondents are: ESSaRel followed by FaultTree+, BlockSim, and RiskSpectrum. The same tools are considered to be the *best tools currently used*. Grouped by working environment, the best tool used in industry is RiskSpectrum, at the university the best tools are ESSaRel, BlockSim, and FaultTree+, and at research institutes the best tools are ESSaRel and FaultTree+.

The estimation of the *time spent for performing MCS analysis* tasks was only provided by four participants (the question was not mandatory). These participants were using FaultTree+, ESSaRel, MagicDraw, and C^2FT. Even if these tools are considered being used, the participants spend a long time (i.e., several days) performing MCS analysis tasks for medium (i.e., order of magnitude: 100 MCSs) and large hazards (i.e., order of magnitude: 1000 MCSs). This highlights the need for better visualizations in order to reduce the time and effort spent on MCS analysis.

5.2 RQ-2: Requirements of the Safety Engineers for Performing MCS Analysis

The *most necessary information* for the participants for performing MCS analysis is: SPoF, MCSs' order, BEs with high FP, the system components related to the BEs, BEs with high number of occurrence, FP of the hazards, FP of the MCSs and BEs, MCSs with a high FP, and the system components properties. Surprisingly, the total number of BEs and the total number of MCSs were not considered to be necessary even though there is an inverse relationship between each of these two and the hazard's safety (if all other factors are fixed).

Four participants reported they *need additional information* for performing MCS analysis. The only two related to MCS analysis are: the information about the architectural components providing the largest amount of basic failures or MCS, and dependency matrices. There are five cases possible for a dependency matrix related to this study. These cases are a dependency matrix between: BEs, BEs and physical components, MCSs, MCSs and physical components, and BEs and MCSs. All other additional information reported either refers to higher level analysis tasks comprising MCS analysis (i.e., fault tree analysis, hazard analysis, risk analysis) or to other types of analysis (i.e., minimal path set (MPS) analysis, sensitivity, and uncertainty analysis) that are not in the focus of this study.

Regarding *MCS analysis tasks*, the participants considered 13 out of 17 tasks (removed duplicates) to be needed during performing MCS analysis. From these four tasks the two: "to prioritize MCSs for detailed analysis" and "to know the BEs with high FP" are not consistent with answers to other questions of the questionnaire where they are considered to be important tasks. The reason is probably that these two points in the question were negated and the participants did not negate their answers accordingly. All other answers are consistent across the different questions . Therefore, the tools used for performing MCS analysis should provide the capability of performing these tasks efficiently.

5.3 RQ-3: Quality of Current Tools for Performing MCS Analysis as Perceived by the Safety Engineers

With regard to the *practitioners' perception of the quality of information representation:* The participants rate the information generated by the best tool they currently use for performing MCS analysis as *understandable*. Since the respondents rate the *quality of the information representation* as fair, where fair is the midpoint of the ratings, improving current representations (currently mostly in textual and tabular form) or adding new ones is highly recommended.

The respondents consider FaultTree+, BlockSim, and ESSaRel being the most powerful tools in their ability of providing the *information in one sight*. However, since the ability of the other tools is limited, there should be improvements on the representations of "BEs' number of occurrence", "safety level of the hazard being analyzed", "MCSs effected by a BE", and "System design", followed by "MCSs with low order" and "MCSs with SPoF".

RQ-3b *Practitioners' Perception of the Quality of Interaction Capabilities:* From the point of view of the respondents, the most powerful tools providing the *interaction capability of sorting* all information other than "both BE's FP and BE's number of occurrence" are RiskSpectrum, FaultTree+, and BlockSim. All other tools provide the sorting for only 2 out of 6 options mentioned in the questionnaire. Therefore, the interaction capability for sorting the BE's FP, BE's number of occurrence, both BE's FP and number of occurrence, and both MCS's FP and order should be added to the tools, according to the requirements.

Surprisingly, from the point of view of the respondents, there is only one tool that can provide the *interaction capability of filtering* all of the 4 options provided in the questionnaire, namely WinNUPRA. According to the requirements, filtering by MCS FP, MCS order, BE FP, and BE number of occurrence are very important interaction features that should be added to the tools for supporting MCS analysis.

Finally, the *amount of interaction regarding scrolling up/down in the same view* is found to be "too much" and should be reduced.

RQ-3c *Practitioners' Perception of Usability:* From the point of view of most of the respondents the best tool they are currently using has a good *usability*. Nevertheless, there were 2 points that the participants were unsure of: the ease of use of the tools and whether the tools support them in communicating with their stakeholders.

Since all other points regarding the ease of use are ranked positively, the reason is probably that the statement of the ease of use was negated and the participants did not negate their answers accordingly.

Thus, the only point that the respondents were unsure of is whether the tools support them in communicating with the stakeholders. Therefore, it is advisable to develop improved visualizations or suitable environments to support this communication.

RQ-3d*: Level of Satisfaction with Current Tools for Performing MCS Analysis:* The *satisfaction level* of most participants is positive regarding the best tool they are currently using for performing MCS analysis.

Finally, most participants prefer *2D over 3D interaction* when analyzing MCSs. One reason of such a preference might be that they are used to working in 2D environments.

6 FINDINGS

The results of this study show, that MCS analysis is performed for a large amount of purposes, and thus the results confirm the importance of MCS analysis for safety analysis. However, we found some unexpected results that are reported next. Moreover, we derive recommendations for improving the safety tools to support MCS analysis.

Unexpected Results: There were two unexpected results of this study. The results show, that classifying MCSs regarding their FPs is a task that is rarely performed by the participants even though it helps in identifying weak points of a top level event of a system (**RQ-1**). Surprisingly, the total number of BEs and the total number of MCSs were not considered to be necessary even though there is an inverse relationship between each of these two and the hazard's safety (fixing all other factors, **RQ-2**).

Recommendations for Tool Improvements: The tools should provide and adequately represent: BEs with high FP, system components related to BEs, BEs with high number of occurrence, properties of system components, and dependency matrices (**RQ-2**). Further, the most important need (**RQ-3**) is to improve the representation for gaining immediate insight regarding the BEs' number of occurrence, the safety level of the hazard being analyzed, the MCSs effected (and their quality) by a BE, MCSs with SPoF, and the system design (i.e., physical parts). Regarding interaction, sorting by BE's FP, BE's number of occurrence, both BE's FP and number of occurrence, and both MCS's FP and order, should be provided by the tools, according to the requirements (**RQ-3**). Moreover, filtering by MCS FP, MCS order, BE FP, and BE number of occurrence is a very important interaction feature lacking in many of the tools, and thus, should be added to the tools for supporting MCS analysis (**RQ-3**). Finally, the amount of interaction regarding scrolling up/down in the same view is found to be too much and should be reduced.

Although the high usability rating of the tools, these tools do not support the communication between the safety analysts and their stakeholders. Therefore, visualizations beyond text and tables should be introduced to support this communication.

Surprisingly, the time estimation for comparing two hazards of an average size (medium) regarding their single points of failure reaches up to two days (**RQ-1**). This is a long time and implies a large effort spent on this task. Additionally, the time spent to perform MCS analysis tasks in general is up to several days, even for hazards having an order of magnitude of 100s MCSs (medium size, **RQ-1**). As most information generated by MCS analysis is represented in text and tables, the recommendation is as well to use new visualizations in order to reduce the amount of time and effort spent in exploring and searching for the relevant information.

As expected, 2D interaction is preferred over 3D interaction for interacting with the generated information from MCS analysis (**RQ-3**).

7 THREATS TO VALIDITY

Content validity: To improve content validity, we performed peer reviews with experts from empirical research, safety analysis, and visualization. Additionally, we used standardized instrument for measuring the practitioners' perception of usability, namely the Technology Acceptance Model [32]: p. 448.

Response rate: To increase the response rate, we followed the recommendations of Cleeland et al. [9], DeMaio at al. [10], and [23]. The final response rate was 45%. This is considered high in case of online surveys. The average time for answering the survey was 26 minutes (SD = 17 minutes). Respondents did not interrupt the questionnaire or had time to talk to colleagues about it. Thus, there is no history of maturation bias possible.

Representative sample: We used a purposive sample. The demographic analysis of the respondents shows that they are representative of the target population. All participants are experienced in MCS analysis and in using tools for performing MCS analysis. However, the generalization of the results requires a replication of the survey with a larger sample of safety engineers. Overall, we conclude that there is no socially desirable responding bias and that the answers mirror the participants' insights.

Section/Question completion rate: We aimed at increasing the sections and questions completion by adding filters. Only one respondent did not complete the survey. Therefore, the corresponding responses were not included in the analysis. The participants answered all mandatory questions (31 from 36 for survey I, 32 from 37 for survey II). Optional questions (5) were intended to give the opportunity to comment mandatory questions. Thus, the non-observation bias is low.

Hawthorne effect: To avoid that participants answered questions randomly, we included duplicated and rephrased questions. The analysis of these questions shows that their answers are mostly consistent. Further, we checked whether the participants were always marking left or right extreme or the middle of the options (controlled integrity). That was not the case.

8 CONCLUSION

We aimed at (1) exploring the state of the practice regarding Minimal CutSet (MCS) analysis from the perspective of practitioners, i.e., safety engineers and (2) identifying gaps in the representation of the important information used for MCS analysis. Furthermore, this research is intended for transferring the knowledge of the safety to the visualization domain to provide a better understanding, and guide the development of MCS analysis tools and visualizations. An online survey was used to gain as many replies as possible from experienced people in MCS analysis from different working environments who use safety tools for performing the analysis. This study determined important tasks performed during MCS analysis, its purposes, the requirements of the analysts, hazard sizes being confronted with, and the stakeholders benefiting from the results of MCS analysis. Moreover, it contributes to identify the improvements needed: integrating new or improved visualizations while preserving the 2D interactions. Finally, we identify the need of replicating this survey with a larger sample in order to get a better understanding of MCS analysis needs in specific safety domains and identifying commonalities and variability across different domains.

Acknowledgements

We would like to thank our participants for their time and effort and the PSAM11/ESREL2012 conference for supporting this study. This work was partially funded by the DFG (IRTG 1311), the DAAD (PhD grant), and the BMBF (project ViERforES (I) [17]).

References

[1] Yasmin I. Al-Zokari, Liliana Guzman, Dirk Zeckzer, and Hans Hagen, 2012. http://www.irtg.uni-kl.de/~alzokari/Demos/MyDemos.html.

[2] Yasmin I. Al-Zokari, Taimur Khan, Daniel Schneider, Dirk Zeckzer, and Hans Hagen. CakES: Cake Metaphor for Analyzing Safety Issues of Embedded Systems. In Hans Hagen, editor, *Scientific Visualization: Interactions, Features, Metaphors*, volume 2 of *Dagstuhl Follow-Ups*, pages 1–16. Schloss Dagstuhl–Leibniz-Zentrum fuer Informatik, 2011.

[3] Yasmin I. Al-Zokari, Daniel Schneider, Dirk Zeckzer, Liliana Guzman, Yarden Livnat, and Hans Hagen. Enhanced CakES representing Safety Analysis results of Embedded Sys-

tems. In *Federated Conference on Computer Science and Information Systems (FedCSIS), RAMSS track*, 2011.

[4] Yasmin I. Al-Zokari, Daniel Schneider, Dirk Zeckzer, and Hans Hagen. An Enhanced Slider for Safety Analysis. In *IEEE VisWeek Poster*, 2011.

[5] Yasmin I. Al-Zokari, Dirk Zeckzer, Peter Dannenmann, Liliana Guzman, and Hans Hagen. Evaluation of "Safety-Domino": a Graphical Metaphor for Supporting Minimal Cut Set Analysis. *IADIS International Journal on Computer Science and Information Systems, online journal*, 7:129–151, 2012.

[6] Marco Bozzano and Adolfo Villafiorita. *Design and Safety Assessment of Critical Systems*. CRC Press (Taylor and Francis), an Auerbach Book, 2010.

[7] Marko Cepin. DEPEND-HRA-A method for consideration of dependency in human reliability analysis. *Reliability Engineering & System Safety*, 93(10):1452–1460, 2008.

[8] Jason Cleeland, Carsten Schmitz, Thibault Le Meur, Marcel Minke, David Olivier, and Thomas White. LimeSurvey: Features. http://www.limesurvey.org/en/about-limesurvey/features; Online; accessed 31-Jan-2012.

[9] Jason Cleeland, Carsten Schmitz, Thibault Le Meur, Marcel Minke, David Olivier, and Thomas White. LimeSurvey: Question Types. http://docs.limesurvey.org/Question+types; Online; accessed 31-Jan-2012.

[10] Theresa J. DeMaio, Jennifer Rothgeb, and Jennifer Hess. Improving Survey Quality Through Pretesting, 1998.

[11] Nasser S. Fard. Determination of minimal cut sets of a complex fault tree. *Comput. Ind. Eng.*, 33(3–4):59–62, December 1997.

[12] N.S. Fard and T.H Lee. Cutset enumeration of network systems with link and node failure. *Reliability Engineering & System Safety*, 65:141–146, 1999.

[13] J. B. Fussell and W. E. Vesely. A New Methodology for Obtaining Cut Sets for Fault Trees. *Transactions American Nuclear Society*, 15(1):262–263, 1972.

[14] Stian Høiset and Eli Glittum. Risk and Safety Training Using Virtual Reality (VR-Safety). In *SPE International Conference on Health, Safety, and Environment in Oil and Gas Exploration and Production*. Society of Petroleum Engineers, April 2008. http://www.cmr.no/cmr_computing/index.cfm?id=179523.

[15] IBM. SPPS Statistics 17. http://www.ibm.com/software/de/analytics/spss/.

[16] IBM. Transform to an on demand business thru IBM Business Integration Solution, 2007. www-07.ibm.com/hk/events/2007/09_csolution/downloads/TransformtoanondemandbusinessthruIBMBusinessIntegrationSolution.pdf.

[17] Fraunhofer Institute. Virtual and Augmented Reality for Maximum Embedded System Safety, Security and Reliability (ViERforES). http://www.iff.fraunhofer.de/en/research-network/vierfores.html.

[18] Taimur Khan, Daniel Schneider, Yasmin Al-Zokari, Dirk Zeckzer, and Hans Hagen. Framework for Comprehensive Size and Resolution Utilization of Arbitrary Displays. In Hans Hagen, editor, *Scientific Visualization: Interactions, Features, Metaphors*, volume 2 of *Dagstuhl Follow-Ups*, pages 144–159. Schloss Dagstuhl–Leibniz-Zentrum fuer Informatik, 2011.

[19] Christoph Lauer, Reinhard German, and Jens Pollmer. Fault tree synthesis from UML models for reliability analysis at early design stages. *SIGSOFT Softw. Eng. Notes*, 36(1):1–8, January 2011.

[20] Hung-Yau Lin, Sy-Yen Kuo, and Fu-Min Yeh. Minimal Cutset Enumeration and Network Reliability Evaluation by Recursive Merge and BDD. In *Proceedings of the Eighth IEEE Symposium on Computers and Communications (ISCC 2003)*, pages 1341–1346, 2003.

[21] Y. Papadopoulos, J. McDermid, R. Sasse, and G. Heiner. Analysis and synthesis of the behaviour of complex programmable electronic systems in conditions of failure. *Reliability Engineering & System Safety*, 71(3):229–247, 2001.

[22] S. A. Patterson and G. E. Apostolakis. Identification of critical locations across multiple infrastructures for terrorist actions. *Reliability Engineering & System Safety*, 92(9):1183–1203, 2007.

[23] QuestionPro Online Survey Software. 10 Easy Ways to Increase Response Rates for your Online Survey, 2012. http://www.questionpro.com/a/showArticle.do?articleID=deploy01.

[24] J. Russell. *Exploring Psychology: AS Student Book for AQA A*. Exploring Psychology. OUP Oxford, 2008. http://books.google.de/books?id=4ereiHom3d8C.

[25] E. O. Schweitzer, Bill Fleming, Tony J. Lee, and Paul M. Anderson. Reliability Analysis of Transmission Protection Using Fault Tree Methods. In *Proceedings of the 24th Annual Western Protective Relay*, 1997.

[26] Robert F. Szafran. *Answering Questions With Statistics*. SAGE Publications, 2011. http://books.google.de/books?id=GEJTD6kr08oC.

[27] A. Thums and G. Schellhorn. Formal Safety Analysis in Transportation Control. In *Proceedings of the Workshop on Software Specification of Safety Relevant Transportation Control Tasks*, 2002.

[28] Edward R. Tufte. *Visual explanations: images and quantities, evidence and narrative*. Graphics Press, Cheshire, CT, USA, 1997.

[29] Edward Rolf Tufte. *The visual display of quantitative information*. Graphics Press, 1992.

[30] J. Utts and R. Heckard. *Statistical Ideas and Methods*. Cengage Learning, 2005.

[31] Jørn Vatn. Finding minimal cut sets in a fault tree. *Reliability Engineering & System Safety*, 36(1):59–62, January 1992.

[32] Viswanath Venkatesh, Michael G. Morris, Gordon B. Davis, and Fred D. Davis. User acceptance of information technology: Toward a unified view. *MIS Quarterly*, 27(3):425–478, 2003. http://www.jstor.org/stable/30036540.

[33] Matthew Ward, Georges Grinstein, and Daniel A. Keim. *Interactive Data Visualization: Foundations, Techniques, and Application*. A.K. Peters, Ltd, May 2010.

[34] Claes Wohlin, Per Runeson, Martin Höst, Magnus C. Ohlsson, Björn Regnell, and Anders Wesslén. *Experimentation in Software Engineering*. Computer Science. Springer, 2012.

[35] Yi Yang, Patric Keller, and Peter Liggesmeyer. Visual Approach Facilitating the Importance Analysis of Component Fault Trees. In Frank Ortmeier and Peter Daniel, editors, *Computer Safety, Reliability, and Security - SAFECOMP 2012*, volume 7613 of *Lecture Notes in Computer Science*, pages 486–497, Berlin/Heidelberg, Germany, 2012. Springer.

[36] Yi Yang, Patric Keller, Yarden Livnat, and Peter Liggesmeyer. Improving Safety-Critical Systems by Visual Analysis. In Christoph Garth, Ariane Middel, and Hans Hagen, editors, *Proceedings of IRTG 1131 Workshop 2011*, volume 27 of *OpenAccess Series in Informatics (OASIcs)*, pages 43–58, 2012.

[37] Yi Yang, Dirk Zeckzer, Peter Liggesmeyer, and Hans Hagen. ViSSaAn: Visual Support for Safety Analysis. In Hans Hagen, editor, *Scientific Visualization: Interactions, Features, Metaphors*, volume 2 of *Dagstuhl Follow-Ups*, pages 378–395. Schloss Dagstuhl–Leibniz-Zentrum fuer Informatik, 2011.

[38] Wei-Chang Yeh. An improved algorithm for searching all minimal cuts in modified networks. *Reliability Engineering & System Safety*, 93(7):1018–1024, 2008.

[39] Wei-Chang Yeh. A simple minimal path method for estimating the weighted multi-commodity multistate unreliable networks reliability. *Reliability Engineering & System Safety*, 93(1):125–136, 2008.

[40] T. Yuge and S. Yanagi. Quantitative analysis of a fault tree with priority AND gates. *Reliability Engineering & System Safety*, 93(11):1577–1583, 2008.

[41] Tan Z. Minimal cut sets of s-t networks with k-out-of-n nodes. *Reliability Engineering & System Safety*, 82(1):49–54, 2003.

[42] E. Zio, M. Librizzi, and G. Sansavini. Determining the minimal cut sets and fussell-vesely importance measures in binary network systems. *Reliability Engineering and System Safety*, pages 723–729, 2006.

Risk Analysis and Decision Theory: An Extended Summary

E. Borgonovo[a], V. Cappelli[*a], F. Maccheroni[a], M. Marinacci[a], and C. Smith[b]

[a] Department of Decision Sciences and IGIER, Università Bocconi, Milan, Italy
[b] Idaho National Laboratory, Idaho Falls, Idaho, USA.

Abstract: We reconcile Kaplan and Garrick's seminal definition of risk with classical subjective expected utility, filling in the relevant gaps and providing a framework that is ready-to-use in applications. We show that Kaplan and Garrick's "frequency" format can be set in one-to-one correspondence with [26]'s utility theory. Kaplan and Garrick's "probability" format corresponds to the framework of [22] in which epistemic uncertainty is captured by a subjective probability over uncertain events. Finally, Kaplan and Garrick's "probability of frequency" format, the most general one, corresponds to the recently proposed framework of [13], which distinguishes aleatory and epistemic uncertainty in a Bayesian perspective. The classic Kaplan and Garrick's risk triplets are then cast in the powerful setting of axiomatic Decision Theory, with its solid behavioral foundations, allowing one to make explicit the often implicit decisions of a Risk Analysis.

Keywords: Decision Analysis, Risk Analysis, Decision Theory, Uncertainty Analysis.

1. INTRODUCTION

In the management of complex technological systems, the term risk analysis refers to the part of the policy-making process associated with the identification of scenarios and their likelihoods [14,20]. The location of a nuclear waste repository [17], the programming of a space mission [10,16,24], and the evaluation of design-changes in chemical and nuclear plants [11,12] are a few examples in which decision-making is informed by a risk analysis, in the so-called risk-informed decision-making [3]. This discipline has gained a significant amount of attention from both policymakers and the public over the past 30 years, as the interaction of technology and policy choices has become more predominant in the evaluation of trade-offs in a democratic society [4, p. 621].[1]

Over the years, Kaplan and Garrick's definition of risk [19] has become one of the pillars of risk analysis, guiding several key studies performed by national and international agencies and laboratories (for instance, Kaplan and Garrick's risk triplets are a structural part of NASA's recent risk management handbook, see, for example, [15]). From a theoretical viewpoint, the triplet structure introduced by Kaplan and Garrick remains in recent generalizations of the risk concept [1,6,7,8]. These works signal a common trait of risk analysis, that is, the consideration of risk as a self-standing concept, apart from an underlying decision-analytical background. This separation is considered attractive by some researchers (see the debates reported in [5]), insofar it permits the extension of Kaplan and Garrick's definition of risk to non-probabilistic approaches.[2] However, it has the drawback of vanishing the normative support that an underlying decision-analytical rationale brings to a risk analysis.

Indeed, in their seminal 1981's article Kaplan and Garrick maintain that risk must thus be considered always within a decision theory context [19, p. 25]. Several subsequent works discuss risk analysis from a decision-making viewpoint [2,4,18,28]. Both [20] and [14] underline that risk analysis and

---

[*] Corresponding author: veronica.cappelli@unibocconi.it
[1] For an early and critical review about risks and benefits of technological systems, we refer to [25]).
[2] The problem was already clear at the time of Kaplan and Garrick. As they state: *one often hears people say that we cannot use probability because we have insufficient data, in light of our current definitions, we see that this is a misunderstanding. When one has insufficient data, there is nothing else one can do but use probability* [19].

decision analysis are intertwined: a decision analysis can include a risk analysis component [20, p. 220]. Nonetheless, the decision-analytical background upholding Kaplan and Garrick's definition itself has not been investigated in depth to date.

This gap in risk analysis leads us to the decision side. Subjective expected utility originates in the seminal works of [26], [27], and [22], which have become the pillars of modern decision analysis [21,23]. This theory features a decision maker (DM) that evaluates acts whose consequences depend on states of the environment generated by mechanisms that are only partially known or understood. Each such mechanism corresponds to a probabilistic model that describes the frequency of the various states inherent to the phenomenon at hand. The information available to the DM allows her to posit a set of possible mechanisms, that is, of possible probabilistic models. In general, such set is not a singleton because information is not sufficiently accurate to pin down a single mechanism. In other words, the DM is uncertain about the true probability model. Thus, one considers two layers of uncertainty as follows [2]: the irreducible aleatory uncertainty (physical risk) about states, described by probabilistic models, and epistemic uncertainty about such models characterized by a prior probability over them. [13] show that, if the DM's preferences satisfy [22]'s axioms plus a consistency condition, then one obtains a subjective expected utility functional where the distinction of the two layers becomes meaningful.

The heart of the present work is the reconciliation of Kaplan and Garrick's definition of risk in its various aspects with the corresponding decision-theoretical rationales. We show that the notion of scenario in Kaplan and Garrick's risk triplets is in one-to-one correspondence with the decision-theoretical notion of event. We then show that Kaplan and Garrick's "frequency" format can be embedded in the [26] decision-theoretic framework and that Kaplan and Garrick's "probability" format is in correspondence with [22]'s expected utility framework, in which uncertainty is described by a subjective probability. Finally, we show that the "probability of frequency" format, where the aforementioned two layer distinction is applied, finds its natural axiomatic collocation within the recently proposed extension of subjective expected utility of [13]. As a side finding, [2]'s concept of unconditional model of the world finds its correspondence in the decision-theoretical notion of probabilistic reduction.

We illustrate the discussion through several examples. In particular, starting from seismic probabilistic risk assessment, we embed the three levels of Probabilistic Safety Assessment (PSA) in the decision-theoretical setup. Our numerical application concerns decision-making in space PSA. Since the late '90s, NASA uses probabilistic risk assessment in all of its programs and projects to support optimal management decision for the improvement of safety and program performance [24, p. 11]. In evaluating space missions, two consequences are typically considered, loss of crew and loss of mission. We address the decision-analytical rationale supporting the formulation of the problem in terms of an acceptability threshold on the probability of loss of crew (loss of mission), providing the overarching framework to the approach used in current practice.

For more details, we refer the interested reader to the working paper version of this paper [9], which will be soon available on the IGIER website.

2. CONCLUSION

Several applications foresee a risk analysis that supports an overall decision analysis problem. One of the cornerstones of risk analysis is represented by Kaplan and Garrick's risk triplets and their quantitative definition of risk. In this work, we have addressed what is the decision-theoretical rationale that supports the use of these risk triplets, in consideration of the well-known distinction between aleatory and epistemic uncertainty. We have seen that Kaplan and Garrick's framework can be set in one-to-one correspondence with suitable decision-theoretic rationales and their more general format is encompassed by the Classical Subjective Expected Utility framework. The work provides an approach ready to use in application, which is illustrated in [9] through the use of several case studies, among which the decision-process associated with the planning of a space mission.

Acknowledgements

The authors at the Department of Decision Sciences and IGIER, Università Bocconi, Milano, Italy, gratefully acknowledge the financial support of the European Research Council (BRSCDP-TEA) and of the AXA Research Fund.

References

[1] C. E. Althaus. "*A disciplinary perspective on the epistemological status of risk*", Risk Analysis, 25, pp. 567-588, (2005).
[2] G. E. Apostolakis. "*The concept of probability in safety assessments of technological systems*", Science, 250, pp. 1359-1364, (1990).
[3] G. E. Apostolakis. "*How useful is quantitative risk assessment?*", Risk Analysis, 24, pp. 515-520, (2004).
[4] G. E. Apostolakis and S. E. Pickett. "*Deliberation: Integrating analytical results into environmental decisions involving multiple stakeholders*", Risk Analysis, 18, pp. 621-634, (1998).
[5] T. Aven. "*Foundational issues in risk assessment and risk management*", Risk Analysis 32, pp. 1647-1656, (2012a).
[6] T. Aven. "*The risk concept - historical and recent development trends*", Reliability Engineering & System Safety, 99, pp. 33-44, (2012b).
[7] T. Aven. "*Practical implications of the new risk perspectives*", Reliability Engineering and System Safety, 11, pp. 136-145, (2013).
[8] T. Aven, O. Renn, and E. A. Rosa. "*On the ontological status of the concept of risk*", Safety Science, 49, pp. 1074-1079, (2011).
[9] E. Borgonovo, V. Cappelli, F. Maccheroni, M. Marinacci, and C. L. Smith. "*Risk analysis and decision theory*", submitted, (2014).
[10] E. Borgonovo and C. L. Smith. "*A study of interactions in the risk assessment of complex engineering systems: An application to space psa*", Operations Research, forthcoming, (2011).
[11] R. F. Boykin, R. A. Freeman, and R. R. Levary. "*Risk assessment in a chemical storage facility*", Management Science, 30, pp. 512-517, (1984).
[12] M. A. Caruso, M.G. Cheok, M.A. Cunningham, G. M. Holahan, T.L. King, G.W. Parry, A. M. Ramey-Smith, M.P. Rubin, and A.C. Thadani. "*An approach for using risk assessment in risk-informed decisions on plant-specic changes to the licensing basis*", Reliability Engineering & System Safety, 63, pp. 231-242, (1999).
[13] S. Cerreia-Vioglio, F. Maccheroni, M. Marinacci, and L. Montrucchio. "*Classical subjective expected utility*", PNAS, 110, pp. 6754-6759, (2013).
[14] R. T. Clemen, and T. Reilly. "*Correlations and copulas for decision and risk analysis*", Management Science, 45, pp. 208-224, (1999).
[15] H. Dezfuli, A. Benjamin, C. Everett, G. Maggio, M. G. Stamatelatos, and R. Youngblood. "*NASA risk management handbook*", NASA SP-2011-3422, (2011).
[16] R. L. Dillon, M. E. Paté-Cornell, S. D. Guikema. "*Programmatic risk analysis for critical engineering systems under tight resource constraints*", Operations Research, 51, pp. 354-367, (2003).
[17] J. B. Garrick and S. Kaplan. "*A decision high- level theory perspective on the disposal of radioactive waste*", Risk Analysis, 19, pp. 903-913, (1999).
[18] R.A. Howard. "*Decision analysis: Practice and promise*", Management Science, 34, pp. 679-695, (1988).
[19] S. Kaplan and B. J. Garrick. "*On the quantitative definition of risk*", Risk Analysis, 1, pp.1-28, (1981).
[20] M. E. Paté-Cornell and R.L. Dillon. "*The respective roles of risk and decision analyses in decision support*". Decision Analysis, 3, pp. 220-232, (2006).

[21] J. W. Pratt, H. Raiffa, and R. Schlaifer. *"Statistical Decision Theory"*, MIT Press, 1995, Cambridge (MA), USA.
[22] L. J. Savage *"The foundations of Statistics*, Wiley and Sons, 1954, New York (NY), USA.
[23] J. E. Smith and D. von Winterfeldt. *"Decision analysis in management science"*, Management Science, 50, pp. 561-574, (2004).
[24] M. Stamatelatos, M., H. Dezfuli, G. Apostolakis, C. Everline, S. Guarro, D. Mathias, A. Mosleh, P. Todd, R. David, S. Curtis, V. William, and R. Youngblood. *"NASA probabilistic risk assessment procedures guide for NASA managers and practitioners"*, NASA/SP-2011-3421, (2011).
[25] C. Starr. *"Social benefit versus technological risk"*, Science, 165, pp. 1232-1238, (1969).
[26] J. von Neumann and O. Morgenstern. *"Theory of Games and Economic Behavior"*, Princeton University Press, 1947, Princeton (NJ).
[27] A. Wald. *"Statistical decision functions"*, Wiley and Sons, 1950, New York (NY), USA.
[28] R. Winkler. *"Uncertainty in probabilistic risk assessment"*, Reliability Engineering & System Safety, 54, pp. 127-132, (1996).

Maritime oil spill risk assessment for Hanhikivi nuclear power plant

Juho Helander[*]
Fennovoima, Helsinki, Finland
juho.helander@fennovoima.fi

Abstract:
Fennovoima is planning to build a new nuclear power plant unit, Hanhikivi 1, on a greenfield site in Pyhäjoki in Northern Finland. A nearby maritime oil spill accident is one of the external events analysed in the probabilistic risk assessment (PRA) of the plant. The oil effects on a nuclear power plant are not well-known, but in the worst case the oil could cause a loss of the ultimate heat sink by blocking the sea water intake screens.

By considering the maritime traffic, oil transport and oil spill accident data in the Baltic Sea area, it is evaluated that a nearby medium oil spill (100 - 1000 tonnes) occurs with a frequency of $1{,}0 \cdot 10^{-2}$ /a and a large spill (> 1000 tonnes) with a frequency of $3{,}0 \cdot 10^{-3}$ /a.

The probability that the spill drifts to Hanhikivi and oil combat measures fail is assessed by using event tree analysis. The spill behaviour is considered, including oil spreading, dissolution, dispersion and movement due to wind and currents. In addition, oil combat measures including the use of oil booms and skimmers are evaluated. According to the results, significant amount of oil could enter the plant intake tunnel with a frequency of $4{,}2 \cdot 10^{-5}$ /a.

Keywords: PRA, oil spill, oil risk, event tree

1. INTRODUCTION

An accidental oil spill at the sea near a nuclear power plant could affect the safe operation of the plant. The oil could enter the sea water intake, block the fine screens and basket chain filters and lead to loss of sea water cooling. This report presents an evaluation of the oil risk regarding the Hanhikivi nuclear power plant, which will be constructed in northern Finland in Pyhäjoki at the coast of the Bothnian Bay. Dozens of small oil spills are observed annually in the Finnish sea areas of the Baltic Sea. However, only large scale accidents related to oil tankers or other large ships containing large amounts of fuel oil could have an impact on the operation of a nuclear power plant.

In this report, a method for oil risk evaluation is presented. The accident frequencies are evaluated by using the accident data and maritime traffic volumes in the Baltic Sea and specifically in the Bothnian Bay. The accident propagation is modelled and quantified by using event tree analysis. The drifting direction of the oil spill, the reporting of the accident and oil combat measures at the open sea and near Hanhikivi are considered in the event trees. Also sensitivity and uncertainty analysis is performed.

2. EVALUATION METHOD

Let us give the following definition for the oil initiating event: "significant amount of oil enters the sea water intake tunnel of the plant". The evaluation of the oil initiating event frequency includes the following evaluation phases:
  1. Frequency of a nearby maritime oil spill accident
  2. Impact probability (the oil spill travels to Hanhikivi)
  3. Probability that oil combat measures fail

According to expert judgement, the operation of a nuclear power plant could be affected if 1 tonne of oil enters the intake tunnel, although no detailed studies have been performed. Furthermore, it can be assumed that at most 1 % of the spill total volume enters the intake tunnel. According to these assumptions, all oil spills that reach Hanhikivi and contain at least 100 tonnes of oil should be considered. Accident frequencies are calculated by using the compromise version of the PREB evaluation method described in [17].

The spill impact probability is assessed by considering oil product properties and prevailing wind and current conditions. The spill propagation can be prevented by using oil combat measures at the open sea and near Hanhikivi. Event tree analysis is used to determine the annual probability that significant amount of oil enters the plant intake tunnel.

The oil spill risks for the existing Finnish nuclear power plants (Loviisa and Olkiluoto) have been evaluated by using similar methods described in [1] and [15].

3. OIL ACCIDENT PROBABILITY

3.1 Location of the Hanhikivi site

The Baltic Sea is a mediterranean sea in northern Europe bounded by Finland, Sweden, Denmark, Germany, Poland, Lithuania, Latvia, Estonia and Russia. Fennovoima's nuclear power plant site is located at the coast of Bothnian Bay, which is the northernmost part of the Baltic Sea. The Bothnian Bay is relatively shallow with an average depth of approximately 40 meters and a maximum depth of 147 meters. The surface area of Bothnian Bay is roughly 37 000 km^2, the salinity ranges between 0.2 and 0.4 % and the sea is covered by an ice sheet typically from December to May.

3.2 Maritime traffic in the Baltic Sea and the Bothnian Bay

The ship traffic volumes in the Bothnian Bay are relatively low when compared to other parts of the Baltic Sea. Figure 1 illustrates the location of the Baltic Sea and its subregions (left picture), the ship movements in the Baltic Sea area during one week in 2008 (middle picture) and all the significant ports at the Bothnian Bay coast (right picture). The Hanhikivi site is located between Kalajoki and Raahe.

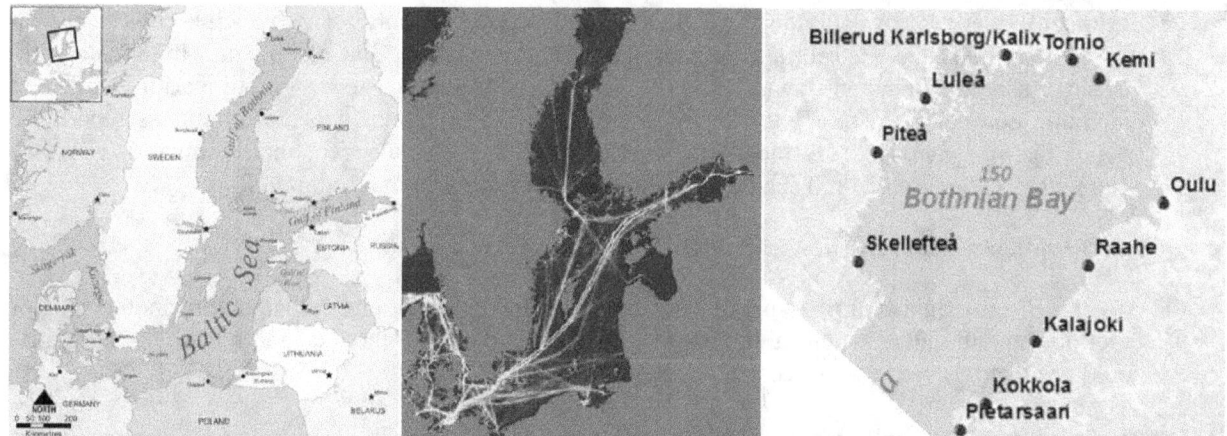

Figure 1. Baltic Sea location and subregions (left); ship movements in the Baltic Sea during one week in 2008 (middle); main ports in the Bothnian Bay coast (right). [3]

Most of the sea transports in the Bothnian Bay are related to other than oil products. No crude oil is being transported in the whole Gulf of Bothnia. The oil products transported include for example: coal tar, benzene, propane, butane, LNG, gasoline, kerosene and diesel fuel. Only relatively small ships can operate in the Bothnian Bay because the quays in the ports are typically less than 10 m deep. A typical cargo ship or oil tanker operating in the Bothnian Bay has a DWT (deadweight tonnage; a measure of how much weight a ship can carry, including cargo, fuel, fresh water, ballast water, provisions, passengers and crew) of roughly 10 000 and draught between 6 to 8 m. This also limits the size of the maximum possible oil spill. [19]

Statistics of the Bothnian Bay ports are presented in Table 1. The first figure (cargo) presents the sum of all cargo handled in the port annually. In addition, the sum of all liquid bulk cargo (including oil products, liquid chemicals and other liquid bulk) handled in the port is presented to give an insight on how much oil products are transported. Also the annual number of all ships and tankers visiting the port are presented. The target is to estimate which share of the Baltic Sea maritime traffic takes place in the Bothnian Bay, and by using this information, to deduce which share of the Baltic Sea oil spill accidents occurs in the Bothnian Bay. The Bothnian Bay accident data alone is insufficient for reliable estimation.

Table 1. Bothnian Bay port statistics in 2006. [2, 10, 11, 14]

Port name	Country	Cargo (million tonnes)	Liquid bulk (million tonnes)	Ships	Tankers
Pietarsaari	Finland	1,54	0,16	393	25
Kokkola	Finland	5,32	0,86	600	161
Kalajoki Rahja	Finland	0,29	0,00	74	0
Raahe	Finland	6,09	0,05	703	66
Oulu	Finland	2,99	0,97	586	144
Kemi	Finland	2,71	0,13	740	67
Tornio	Finland	1,89	0,09	425	*10*
Kalix	Sweden	0,17	0,00	36	0
Billerud Karlsborg	Sweden	0,10	0,02	33	*2*
Luleå	Sweden	7,49	0,35	603	31
Piteå	Sweden	1,53	0,21	336	21
Skellefteå	Sweden	1,83	0,57	371	19
Bothnian Bay SUM		32,0	3,4	4 900	546
Baltic Sea SUM		798	296	379 000	
Bothnian Bay / Baltic Sea		4,0 %	1,1 %	1,3 %	

estimate (no detailed data available)

The figures indicate that roughly 4 % of all cargo handled in the Baltic Sea area is handled in the Bothnian Bay ports. However, the share in liquid bulk products is only 1,1 %, which is consistent with the fact that the Baltic Sea oil transportation is concentrated in the Gulf of Finland. In terms of ship visits, the share of Bothnian Bay is 1,3 %. It can be concluded that 1 to 4 % of all maritime traffic in the Baltic Sea occurs in the Bothnian Bay, depending on the factor considered. In this report, we assume that the share is 2 %, which is the average from the three different share estimates. Thus, we assume that 2 % of the oil spill accidents in the Baltic Sea occur in the Bothnian Bay.

3.3 Oil spill accident statistics

An oil spill can originate from a ship that is leaking oil due to grounding, collision with another ship or other object, hull failure, technical failure, fire, explosions or some other reason. In addition, the spill could originate from a leak in a coastal oil depot.

The most severe oil spill accidents are related to tankers that could carry crude oil volumes up to 500 000 tonnes. One of the most recent major accident occurred to Prestige off the coast of Spain and Portugal in 2002 resulting in a 63 000 tonne spill of heavy oil. Large recovery operations were initiated and reportedly 50 000 tonnes of oil-water mixture was removed at the open sea. Nevertheless, roughly 1900 km of coastline was contaminated, and massive clean-up operations were necessary. [7]

Despite the steady global increase in the maritime oil transport since 1980's, the number of oil spills has decreased due to improved marine safety and the adoption of two-hull tankers. The number of spills larger than > 7 tonnes compared to the seaborne oil trade development since 1969 is illustrated in Figure 2. Similar trend can be seen in the Baltic Sea statistics. The total cargo volumes in the Finnish ports have increased from roughly 20 million tonnes in 1970 to more than 100 million tonnes in 2012. Also the oil transport volumes in the Baltic Sea have increased rapidly from roughly 120 million tonnes in 2000 to 290 million tonnes in 2010. All registered oil spills larger than 100 tonnes in the Baltic Sea area between 1969 and 2011 are listed in Table 2. Despite the increase in maritime traffic volumes, the number of accidents has remained roughly the same during different decades. [3, 7, 9, 11]

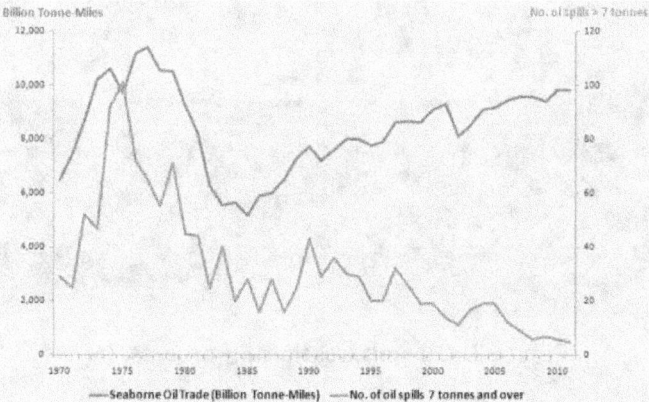

Figure 2. The seaborne oil trade and number of oil spills since 1969. [7]

Table 2. Registered oil spill accidents (≥ 100 tonnes) in the Baltic Sea between 1969-2011. [9]

Year, ship name, location and spill size (tonnes)			
1969, Raphael, Finland (Emäsalo)	250	1990, Volgoneft, Sweden (Karlskrona)	1000
1969, Palva, Finland (Utö)	200	1992, Unknown, Sweden (Västra Götaland and Halland)	200
1970, Esso Nordica, Finland (Pellinki)	600	1993, Kihnu, Estonia (Kopli Peninsula)	100
1970, Pensa, Finland (Hailuoto)	500	1995, Hual Trooper, Sweden (Öresund)	180
1977, Tsesis, Sweden (Stockholm)	1000	1998, Weston, Sweden (Västra Götaland)	4000
1979, Antonio Gramsci, Finland (Åland)	5500	1998, Pallas, Germany (Wadden Sea)	244
1979, Lloyd Bage, Finland (Harmaja)	100	1998, Nunki, Denmark (Kalundborg fjord)	100
1981, Globe Asimi, Lithuania (Klaipeda)	16000	2000, Alambra, Estonia (Muuga)	250
1984, Eira, Finland (Merenkurkku)	300	2001, North Pacific, Lithuania (Klaipeda)	3427
1985, Sotka, Finland (Märket)	370	2001, Baltic Carrier, Denmark (Kadetrenden)	2700
1987, Antonio Gramsci, Finland (Vaarlahti)	650	2003, Fu Shan Hai, Sweden (Ystad)	1200
1987, Tolmiros, Sweden (Västra Götaland)	400	2004, Herakles, Sweden (Grundkallen)	200
1987, Thuntank 5, Sweden (Bay of Gävle)	230	2006, Runner 4, (Gulf of Finland)	100
1988, Unknown, Sweden (Torekov)	287		

Let us calculate the Baltic Sea oil spill accident frequencies by using the compromise version of the PREB estimation method described in [17]: all spills ≥ 100 tonnes 0,64 /a; spills 100 - 1000 tonnes 0,50 /a; spills > 1000 tonnes 0,15 /a.

4. OIL SPILL BEHAVIOUR

4.1 The weathering process

Quite sophisticated models (e.g. Spillmod by Russian Sergey Ovsienko, State Oceanographic Institute, Russia), have been developed to evaluate oil spill behaviour after an oil spill accident. Spill movement and volume development is affected by several factors, such as accident location, oil properties, local hydrometeorological conditions (mostly wind and currents) and volume and intensity of the leakage. After oil is released in the sea, a process called weathering (Figure 3) begins, including rapid loss of volatile materials and mixing with water due to wind and waves. The oil becomes denser, more viscous and often forms emulsions. [5, 6]

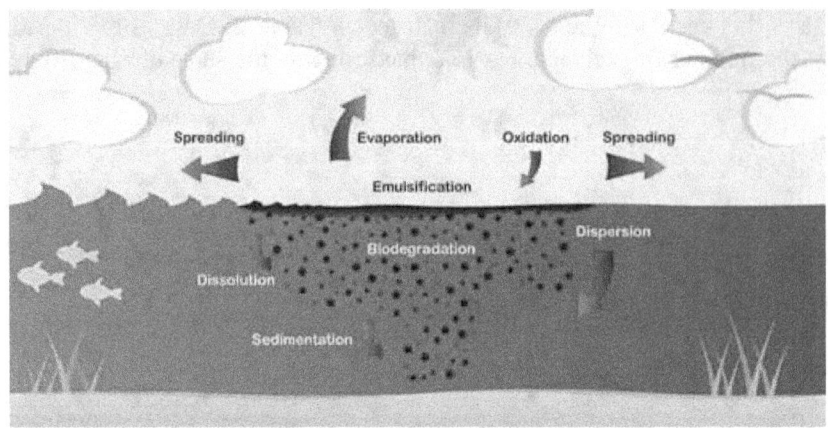

Figure 3. Oil spill weathering process. [6]

4.2. Crude oil and oil products

Crude oil (also called petroleum) is a naturally occurring liquid found in geologic formations beneath the Earth's surface. The densities of crude oils vary between 0,7 - 1,0 kg/dm^3 (typically around 0,85 kg/dm^3) and also the viscosity, color and other properties are different depending on the crude oil product in question.

Crude oils are mixtures of different hydrocarbons with different chemical properties. The different components can be separated by fractional distillation that allows the separation of chemical components with different boiling points one by one. The lightest distillates, such as liquified petroleum gas, propane and butane, have boiling temperatures below zero degrees (^{o}C) and are thus gases in room temperature. Some of the most important medium and heavy distillates are listed in Table 3. The terminology used is not completely established and products are called with different names in different countries. Some of the products listed (e.g. petrol) are not pure distillation products because they are enhanced with different additives. The products with a higher density are generally more viscose, less volatile and have higher boiling points. Residual fuel oil is actually not a distillate but a more impure residual product that is left when lighter components have been distilled away. Other significant raw oil distillate products not included in the table include for example: alkanes, lubricants, tar, asphalt, petroleum coke and aromatic petrochemicals.

Table 3. Some important oil products.

Fuel product	Density (kg/dm^3)	Primary use
Petrol (gasoline)	0,71 to 0,77	Vehicle internal combustion engines
Kerosene, jet fuel (paraffin)	0,78 to 0,81	Aircraft jet and gas turbine engines, cooking, lighting
Diesel fuel	~0,83	Diesel engines in cars, buses, trucks and ships, electricity production
Light fuel oil	0,80 to 0,90	Heating, diesel engines
Crude oil (petroleum)	~0,85 (0,7 to 1,0)	Raw product of different oil products
Distillate fuel oil (heavy fuel oil, marine gas oil)	0,89 to 0,92	Ships, other large diesel engines
Residual fuel oil (heavy fuel oil, marine fuel oil, bunker fuel)	0,96 to 1,01	Ships, heating plants

4.3 Oil spill volume development

Oil dissipation and persistence depend on its properties. Light crude oil distillates, such as petrol, kerosene and diesel oil evaporate quickly and do not require any cleaning-up measures. Lighter products also have a smaller effect on the intake screens and heat exchangers of the power plant. Heavier products, such as crude oils, fuel oils and lubricating oils, are much more persistent.

The development of the oil-water mixture volume as a function of time has been assessed by categorizing the oil products into 4 different categories according to their density. The volume development and the categorization of oil products according to [6] is presented in Figure 4. At an early phase of the accident, the volume of oil waste increases in groups II, III and IV because of the viscous emulsions formed with water.

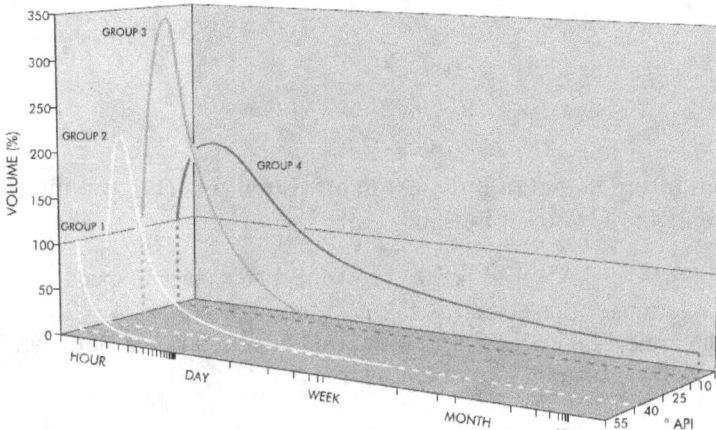

	Density (kg/dm^3)	Examples
I	< 0,80	Petrol, Kerosene
II	0,80 - 0,85	Diesel fuel, Light fuel oil, Abu Dhabi Crude Oil
III	0,85 - 0,95	Distillate fuel oil, Arabian Light Crude Oil, North Sea Crude Oil
IV	> 0,95	Residual fuel oil, Venezuelan Crude Oil

Figure 4. The volume of oil-water mixture remaining on the sea surface as a function of time for different oil product categories. [6]

4.4 Oil spill movement

Spreading of oil is mainly affected by oil viscosity and also by environmental conditions. If a 10 000 tonne oil spill occurs, the estimated spill area after 1 day is roughly 10 km^2 and the oil layer thickness roughly 1 mm. After 3 days the estimated area is 15-20 km^2 and thickness 0,5 mm and after 5 days 20-30 km^2 and 0,3 mm. If the oil accident occurs during winter time, ice partially prevents the spill from spreading. Part of the oil could, however, spread under the ice sheet. Also oil combating is more difficult during winter. It has been estimated that if more than 10 000 tonnes of oil come ashore, more than 100 km of coast is contaminated. Respectively, 10-100 km of shore is assumed to be affected if the amount of oil is 1000-10000 tonnes. If the oil amount is less than 1000 tonnes, less than 10 km of shoreline is affected. [19]

The oil spill is moved by wind and water currents. There are no strong and stable currents in the Bothnian Bay, and thus the water movements are fluctuating and unpredictable. The wind only affects the water movement in the summer. When the sea is covered by an ice sheet (on average from November to May), the water movement is mostly affected by river flow rates, water level fluctuations and temperature and salinity differences. [13]

The average wind speed in the Bothnian Bay is roughly 7 m/s and the probabilities of different wind directions are presented in Figure 5 (left picture). Roughly half of the time, the wind direction is between south and west. [16]

In the Gulf of Bothnia, two main currents can be recognized: one in the Bothnian Bay and another one in the Bothnian Sea (Figure 5, right picture). The Coriolis effect causes the water to flow anticlockwise and thus the water flows northwards in the Finnish coast and southwards in the Swedish coast. The Quark is an intermediate zone between these two main currents. The average current speeds in the Baltic Sea are typically between 5-10 cm/s. The measurements performed around Hanhikivi confirm that the prevailing current direction is from southwest to northeast and the average current speed near the surface is less than 10 cm/s, whereas the highest speeds measured are 50 cm/s. [13]

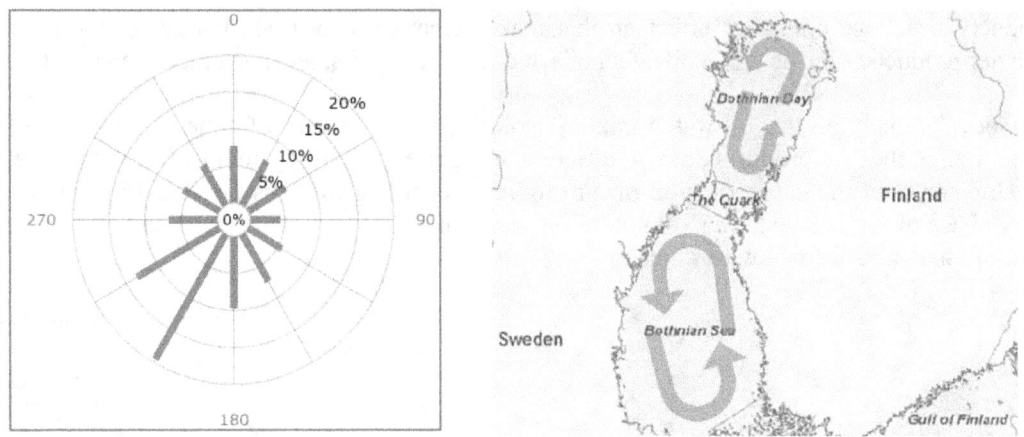

Figure 5. The average wind directions in the Bothnian Bay (40 km off Hanhikivi) [16] and the main currents in the Gulf of Bothnia. [13]

It has been roughly estimated that the oil spill travel speed is 100 % of the water current speed and 3 % of the wind speed [5]. By using the estimates presented (5-10 cm/s current to northeast + 7 m/s wind to northeast), we can assess that the typical oil spill drifting speed and direction in the Bothnian Bay is approximately 30 cm/s to northeast. This equals 1,1 km/h or 26 km/d.

5. OIL COMBAT

5.1 Oil combat techniques

Several maritime oil combat measures are available and have been discussed for example in [5]. When assessing the oil spill risk for Hanhikivi 1, it is important to evaluate the open sea combat measures that can be used to prevent the oil spill from reaching the plant sea water intake.

Oil booms are used to surround and concentrate an oil spill in the open sea to enable oil recovery. Booms are also effective in protecting small and important targets in the coast, such as power plants. Booms are typically deployed by using boats, but deployment from shore or helicopter is also possible. The effectiveness of booming depends on the oil properties and weather conditions. If the spill is already spread out on a large area, it might be impossible to surround it. Heavy oils may go under the booms, and strong wind and high waves could cause the oil to splash over the boom. The booms could also be carried away or be completely broken up by wind, wave and currents. Figure 6 illustrates a deployed oil boom in the Kalajoki harbour. This is a typical boom used in Finland with an approximate 1 meter height (1/3 of which above surface) and consisting of consecutive sections chained together. There are also single-use booms that can be used to absorb oil.

Figure 6. A deployed oil boom in the Kalajoki harbour. [8]

Specialized ships equipped with skimmers are used for collecting oil from the sea surface. Different types of skimmers (mechanical, weir, oleophilic, vacuum) can be used for collecting oil products with different characteristics. Oil collection is usually only effective immediately after the spill because the weathering process causes oil to spread, fragment and disperse under the influence of wind, waves and currents. It has been estimated that the effective gathering time after the spill is 3 days in the case of crude oil and 10 days in the case of heavier and more persistent fuel oil. In the winter, the ice sheet prevents oil spreading and the effective gathering times could be even three times longer. [19]

Other possible measures that can be used for oil combating are for example bubble barriers, burning of oil and dispersants. These are discussed in more detail for example in [6].

5.2 Preparedness in the Bothnian Bay

The Finnish Environment Institute (SYKE) has set the following targets for oil combat measures in the Gulf of Bothnia [12, 19]:
- Capacity to gather 5000 tonnes of oil at the open sea (during three days in summertime and during 10 days in wintertime)
- Capacity to deploy coastal oil boom: 2 km within 12 hours of the accident, 12 km within 24 hours, 42 km within 48 hours, 80-90 km within 72 hours.

SYKE maintains 13 oil combating stations equipped with oil combating boats and ships, oil booms and other material. Two of the stations are located in the Bothnian Bay (Oulu and Rahja). There are 16 oil combat ships that are ready to depart for an oil combat mission after a delay of 0 to 48 hours. The ships are equipped with extendable oil gathering arms that allow sweeping widths from 19 to 42 meters and oil gathering capacities ranging from 35 to 78 m^3/h. In addition to oil combat ships, there are roughly 100 smaller oil combat boats that can be used for oil booming and other support operations. [18]

If an oil spill occurs, the ship captain is obliged to immediately report the accident to the Maritime Rescue Coordination Centre. In the case of an oil spill, the Finnish oil combat boats and ships stationed in the Bothnian Bay coast would arrive to the accident location first. The local vessels would be supported by Finnish oil combat vessels from other areas and by vessels from other Baltic sea countries, specifically from Sweden. These additional ships would arrive gradually within 1-4 days.

5.3 Plant protection against oil spills

Oil booms are the most effective measure in protecting a nuclear power plant against oil spills. Booms can be permanent or they can be placed on demand. In the the latter case, there shall be a sufficient preparedness (boats and personnel) for rapid boom deployment. Booms can be placed in the immediate vicinity of the intake and discharge side and further away on the sea. Also the intake structure design has some effect on the oil spill consequences. A deep intake collects less oil than a surface intake. It shall also be possible to clean the fine screens and basket filters effectively if they are contaminated by oil. Furthermore, the plant personnel shall be informed without delay from any nearby spills and there shall be operating procedures to guide the actions of the plant personnel.

The risk related to oil spills is decreased if it is possible to take cooling water via an auxiliary intake or from the discharge side. The oil risk is also decreased by enabling sufficient measures for cooling also in the case of loss of primary ultimate heat sink (i.e. the sea). The Finnish YVL guide B.7 requires that it shall be possible to remove residual heat also by using an alternative, independent heat sink.

6. RESULTS

The frequency of the oil initiating event (significant amount of oil enters the intake tunnel) in the Hanhikivi plant is estimated by considering two different accident scenarios: **a medium spill** (100-1000 tonnes, mean size roughly 550 tonnes) and **a large spill** (> 1000 tonnes, mean size roughly 4000 tonnes). The risk related to spills smaller than 100 tonnes is assumed insignificant. A medium spill is possible if any type of ship carrying sufficient amount of fuel oil suffers an accident. For example, the fuel capacity of a large passenger cruise ship exceeds 1000 tonnes. Ships need fuel to produce the propulsion force, but also for producing all the electricity needed in the ship. A large spill can be caused by an accident of a tanker carrying various oil products that form a floating spill on the sea surface.

6.1 Spill frequencies

Ship fuel oils and light and heavy fuel oils transported in tankers in the Bothnian Bay are relatively heavy and persistent oil products belonging mainly to groups III and IV in the categorization of Figure 4. Due to the high persistence of these oil products, the whole Bothian Bay is considered as a potential accident area. However, because of the prevailing currents (see Figure 5), accidents occurring further away in the Baltic Sea are not considered a threat to the Hanhikivi plant.

The frequency for an oil spill accident with a volume between 100 and 1000 tonnes is received by assuming 21 events in the Baltic Sea during 43 years (see Table 2), and assuming that 2 % of the Baltic Sea accidents occur in the Bothnian Bay (see Table 1): $21,5 / 43 \text{ a} \cdot 0,02 = 1,0 \cdot 10^{-2}$ /a. Similarly, the large spill (> 1000 tonnes) frequency estimate is: $6,5 / 43 \text{ a} \cdot 0,02 = 3,0 \cdot 10^{-3}$ /a.

6.2 Impact probability

If an oil accident occurs in the Bothnian Bay, the spill should travel to Hanhikivi to have any effect on the plant operation. According to section 4.4, we assume that a medium spill contaminates 10 km of coastline and a large spill 40 km. The Bothnian Bay coastline length is roughly 700 km and the Finnish coastline length in the Bothnian Bay is 350 km. Based on the wind rose presented in Figure 5, we estimate that the wind blows between southwest and north with a 60 % probability causing the spill to land on the Finnish coast. By using the assumptions presented, we can calculate that the medium spill impact probabilities on Hanhikivi is: $0,60 \cdot 10 \text{ km} / 350 \text{ km} \approx 0,017$. Similarly, the large spill impact probability is: $0,60 \cdot 40 \text{ km} / 350 \text{ km} \approx 0,069$.

6.3 Spill warning

In the case of a significant oil spill, the ship captain is obliged to report to the Finnish maritime rescue coordination centre. The warning could fail if the ship captain would, for some reason, intentionally decide not to report the event. Also the communication equipment in the ship could fail, although there are probably alternative methods for communication. In the most drastic scenario, the ship could sink so quickly that there would be no time to report the event. Even if the ship itself could not give the warning, other bypassing ships or airplanes could do that. There is no data available to assess how often large oil spills are left unreported. According to the information presented above, it can be assumed that a duly warning is received with a high probability. By using expert judgement, we assume that no early warning is obtained with a probability of 0,01.

6.4 Open sea oil combat

The most effective oil collection time is immediately after the spill because the spill spreads out quickly. Thus, we assume that the open sea oil combat fails if no early oil spill warning is received.

SYKE is prepared to collect 5000 tonnes of oil in the Gulf of Bothnia area during three days. Thus we can assume that the open sea oil combat succeeds with a high probability if an early warning is received and if the weather and sea conditions are not very harsh. Strong wind, high waves and strong currents cause the oil to spread out quickly, which makes the oil booming and collection very difficult.

We assume that a medium oil spill combat fails if the wind speed exceeds 14 m/s (corresponding average wave height is roughly 3 m). According to measurement data, these conditions prevail in the

Bothnian Bay area with a probability of 0,09 [4]. Similarly, we assume the large oil spill combat to fail if the wind speeds exceeds 10 m/s (wave height 1,5 m), the probability of which is 0,35 [4].

6.5 Near Hanhikivi oil combat

If strong wind causes the failure of open sea oil combat, there is also a relatively high probability that also the oil booms near Hanhikivi fail. However, the success probability of oil combat is somewhat larger near Hanhikivi than in the open sea. The open sea oil combat aims at both booming and collecting the spilled oil, whereas near Hanhikivi the only aim is to prevent the oil spill from drifting to the intake. It could also occur that the strong wind settles before the spill reaches Hanhikivi several days after the accident. By taking these factors into account, we use expert judgement and estimate the following failure probabilities for near Hanhikivi oil combat if the open sea oil combat has failed due to strong wind: medium spill 33 % and large spill 50 %.

If the open sea oil combat failed because no early warning was received, the spill might be detected so late that it reaches the Hanhikivi intake harbour before the oil boom is placed. If the boom is placed on time, the oil combat success probability can be estimated to be relatively high. We assume the same failure probabilities for near Hanhikivi oil combat also in this case (medium 33 %, large 50 %).

6.5 Event trees

The event trees for medium and large spills are presented in Figure 7 and Figure 8. The estimated medium spill frequency is $1,0 \cdot 10^{-2}$ /a, and the spill reaches the plant intake with a 0,06 % probability (frequency $5,6 \cdot 10^{-6}$ /a). The estimated large spill frequency is $3,0 \cdot 10^{-3}$ /a, and the spill reaches the plant intake with a 1,2 % probability (frequency $3,7 \cdot 10^{-5}$ /a).

The oil initiating event for the PRA was defined to be an event where significant amount of oil enters the intake tunnel. The frequency for this initiating event is obtained by summing up the two scenarios: $5,6 \cdot 10^{-6}$ /a + $3,7 \cdot 10^{-5}$ /a = **$4,2 \cdot 10^{-5}$ /a**

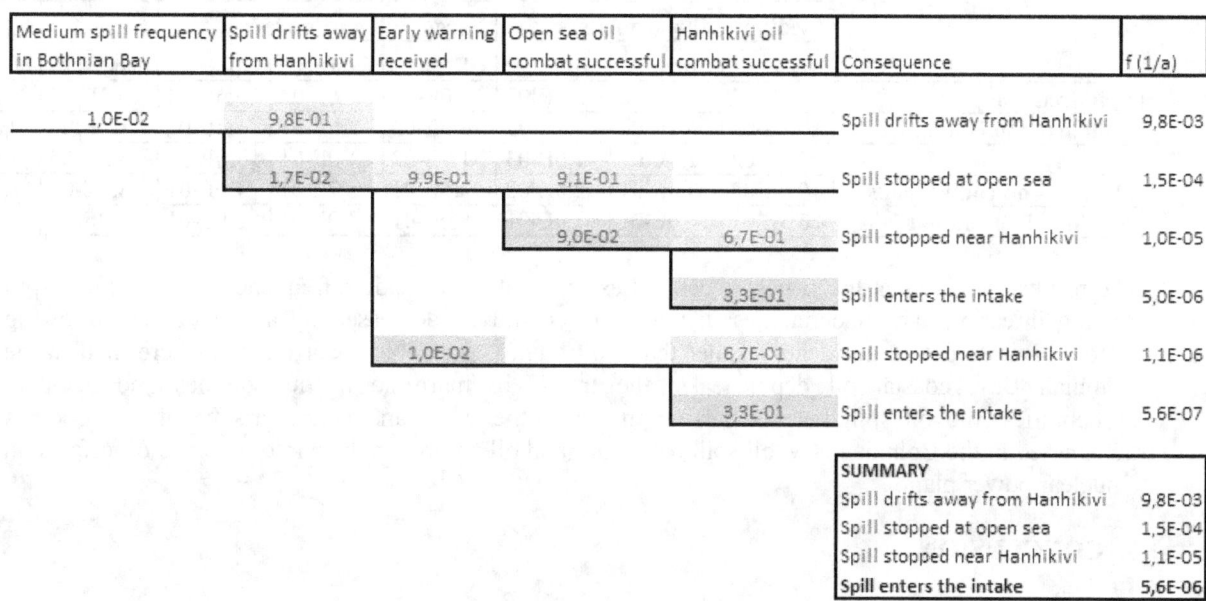

Figure 7. Event tree for a medium oil spill (100 - 1000 tonnes).

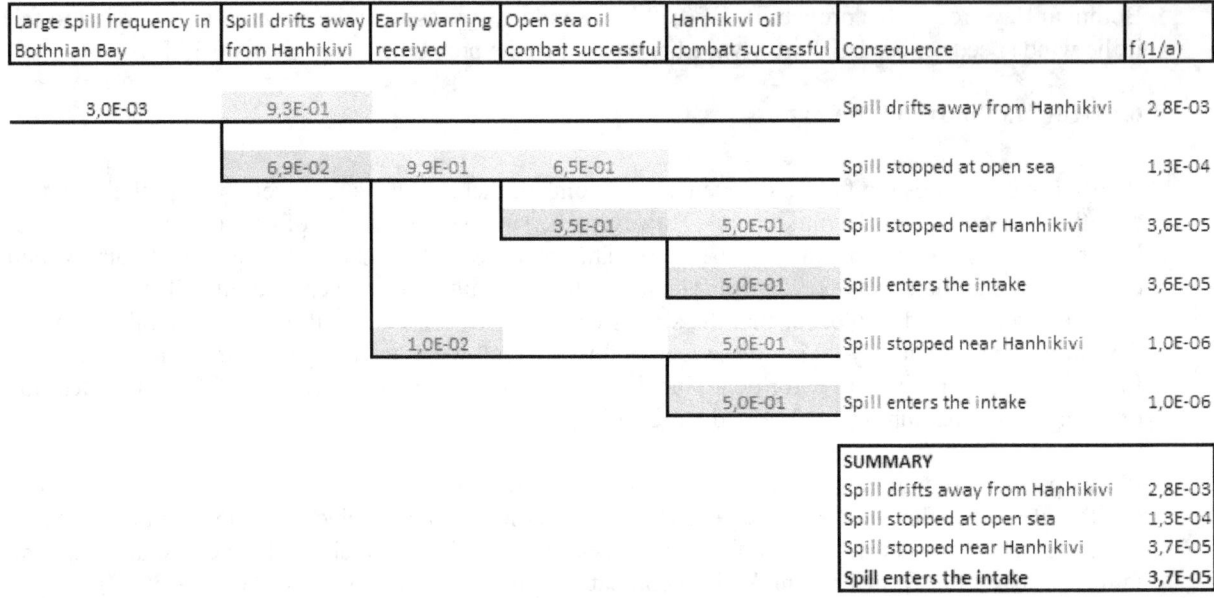

Figure 8. Event tree for a large oil spill (> 1000 t).

6.5 Sensitivity and uncertainty analysis

The event tree probabilities include somewhat large uncertainties. Thus, sensitivity analysis was performed by recalculating the results by using both conservative and optimistic estimates for each parameter. The parameter values and the effect on the results is presented in Table 4.

Table 4. Sensitivity analysis for the oil initiating event frequency by using both conservative and optimistic assumptions for different parameters.

	Medium spill parameters			Large spill parameters			f_{CONS}	f_{OPT}
	BE	Cons.	Opt.	BE	Cons.	Opt.	(1/a)	(1/a)
Spill frequency	1,0E-02	5,3E-02	2,5E-03	3,0E-03	1,6E-02	7,6E-04	2,3E-04	1,1E-05
Spill drifts to Hanhikivi	1,7E-02	5,1E-02	7,1E-03	6,9E-02	2,6E-01	1,4E-02	1,6E-04	9,8E-06
No early warning	1,0E-02	1,0E-01	1,0E-03	1,0E-02	1,0E-01	1,0E-03	5,3E-05	4,1E-05
Open sea oil combat fails	9,0E-02	3,3E-01	2,0E-02	3,5E-01	6,7E-01	1,0E-01	8,9E-05	1,3E-05
Hanhikivi oil combat fails	3,3E-01	6,6E-01	1,0E-01	5,0E-01	8,0E-01	2,0E-01	7,0E-05	1,6E-05

The results indicate that the largest uncertainties are related to accident frequency estimates and spill drifting direction. The uncertainties in the results could be decreased by analysing the following factors in more detail: accidents smaller than 100 tonnes, accidents occurring elsewhere than in the Bothnian Bay, coastal oil depot leaks, the trends in maritime traffic volumes and accident probabilities, the oil spill behaviour after the leak, the types and characteristics of oil products transported in the Bothnian Bay, oil spill behaviour and oil combat in the wintertime and oil effects on a nuclear power plant.

7. CONCLUSION

In this report, the risk of a maritime oil spill accident near the Hanhikivi nuclear power plant was evaluated. If the floating oil spill drifts to Hanhikivi and oil combat measures fail, oil could enter the sea water intake, block the fine screens and basket chain filters and lead to loss of sea water cooling.

By considering the maritime traffic, oil transport and oil spill accident data in the Baltic Sea area, it was evaluated that a nearby oil spill of medium size (100 - 1000 tonnes) occurs with a frequency of

$1,0 \cdot 10^{-2}$ /a and an oil spill of larger size (> 1000 tonnes) with a frequency of $3,0 \cdot 10^{-3}$ /a. The probability of a major oil spill accident is relatively low in the Bothnian Bay due to the small oil product transportation volumes and the lack of crude oil transports.

The probability that the spill drifts to Hanhikivi and oil combat measures fail was assessed by using event tree analysis. The typical wind and current conditions in the Bothnian Bay cause the spill to float towards northeast with an approximate speed of 30 cm/s, which equals 1,1 km/h or 26 km in one day. Light oil products, such as petrol, evaporate quickly, but heavy and persistent ship fuel oils could travel over long distances. Open sea oil combat measures are coordinated by the Finnish Environment Institute and include booming of the spill and gathering oil with specialized ships. There is a capacity to gather 5000 tonnes of oil in the open sea during three days, provided that the wind and sea conditions are not specifically harsh. Also the Hanhikivi plant can be protected by using deployable oil booms. According to the results of the event tree analysis, significant amount of oil could enter the Hanhikivi plant intake tunnel with a frequency of $4,2 \cdot 10^{-5}$ /a.

The most significant uncertainties are related to the accident frequency estimation and oil spill drifting direction. The uncertainties could be decreased by performing more detailed analysis. However, due to the low risk of the event, further analysis is not considered necessary.

References

[1] N. Bergroth, K. Jänkälä and S. Ovsienko. 2006. Oil spill risk assessment for Loviisa nuclear power plant. Fortum Nuclear Services Ltd. PSAM-0388. *PSAM8 conference*. New Orleans, Louisiana, USA.
[2] European Commission. 2006. *Eurostat*. Retrieved August 2013, from Statistics: epp.eurostat.ec.europa.eu/portal/page/portal/transport/data/database
[3] HELCOM. 2010. *Maritime Activities in the Baltic Sea – An integrated thematic assessment on maritime activities and response to pollution at sea in the Baltic Sea Region. Baltic Sea Environment Proceedings No. 123*. Helsinki Commission.
[4] Ilmatieteen laitos. 2013. Retrieved September 2013, from Merialueiden tuulipäivät: ilmatieteenlaitos.fi/tuulitilastot. The Finnish Meteorological Institute.
[5] ITOPF. 1987. *Effects of oil pollution on major municipal and industrial sea water abstractions*. The International Tanker Owners Pollution Federation.
[6] ITOPF. 2013a. *Handbook 2013/14*. The International Tanker Owners Pollution Federation.
[7] ITOPF. 2013b. *Oil tanker spill statistics 2012*. The International Tanker Owners Pollution Federation.
[8] Port of Kalajoki. 2013. *Kalajoen Satama - Port of Kalajoki*. Retrieved August 2013, from Öljyntorjuntaa: portofkalajoki.com
[9] Pålsson, J. 2012. *Oil spill preparedness in the Baltic Sea countries*. World Maritime University.
[10] A. Saurama, E. Holma and K. Tammi. 2008. *Baltic Port List 2006. Annual cargo statistics of ports in the Baltic Sea region*. Centre for maritime studies, University of Turku.
[11] Suomen satamaliitto. 2013. Retrieved July 2013, from Statistics: www.finnports.com
[12] SYKE. 2009. *Kokonaisselvitys valtion ja kuntien öljyntorjuntavalmiuden kehittämisestä 2009-2018*. The Finnish Environment Institute.
[13] SYKE. 2013. *Itämeriportaali. The Finnish Environment Institute, Ilmatieteen laitos, Ympäristöministeriö*. Retrieved August 2013, from www.itameriportaali.fi
[14] Transportgruppen. 2006. *Sveriges hamnar*. Retrieved July 2013, from Hamnstatistik: www.transportgruppen.se/ForbundContainer/Svenskahamnar/Branschfragor/Hamnstatistik/Trafik/
[15] L. Tunturivuori. 2013. External hazards in the PRA of Olkiluoto 1 and 2 NPP units - Accidental oil spills. *OECD NEA Workshop on PSA of natural external hazards including earthquakes*. Prague.
[16] Tuuliatlas. 2009. *The new Finnish Wind Atlas*. Retrieved August 2013, from www.tuuliatlas.fi
[17] J. Vaurio and K. Jänkälä. 2006. Evaluation and comparison of estimation methods for failure rates and probabilities. *Reliability Engineering and System Safety. 91*, 209-221.
[18] Ympäristöhallinto. 2014. *Ymparisto.fi*. Retrieved February 2014, from www.ymparisto.fi
[19] Ympäristöministeriö. 2011. *Toiminta isoissa öljyvahingoissa. Torjunnan järjestäminen, johtaminen ja viestintä*. Ympäristöministeriön raportteja 26/2011 .

Using bond graphs for identifying and analyzing technical and operational hazards in complex systems

Ingrid Bouwer Utne[*][a], Eilif Pedersen[a] and Ingrid Schjølberg[a]

[a]Department of Marine Technology, Norwegian University of Science and Technology, 7491 Trondheim, Norway

Abstract: Oil and gas exploration and production is moving into harsher environments, such as the Arctic, which increases the complexity of operations. Technology development introduces more advanced functionality and operators may not have the full overview or knowledge to handle deviations that propagate in the systems. The increasing complexity and couplings in systems and operations means that a systems approach is necessary to ensure sufficient risk management for accident prevention. Current risk analysis methods, however, have limitations. This paper investigates the use of bond graphs as a systemic method for analyzing risk in dynamic systems, for example, as a supplement to hazard and operability analysis (HAZOP) and system theoretic process analysis (STPA). The article uses a remote operated vehicle (ROV) for subsea operation as a case study.

Keywords: Risk assessment, Systems theory, Bond graphs, Remote operated vehicle (ROV)

1. INTRODUCTION

Exploration and production of oil and gas resources is moving into harsher environments, such as the Arctic. To obtain permission to start up and operate oil and gas installations in Arctic areas it is necessary to focus resources specifically on preventing accidents and damages to human beings and the environment. Even though this is already very important in the North Sea having sufficient and significant risk management systems in place for accident prevention in Arctic areas will become increasingly important [1-2].

Technology is changing faster than ever and the time to market for new solutions is decreasing. Complexity is increasing in the new systems, particularly in terms of coupling between sub-systems and components. These couplings may be very difficult to understand during planning and design. In addition, operation and the surroundings may be so complex and demanding that most operators do not have the full overview of potential behavior of the systems, and may not have the ability to handle all arising situations and disturbances safely.

The added complexity of systems and challenges related to their operating surroundings put increased demands on risk analysis as input to decision-making for both industry and authorities. Current risk analysis methods are feasible for addressing technical risk, whereas integration of operational risk remains a challenge [3]. Human operators and their role with respect to safety have traditionally been simplified. As systems become more technological advanced and complex, operators increasingly assume supervisory roles over automation, which requires cognitively complex decision-making where mistakes can no longer be effectively considered as simple random failures. The design of systems today leads to new types of operator errors, such as confusion and lack of situation awareness. Accidents occur due to interactions among system components and are not a result of one single failure [4].

According to [4], safety is controlled or enforced in terms of constraints on the system behavior. Hence, safety can be considered a control problem and accidents happen when safety is not controlled sufficiently. Hazardous events reflect inadequate control, for example, in the design process or during operation. As such, the control structure has to be understood in order to gain understanding of what went wrong in an accident. Enforcing control in terms of safety constraints on component behavior can

---

[*] ingrid.b.utne@ntnu.no

be performed through design, production or the maintenance process, or through social controls, such as management, regulations, culture or societal, and policies.

Hazard and operability analysis (HAZOP) is a method for risk analysis commonly used during the design process, both before commissioning of the system and during modification of an existing system [5]. [4] states that there is a need for "new types of hazard analyses and risk assessments that go beyond component failures and can deal with the complex role software and humans are assuming in high-tech systems". The basis for such new methods should be constituted by systems theory. Systems theory assumes that some systems cannot be separated into subsystems without losing information about systems' interactions and the relationships between technology and social aspects. The system theoretic process analysis (STPA) is a recently proposed risk analysis method based on systems theory, with some similarities to HAZOP, but it focuses more on system functions and control [4].

In the systems approach understanding functions is decisive, which can be facilitated through system modeling, especially for systems whose behavior as a function of time is important. Hazardous events imply that systems are in a deviating state reaching or transgressing the limits of its operating envelope. Hence, dynamic system analysis for hazard identification and analysis, even though it is resource demanding, seems preferable compared to assuming steady state and constant system performance.

The bond graph language is a common conceptual and operational framework for the analysis of a very wide variety of system types. Bond graphs is proven an efficient methodology for modeling of especially hybrid dynamic systems. Advantages are that the same use of symbolism can represent power interaction in a large selection of physical systems, and causality can be shown in terms of input and output variables. In the method of bond graphs it is possible to separate and concentrate properties of a component and then describe it as a system of interconnected basic ideal elements. Hence, it is possible to predict the behavior of systems with an acceptable limit of accuracy [6]. Bond graphs may contribute to improved knowledge about systems, including impacts from human operators, and hence possibly reveal additional hazards than those identified and analyzed in traditional risk analysis.

The objective of this paper is to investigate the use of bond-graphs as part of a systemic approach for assessing technical and operational hazards in complex systems for challenging environments. There is hardly any research on application of bond graphs for use in risk analysis. One exception is [7] who uses bond graphs to analyze disturbance paths and compares with fault tree analysis (FTA). The current paper introduces bond graphs as means for providing useful input to risk analysis, and specifically for HAZOP and STPA. A case study illustrates the approach focusing on subsea operations and a remote operated vehicle (ROV). ROVs are increasingly used and their development towards more autonomy is especially desirable with the expansion of industrial activities in the Arctic [8].

The structure of the paper is as follows: Section 2 briefly presents the risk analysis methods HAZOP and STPA. Bond graphs is introduced in Section 3 and Section 4 illustrates how bond graphs may be integrated into risk analysis through a case study on ROVs, providing additional information concerning operational hazards. Last, conclusions are given.

2. RISK ANALYSIS

2.1. HAZOP
HAZOP considers the purpose of a system, operational modes, and possible deviations during the modes. Causes to any deviations, possible consequences, and the need for risk-reducing measures are considered. HAZOP typically starts with dividing the system into study nodes that are analyzed one by one. Physical component diagrams, such as a piping and instrument diagram (P&ID), are used for understanding the system and separating the study nodes. Various guidewords, shown in Table 1, are used to analyze deviations in operation and any hazards that these deviations may cause. Existing safeguards or barriers are evaluated and a decision is made about the need to follow up and implement improvements. In some cases, more detailed analysis about the risk may be required [5]. When all study nodes have been considered, the guidewords should be used for reviewing the whole system [9].

Currently, HAZOP studies are used in many different industries, and have become a standard activity in the design of the process systems on offshore oil and gas platforms in the North Sea. Initially HAZOP was developed for use in the design phase, but may also be applied to systems in operation [5].

Table 1. Typical generic HAZOP guidewords, adapted from [5, 10-12].

Guideword	Property	Deviation (examples)
No/none	Flow	No flow
Less	Pressure	Less pressure
More	Temperature	High temperature
Reverse/opposite	Time	Delayed control
Threshold	Level	Insufficient communication
Insufficient	Control and monitoring	As well as composition – high concentration of impurities
Faster	Communication	...
Slower	Cleaning and emptying	
High	Function	
Low	Particles/composition:	
Early	• Properties, e.g., viscosity, sand, concentration, reactions, impurities, corrosion materials	
Late/delay		
Malfunction		
Before		
After	Junctions and couplings:	
Other than	• Drainage, ventilation, auxilliary systems	
Part of		
As well as		
	Health, Safety, Environment:	
	• Explosive gases, toxicity, leakages rotating parts, pressure relief, environmental impacts	
	Interface/instruments:	
	• Signal/data processing, capacity, prosess control	
	Maintenance:	
	• Access, isolation	
	Operation mode:	
	• Start-up, stop, shutdown, breakdown, emergency stop	
	Other:	
	• Deviations in drawings, notes	

2.2. System Theoretic Process Analysis (STPA)

STPA was introduced by [4]. Similarly to HAZOP, STPA uses guidewords to assist the analysis, but a functional diagram is used, rather than a physical diagram. The focus is on identifying and analyzing hazards to mitigate or reduce risk in design and operations. In general, an unsafe control action may be unsafe if (i) it creates a hazard, (ii) a required control action is not provided, (iii) it is provided too late, too early or in wrong order, (iv) it is provided too long or stopped too soon, and (v) it is provided, but not followed.

A hierarchical safety control structure includes higher-level controls, such as safety policy, standards and procedures, and lower-level controllers that implement theses. Between each level there is a feed back control loop that provides learning and a possibility to improve effectiveness of the safety controls. Every controller has an algorithm for deciding what control actions to provide, which uses a model of the current system state to help making the decision. This means that the controller issues control actions to alter the state of the controlled process based on assigned requirements that ensure that safety constraints are maintained. For a human controller the process model is a mental model.

STPA has two main steps [4]:
1) Identify potential for inadequate control and hazardous control action
2) Determine how each potentially hazardous control action identified in Step 1 may occur

The causal factors can be related to three general categories; (i) the controller operation, (ii) the behavior of actuators and controlled processes, and (iii) the communication and coordination among controllers and decision – makers. Input to controllers may be wrong; algorithms (procedures for hardware or human operators) can be wrongly implemented. Further, time delays must be considered, and insufficient process models may lead to unsafe actions. Multiple controllers need coordination and responsibilities must be defined [4].

Supervising an automated system requires extra training and skill of a human operator, and not the opposite as sometimes is assumed. Human controllers need additional information not only about the process they are directly controlling, but also of processes they control indirectly. Hence, hazard analysis has to identify the specific human behavior that causes hazards. Humans will update their control algorithms according to goals and feedback [4].

STPA is a qualitative risk analysis method, as it does not quantify probabilities to avoid omitting important causal factors for which probabilistic information does not exist. It is a top-down method, which means that the initial list of hazards should be small and at an overall level, i.e., contain less than 20 and preferably less than 10. When the high level system hazards have been identified safety requirements or constraints have to be derived. At this point the system control structure (which is not part of the STPA method) is needed to perform STPA, reflecting the functional design of the system. The control loop (controlled process, sensor, controller, actuators) has to be examined to see if they contribute to hazards and as details may be added, such as responsibilities, the control actions, and feedback, the control structure is refined and expanded [4].

Bond graphs can be used to model any energy dynamic system and are easily connected to control systems using standard block diagram representation. Hence, it may work as a functional diagram. The bond graph method can be said to represent an object-oriented approach to mathematical modeling.

3. BOND GRAPHS

The bond graph method is methodology for modeling of dynamic systems based on energy flow, i.e. deriving mathematical state space equations for a dynamic system. The method, originally invented by [13] and brought into diverse applications by [6] is a graphical visualization of the energy flow, storage and dissipation of interacting dynamic systems. The bond graph method introduces a unified approach to modeling and is widely recognized as a powerful approach for mathematical modeling of engineering systems combining mechanical, electric, hydraulic, pneumatic, magnetics, chemical and electrochemical physics, i.e., strongly supporting a mechatronic approach.

The flow of energy between two systems are defined using two basic elements of the bond graph language; i.e., ports and power bonds. The ports represent energy points where there is energetic interaction between systems or a system and its surroundings. The power bonds represent the power flow between two systems. There is no loss of power along the power bond. The bonds can be included with a half arrow on one end, indicating the direction of power flow when it is positive. A full arrow on a bond is called an active bond and indicates a signal flow at very low power [6] and is often used to link the power system to control actions or events.

Figure 1. Power bond visualizing the energy flow.

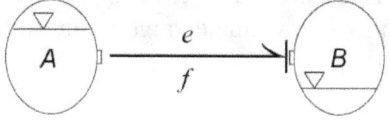

It is possible to predict the behavior of any energetic system knowing the energy level at each system, i.e., the potential or effort, and the energy flow between each system, i.e., the rate or flow, as visualized in Figure 1. Only four variables are needed to represent the dynamics of physical systems, i.e., the power variables e, f and the energy variables p, q. Effort is represented by e, flow by f, the generalized momentum by p, and q is the generalized displacement. A bond graph can represent the variables effort $e(t)$ and flow $f(t)$, for example for mechanical systems, force and velocity can be considered as $e(t)$ and $f(t)$, which multiplied represent power

$$P(t) = e(t) \cdot f(t) \tag{1}$$

The generalized momentum for a system can be defined as the time integral of effort

$$p(t) = \int_0^t e(t)dt + p(0) \tag{2}$$

where $p(0)$ is the initial generalized momentum at $t = 0$.

The generalized displacement can be defined as the time integral of flow

$$q(t) = \int_0^t f(t)dt + q(0) \tag{3}$$

where $q(0)$ is the initial generalized displacement at $t = 0$.

Setting up a mathematical model using the bond graph approach is often supported following simple procedures in most disciplines [6]. Table 2 shows corresponding energy domains and power- and energy variables [6].

Table 2. Variables and energy domains in bond graph modelling [6].

Energy domain	Effort	Flow	Momentum	Displacement
Electrical	Voltage [V]	Current [A]	Flux linkage [Vs]	Charge [As] or [C]
Mechanical translation	Force [N]	Velocity [m/s]	Linear momentum [kgm/s]	Distance [m]
Mechanical rotation	Torque [Nm]	Angular velocity [rad/s]	Angular momentum [Nms]	Angle [rad]
Hydraulic	Pressure [Pa]	Volume flow rate [m³/s]	Pressure momentum [N/m²s]	Volume [m³]
Thermal	Temperature [K]	Entropy flow [J/s]	N/A	Entropy [J]
Magnetic	Magneto-motive force [A]	Flux rate [Wb/s]	N/A	Flux [Wb]
Chemical	Chemical potential [J/mol]	Rate of reaction [mol/s]	N/A	Advancement of reaction [mol]

There are nine basic bond graph elements, shown in Table 3, which can be used to represent the energetic structure of the system. The elements are characterized by the number of energy ports they have (1-port, multiports) and by how they handle the energy. In the bond graph language the effort variable e is written above or to the left of a bond, whereas the flow f variable is put below or to the right of the variable [6].

Basic 1-port elements represent energy supply, storage and dissipation. To model energy supply to a system, energy sources are needed. An energy source provides power continuously by an external energy reservoir to the system. Two idealized sources are defined, i.e., source of effort S_e, and source of flow, S_f. An ideal effort source is capable of providing enough effort on the system regardless of flow, and vice versa for the ideal flow source [6].

Most systems store energy either in the form of potential energy or kinetic energy. This can be represented by 1-port capacitor element or a 1-port inertia element. The capacitor element uses the mnemonic code C and can generally be defined as $q = \Phi_C(e)$. In the linear case it can we written $q = C \cdot e$ where C is the constant capacitance parameter. The capacitor element is the idealization of physical components, such as springs, capacitors, and liquid storage tanks [6].

The energy stored in a capacitor element at any time t can be represented by

$$E(t) = \int_0^t (e \cdot f) dt + E(0) \qquad (4)$$

where $E(0)$ is the initial stored energy at $t = 0$.

The inertia element uses the mnemonic code I and the corresponding constitutive law can be defined as $p = \Phi_I(f)$. In the linear case this can we written $p = I \cdot f$, where I is the constant inertia parameter. The inertia element is an idealization of physical objects, such as mass or inertias, inductance in electrical systems or inertia effects in hydraulic systems.

Power dissipation, energy lost from the system as a function of internal relations between effort and flows, is handled by the 1-port resistor element. The resistor element uses the mnemonic code R and is characterized by a static relationship between e and f: $e = \Phi_R(f)$ in the general case and $e = R \cdot f$ in the linear case where R is called the constant resistance parameter. The R-element is used to model all types of energy dissipation, such as mechanical and hydraulic friction and electrical resistors [6].

Two-port elements are the transformer element (*TF*) and the gyrator element (*GY*). Both elements transmits power through the system without any loss or accumulation. The former relates an effort on one port into an effort on the other port with a magnitude depending on the modulus, displayed on the graph above the *TF*. The gyrator relates an effort on one bond into a flow on the other bond with a magnitude depending on the gyrator modulus, displayed above the *GY* [6].

Table 3. Bond graph elements, based on [6].

Name	Bond graph element	Constitutive relationship
Effort source	$S_e \xrightarrow{e}$	$e = e(t)$
Flow Source	$S_f \xrightarrow{f}$	$f = f(t)$
Capacitance	$\xrightarrow[f]{e} C$	$\int f(t) dt = \Phi_C(e)$
Inertia	$\xrightarrow[f]{e} I$	$\int e(t) dt = \Phi_I(f)$
Resistance	$\xrightarrow[f]{e} R$	$e = \Phi_R(f)$
Transformer	$\xrightarrow[f_1]{e_1} \overset{m}{TF} \xrightarrow[f_2]{e_2}$	$e_1 = m e_2$ $m f_1 = f_2$
Gyrator	$\xrightarrow[f_1]{e_1} \overset{m}{GY} \xrightarrow[f_2]{e_2}$	$e_1 = f_2 r$ $f_1 r = e_2$
0-junction	$\xrightarrow[f_1]{e_1} 0 \xrightarrow[f_2]{e_2}$ with e_3, f_3	$e_1 = e_2 = e_3$ $f_1 - f_2 - f_3 = 0$
1-junction	$\xrightarrow[f_1]{e_1} 1 \xrightarrow[f_2]{e_2}$ with e_3, f_3	$f_1 = f_2 = f_3$ $e_1 - e_2 - e_3 = 0$

The system topology is decisive for the arrangement of the elements. There are two types of junction elements, the *0*-junction and *1*-junction, both that is power continuous and routes energy into the appropriate paths without any delay or loss. The *0*-junction has equal effort on all bonds joining while all flows on connected bonds sums to zero and represents Kirchhoff's current law in the electric domain. The *1*-juntion is a multiport element where the flows are equal and all efforts are summed to zero and as such represents Kirchhoff's voltage law [6].

4. CASE STUDY – REMOTE OPERATED VEHICLE (ROV)

Underwater vehicles can be divided into manned submersibles and unmanned underwater vehicles (UUV), i.e., towed vehicles, remotely operated vehicles (ROV), and autonomous underwater vehicles (AUV). An AUV is not directly controlled by an operator, but is mainly pre-programmed for a given mission. The main power source is integrated and communication with an operator is related to data transmission with limited bandwidth. Contrary to a ROV, an AUV has no permanent connection to an operation centre. There is few fully autonomous underwater vehicles; current systems operates by consent which means that the AUV recommends actions to the operator and the system involves the operator at key points for information or decisions [14].

Increased use of ROVs, and eventually AUVs, is required in the Arctic as new areas become accessible for operation [15]. AUVs can provide easier environmental monitoring, mapping and monitoring of ice [16]. Future AUV capabilities will enable inspection of subsea equipment and pipelines in deep-water. In the following, the focus is on ROV systems as these are most common for subsea inspection, maintenance and repair of equipment and infrastructure.

A typical ROV system, shown in Figure 2, is here defined to consist of six main parts; (i) external sensors, (ii) the guidance and navigation system, (iii) the control system, (iv), thrusters, (v) local sensors, and (vi) power supply.

Figure 2. Main components of a typical ROV.

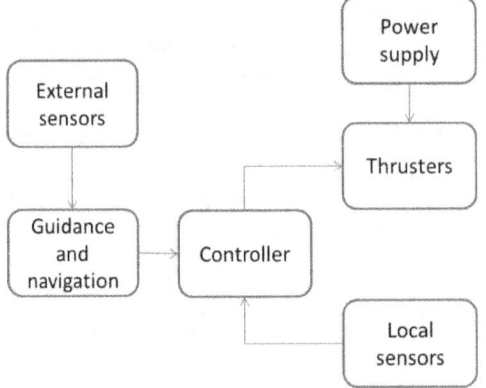

An energy storage system provides the electric power to the motor. The motor supplies mechanical energy to the whole system, i.e., the thrusters, manipulators or tools, the control system/pilot, and the emergency system. The ROV may be equipped with one or several manipulators to carry out different operational functions. The manipulators are usually powered by a hydraulic system connected to the motor and the control system/pilot. The power transmission consists of a gear connected to the motor and the thrusters.

The manoeuvring of the ROV is determined by the control system and/or the operator. The positioning system sends signals to the control system that compares it to pre-set values and mobilizes the propulsion system to minimize the difference. The ROV may have a local control system, i.e., dynamic positioning (DP) system or can be directly controlled by the operator. This is schematically shown in Figure 3.

Figure 3. Control structure of an ROV. The grey box shows the relationship to the bond graph in Figure 4.

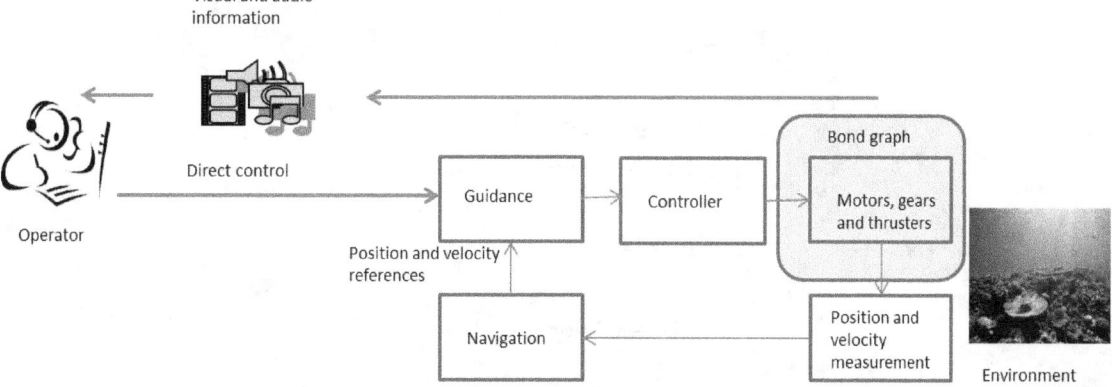

The equations of motion of an ROV with one manipulator arm can be written as

$$M(q)\dot{\varepsilon} + C(q,\varepsilon)\varepsilon + D(q,\varepsilon)\varepsilon + g(q,\theta) = \tau \qquad (5)$$

where q is the joint position of the manipulator arm, ε is the vehicle and manipulator arm velocities ($\dot{\varepsilon}$ is acceleration), and θ is the position and orientation of the vehicle in an inertial frame. M is the mass matrix of the system, C holds the Coriolis and centrifugal terms, D the damping terms and g the gravity terms. Cable dynamics is integrated in this representation, as well as coupling force between vehicle and manipulator. τ is the control input for position, orientation and velocity control [17]. For tracking control a model-based proportional-derivative (PD) controller could be used. The control input is designed to compensate for thruster and motor dynamics.

A bond graph, as shown in Figure 4, can be applied to analyze the energy supply and the functionality of the thruster, gearing and motor system. The mass of the vehicle is described as the inertia with the multiport I-field (I:M), where the translational and angular inertia is represented with corresponding velocities as input. Restoring forces are all forces acting on a vehicle to resist its deviation from its static equilibrium. For an ROV the restoring forces will depend on the location of the centre of gravity and the centre of buoyancy. These forces are efforts in the bond graph model, and can be represented as ideal potential energy accumulation in terms of a C-element. Ocean currents are usually modeled by assuming a constant current velocity. Such currents add to additional Coriolis, centripetal and damping forces and are dependent on vehicle velocities. Thruster forces impacting a vessel is a complex relationship between the vehicle's motion, sea conditions (currents and waves), water flow around the thrusters, the design of the propulsion system and control inputs. The thruster forces can be modelled in terms of a relation between flow and effort, i.e., shown as an R element in the bond graph model in Figure 4. The *TF* element represents the rudder forces due to rudder lift and drag forces generated by the position of the rudder relative to the vehicle main body. A detailed model for one of the thruster systems on a ROV is presented by a bond graph in Figure 4, further developed from [18]. The bond graph receives the control input and calculates the energy flow.

The on-board energy and propulsion system is essential for safe operations and bond graphs is one method to analyze disturbance paths and perform HAZOP or STPA on these systems. The graph supplies a simple framework for fault and consequence analysis of the energy and propulsion system and can (e.g., in Matlab) be linked to the vehicle equations of motion (5) to simulate the behaviour of the total system. This cannot be done with physical layout diagrams, such as the simple layout shown in Figure 2 or even with P&ID. Bond graphs may, as such, be considered an advanced expansion for providing input to STPA.

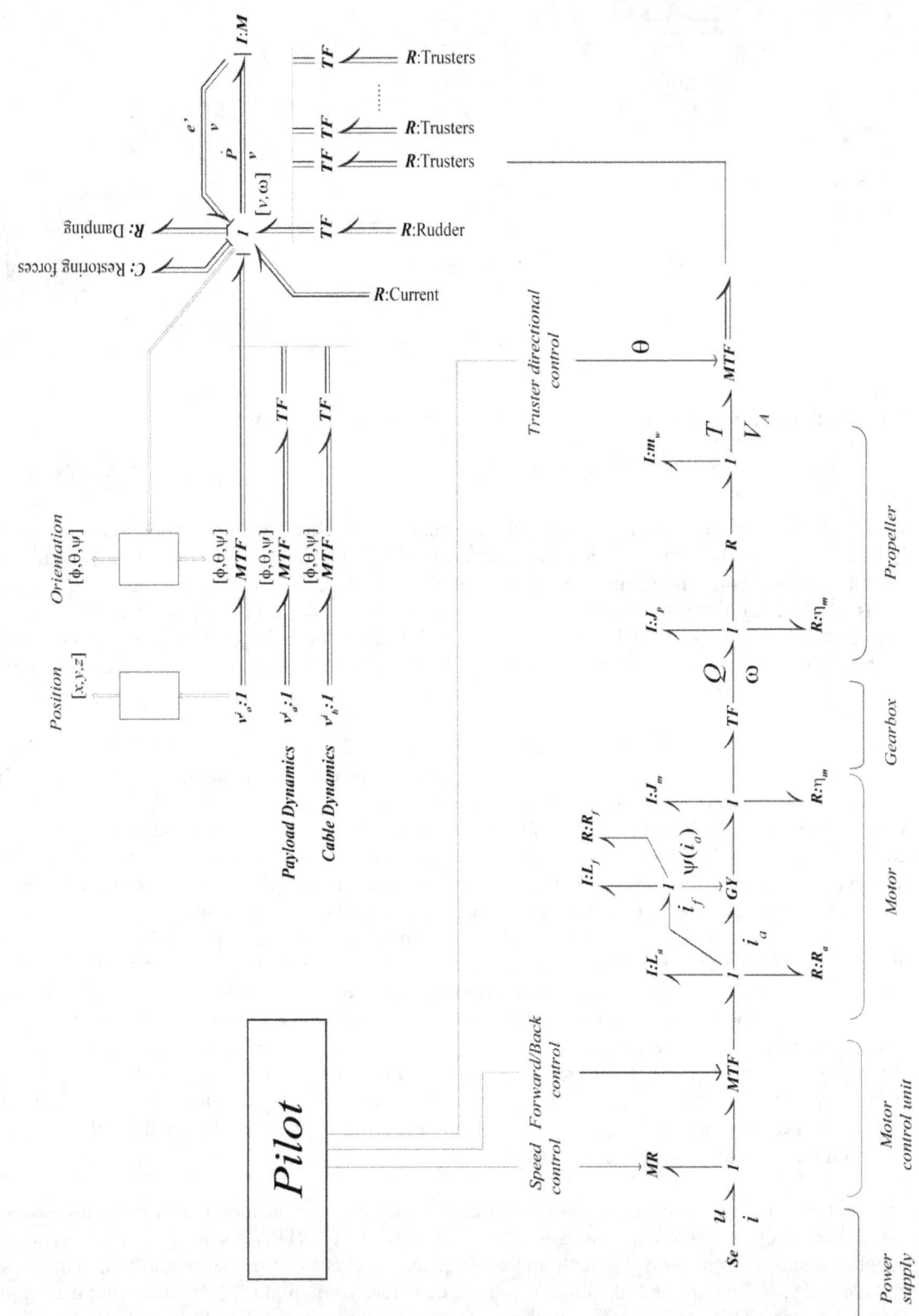

Figure 4. The bond graph model of the ROV.

5. DISCUSSION

To design a robust ROV, requirements to product safety, availability, functional safety, and maintainability have to be fulfilled. These requirements must focus on the technical system, as well as the human operator and interaction when manual control and intervention is needed. A producer of a ROV has to establish a systematic way for ensuring that the safety performance requirements or constraints are addressed throughout the system life cycle. This means that feedback or transfer of experience and knowledge between development and operation is crucial.

Bond graphs can be used to simulate system behavior under various circumstances, for example, to analyze different system configurations and/or operating conditions and impact on system performance. The operator's interaction with the system can be investigated trough the control of the system. This is modeled for the ROV in Figure 3, where the operator controls the speed and position. This means that bond graphs can be used for more detailed and "dynamic" HAZOP analysis. STPA needs a functional diagram rather than a physical diagram, such as a PI&D. Bond graphs can be used to model systems in which the functions of the system and the system control structure can be investigated. This means that several aspects of a system becomes more "visible", such as dynamic forces acting on the system (cf. the motions in Figure 4) and how signals are transmitted through the system in terms of operator interaction and control.

Figure 5. The relationship between bond graph modeling and risk analysis.

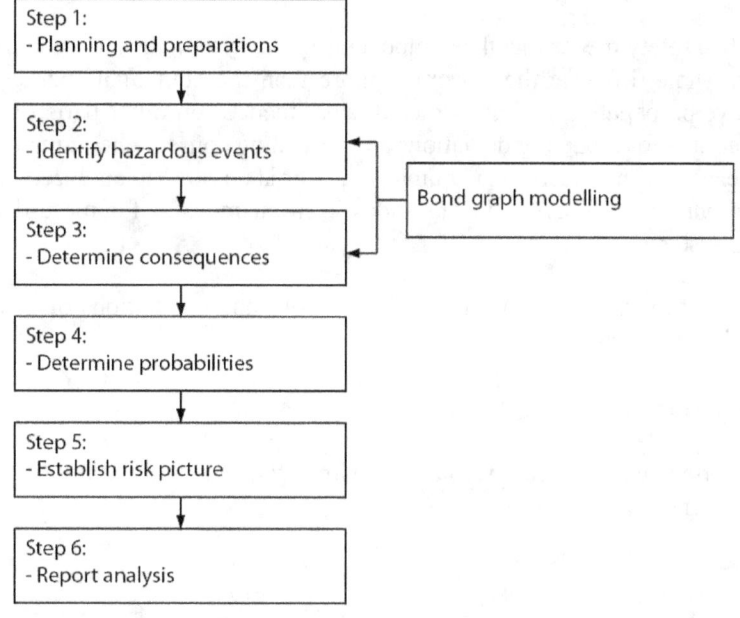

HAZOP is effective for technical failures and to some extent human errors, and develops recommendations for additional or modified barriers. The analysis is, however, dependent on the facilitation of the leader and the knowledge of the team and is optimized for process hazards. Hence, it needs modification to cover other types of hazards and requires development of procedural descriptions, which are often not available in appropriate detail [5]. HAZOP analyzes a system or process using a "section by section" approach and may not identify hazards related to interactions between different nodes. The latter implication is becoming more important as couplings between subsystems and components increase. Analysis of component interdependencies can be facilitated by bond graphs and is one of its main advantages.

Typical guidewords in HAZOP, shown in Table 1, can be used for simulating and investigating deviations in the energy flow throughout a bond graph model. A taxonomy, e.g., from [19], can be used as a starting point for the simulations and more detailed analysis for identifying and analyzing hazards.

For example, the consequences of an operator failing to provide the correct input to the ROV by activating the wrong switch can be investigated. [4] claims that one of the main advantages of STPA is that more hazards are identified and using bond graph supports this view. Figure 5 shows where in the risk analysis process bond graphs can provide useful input, i.e., to hazard identification through modeling and simulation of energy flow and deviations and to determine consequences of such deviations on the system and operation.

Bond graphs is based on modules or "building blocks" which means that a system can be modeled relatively efficiently by reuse and reconfiguration of previously built modules. This means that the initial establishment of the bond graph model is relatively resource demanding, but once it is developed, reuse and modifications are less time consuming.

6. CONCLUSION

In this paper bond graph modelling is presented as a method for identifying and analysing hazards in dynamic systems. A bond graph model of a ROV is provided, including a representation of the control system and the operator. Bond graphs is a unified approach to modeling and visualizing the energy flow, and is widely recognized as a powerful approach for mathematical modeling of engineering systems. The complexity of engineering systems increase with the technological development at the same time as operating surroundings become more demanding, such as in the Arctic. Present types of risk analysis methods are feasible for addressing technical risks, but sufficient analysis and integration of human and organizational factors is still a challenge.

A systems approach to safety means that the components of the system have to be considered in relation to the system as a whole, because the system is more than the sum of its parts. A change in one component may have propagating and unpredicted consequences on other parts of the system. Bond graph models are feasible for modeling deviations and cascading consequences, including in the control structure, which means that impact from the human operator is possible to analyze. Hence, bond graphs support the system approach to safety and provides a promising tool for more detailed analysis of hazards in complex systems.

Further research should focus on more complex applications and simulations of dynamic systems for detailed risk analysis.

ACKNOWLEDGEMENTS

This work was supported by the Research Council of Norway through the Centre of Excellence funding scheme, project number 223254 AMOS.

REFERENCES

[1]   O. T. Gudmestad and C. Quale. *"Technology and operational challenges for the high North."* Report to the Petroleum Safety Authority, IRIS – 2011/166, (2011).
[2]   Petroleum Safety Authority Norway (PSA). *Main priorities 2014. The North.* http://www.ptil.no/the-far-north/category1116.html (Accessed: March 11th, 2014)
[3]   J. E. Skogdalen and J. E. Vinnem. «*Quantitative risk analysis offshore – Human and organizational factors*". Reliability Engineering and System Safety, 96, pp. 468-479 (2011).
[4]   N. Leveson, *"Engineering a safer world. Systems thinking applied to safety"*. The MIT Press, 2011, Cambridge, MA, USA, (2011).
[5]   M. Rausand, *"Risk assessment. Theory, methods and applications"*. Wiley, Hobroken, (2011).
[6]   D. Karnopp, R. Rosenberg, *System Dynamics: A Unified Appraoch*, John Wiley & Sons, Inc., (1975).
[7]   T. Kohda and K. Inoue, *"A simplified risk analysis method of complex systems using the global system model"*. Reliability and Maintainability, 2004 Annual Symposium - RAMS proceedings, IEEE.

[8] R. D. Christ, R. L. Wernli, Sr, *"The ROV Manual: A User Guide for Observation Class Remotely Operated Vehicles"*, Butterworth-Heinemann, (2007).

[9] New South Wales (NSW) Government, *"Hazardous Industry Planning Advisory Paper No 8 HAZOP Guidelines"*, http://www.planning.nsw.gov.au/LinkClick.aspx?fileticket=HT4S8cZpZ3Y%3D&... (Accessed: Sept. 4th, 2013).

[10] NEK IEC 61882. Hazard and operability studies (HAZOP studies) Application guide, (2001).

[11] NEK IEC/ISO31010. Risk management Risk assessment techniques, (2009).

[12] N. Hyatt, *"Guidelines for Process Hazards Analysis, Hazard identification and risk analysis"*, CRC Press, Ontario, Canada, (2003).

[13] H. H. Paynter, "A*nalysis and design of engineering systems"*, The M.I.T. Press, Boston, (1961).

[14] M. Rothgeb. *"Intelligent autonomy for reducing operator workload"*, Intelligent Control Systems Department, Autonomous Control and Intelligent Systems Division, 2007, http://engineering.tamu.edu/media/699818/intelligent_autonomy_2007.pdf (Accessed: Nov. 13th 2013).

[15] P. B. Brito, G. Griffiths and P. Challenor, *"Risk analysis for autonomous underwater vehicle operations in extreme environments"*, Risk Analysis, 30(12), pp. 1771-1788, (2010).

[16] G. Griffiths, M. P. Brito, *"Risk management for autonomous underwater vehicles operating under ice"*. OTC 22162. Offshore Technology Conference, Houston, USA, (2011).

[17] I. Schjolberg, and T. I. Fossen, *"Modelling and Control of Underwater Vehicle-Manipulator Systems,"* Proceedings of the 3rd Conference on Marine Craft Maneuvering and Control, Southampton, UK, pp. 45–57, (1994).

[18] E. Pedersen, *Bond Graph Modeling of Marine Vehicle Dynamics*, 7th Vienna International Conference on Mathematical Modelling (MATHMOD 2012), Vienna, Austria, (2012).

[19] I.B. Utne and I. Schjølberg. *"A systematic approach to risk assessment – focusing on autonomous underwater vehicles and operations in Arctic areas"*. Accepted for proceedings OMAE, San Francisco, USA, (2014).

Estimating Farmer's risk aversion

Patrick Momal [a]
[a]IRSN, Fontenay-aux-Roses, France

Abstract: In the early days of safety, together with his famous diagram, Farmer introduced a form of risk aversion. The first objective of the paper is to propose a general formulation of risk aversion along Farmer's thinking. This theoretical framework is particularly well suited when accident severity cannot be mitigated and prevention efforts are aimed at reducing probabilities, as is the case with nuclear safety. This is shown to go beyond the expected utility theory. The second part of the paper reports on an attempt at estimating Farmer's risk aversion as perceived by a panel of nuclear safety professionals. This tends to confirm Farmer's views.

Keywords: Risk aversion, disasters, prevention behavior, economics of Safety

1. THE VALUE OF EXPECTED VALUE

1.1. Early criticism of expected value

In the early days of probabilities, in the 17-18th century, wealthy people had plenty of time to play card games or engage into lotteries. The moral problem was: how much is it fair to ask for a lottery ticket? The general thinking of scientists of that time was to evaluate lotteries by their expected value. For instance, if a true coin was tossed and the winner would pocket 100 upon heads coming up, the fair value of this lottery was: ½ x 100 = 50. It was honest to ask a potential player to pay 50 for such a lottery ticket both players then having the same probability of winning or losing; the lottery was then changed from a potential swindle to a "pure" game of chance.

Even nowadays, many feel that, faced with a probabilistic outcome, the expected value is the natural and the best quantification with *one* figure. But this was criticized at as early as 1738! Consider the following game:

> Toss a true coin until heads come up; if this happens on the first throw, you win 2 and the game stops; if this happens the second time the coin is tossed, you win 4 and the game stops; in general, if this happens at the nth stage, you win 2^n.

The expected value of this lottery is: $2 \times \frac{1}{2} + 4 \times \frac{1}{4} + \cdots + 2^n \times \left(\frac{1}{2}\right)^n + \cdots$ which is worth $1 + 1 + \ldots + 1 + \ldots = \infty$!

This paradoxical game found a solution when Daniel Bernoulli remarked that:

> *The determination of the value of an item must not be based on the price, but rather on the utility it yields.... There is no doubt that a gain of one thousand ducats is more significant to the pauper than to a rich man though both gain the same amount.*

Bernoulli proposed the logarithm of a sum of money as the utility of the said monies which implies that an "additional ducat" is worth less and less as gains increase. The expected value of the utility of the above lottery is no longer infinite. This solution was published in 1738 in the Commentaries of the Imperial Academy of Science of Saint Petersburg; hence its name as the Saint Petersburg Paradox.

1.2. Expected utility and risk aversion

This approach was later generalized by Non Neumann and Morgenstern as the Expected Utility theory (EU). When utility U(x) of a sum x boils down to x, EU simply judges according to expected values. EU is thus (much) more general.

When judging probabilistic outcomes on the basis of their expected value, if amounts increase by, say, 10%, the value also raises by 10%. (In sophisticated terms, the elasticity of risk with respect to losses is 1). This is no longer the case with EU. It could be lower or larger than 10%. With a decreasing utility of gains, the utility function is concave, such as the logarithm proposed by Bernoulli and the increase in the value of the lottery is lower than 10%. (The elasticity of the value of the lottery with respect to potential gains is lower than 1).

The concavity of the utility function models an important behavior: risk aversion. This gives rise to risk premiums. On financial markets, if you have a wife, two kids and a few loans, you probably wish to limit the risks you bear when investing your savings. The returns you can ask for are then lower than what you might earn if you were prepared to take larger risks; the difference is called the risk premium. A risk premium is earned by those who accept to take larger risks and is paid by those who wish to limit their risk.

The larger your risk aversion, the larger the risk premium you are ready to pay in order to run lower risks. When you have zero risk aversion, your utility function has zero concavity, it is a straight line, the value of risk is the expected value. You are said to be risk neutral, a limit case of risk aversion.

This model is applicable to prevention. With risk neutrality, you only prevent up to the expected value of losses; in contrast, a risk averse decision process results in (much) more prevention; one could use the term prevention premium. Here is a trivial example:

> With risk neutrality, it is equivalent to prevent a grain of sand weighing 1 g falling onto your head with probability 10% and preventing a 1 kg stone falling on your head with probability 10^{-4}. Indeed the expected value is 0.1 g in both cases. The result is minimal prevention.

Or, to paraphrase Bernoulli:

> *The determination of the value of a risk must not be based on the expected losses, but rather on the disutility it yields.... There is no doubt that a loss of one thousand ducats with probability 10^{-3} is more significant to prevention than a certain loss of one ducat.*

More generally, using averages is a profoundly misleading way of approaching risks. It denies any specific character to the disastrous effect of a sudden lump sum loss by simply asserting that it is equivalent to fraction losses incurred more frequently. A fundamental feature of disasters is thus swept under the rug.

1.3. Beyond expected utility

Such risk aversion only relates to losses; in EU theory risk remains linear with respect to probabilities; elasticity of risk with respect to probabilities is identical to 1.

> In formal terms, let w be wealth in the absence of risk and U(w) its utility without risk. Let x be a potential loss incurred with probability p. The existence of such a risk leads the situation to be valued as the expected utility:
>
> $$(1-p) U(w) + p (U(w-x) = U(w) - p [U(w) - u(w-x)]$$

The value of risk is thus $p [U(w) - u(w-x)]$, the expected disutility of the risk materializing. This can be written $px \, \mu(x)$ where μ is a multiplier of expected values which quantifies risk aversion. Conversely, if the value of risk can be expressed as $px \, \mu(x)$, it can be shown that the decision process is EU. Therefore EU is entirely characterized by: "the elasticity of risk with respect to probabilities is identical to 1".

Therefore, despite arguments existing in favor of EU, there is a need to investigate risk aversion with respect to probabilities.

2. FARMER'S RISK AVERSION

As early as 1967, Farmer [1] analyzed risks in a quantitative fashion in order to determine suitable locations for NPPs. At the time, his work was construed as a "probability approach", in contrast with general siting criteria such as "more than 50 km away from any town". His major contribution was seen as a quantification of health risks; however, much progress has been made in risk quantification since this historical paper and what is mainly retained now is the "Farmer diagram".

The idea was first put forward as the left diagram in Figure 1 where consequences appear as Curies released on the x-axis and probabilities appear as return years on the y-axis. The line CD pictures risks (x,p) which are deemed equivalent. In Farmer's words:

> It is seen on Fig. 2 that the area A is one of low risk and area B one of high risk and all parallel lines of equal slope -1 join points of equal risk in terms of curies per year. One such line might be used as a safety criterion by defining an upper boundary of permissible probability for all fault consequences.

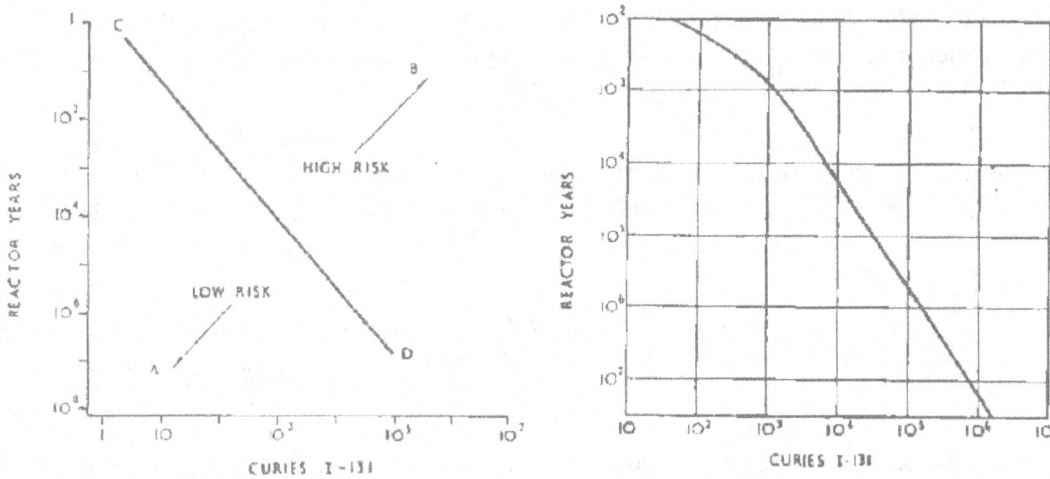

Figure 1 : Farmer's original diagrams

A few pages later, Farmer refines this general idea as follows:

> 7. A BOUNDARY LINE AS A CRITERION
>
> In the introductory section of this paper it was suggested that a useful standard for a reactor might be provided by a boundary line indicated as CD on Fig. 2. Although this line joins points of equal risk in curies per year, it may not represent an equal risk of casualties†. Furthermore, it is likely that most people would apply a relatively heavier penalty against the possibility of a large release than a small release. This would lead to a boundary line of greater negative slope. The one now chosen has a slope of -1.5 so as to reduce by three orders of magnitude the frequency of an event whose severity increases by two orders of magnitude.

The diagram is log-log; therefore the first "line of equal risk" implies px = constant since it has a slope of -1; the expected value of risk is constant along the line. The simplest corresponding value of risk is[1]:

$$V(x,p) = px$$

The lines of equal risk then form a network of parallel lines of slope -1 in log-log coordinates, as shown on the above left diagram. This depicts risk neutrality.

In the second diagram to the right, which represents the actual quantification suggested by Farmer, the equal risk lines can similarly be represented by:

$$V(x,p) = p^{0.66}x$$

Let's generalize by:

$$V(x,p) = p^{1-a}x$$

For a = 0, this formulation boils down to risk neutrality. As a increases, the reduction in probability Δp required to offset an increase in severity Δx increases [(1-a) dp/p + dx/x =0]. For a = 1, the value of risk is independent of its probability of occurrence which is certainly excessive for most prevention specialists. Thus "a" can be conceived as a risk aversion with $0 \leq a < 1$.

This formulation generalizes the well-known Farmer risk diagram in two ways.

1. It is no longer the separation of the risk plan into two zones – the acceptable and unacceptable risk zones[2] – which is a rather gross decision aid in the view of economists. It now gives rise to a risk value formula, a choice function or a (dis)utility function which allows much finer distinctions.
2. The formula for the risk value involves parameter "a" which I propose to call the Farmer Risk Aversion (FRA).

Let me address a paradox. The above value of risk, $V = px.p^{-a}$, is the expected value of losses multiplied by p^{-a}. For standard probabilities of disaster such as 10^{-n}, the multiplier is 10na and may rise to impressively high values; for instance, if a = 0.33 as suggested by Farmer, and $p = 10^{-6}$, the value of risk is 100 times the expected value of losses. For $p = 10^{-9}$, this multiplier is 1000 while it is only 10 for $p = 10^{-3}$. The lower the probability the higher the multiplier. In other words, as safety measures reduce risk (for disasters with fixed damages $x = x_0$), the decision process allocates proportionally more and more funds for safety which may appear odd.

However, if one wanted to have multipliers decreasing with decreasing values of probabilities, the FRA would have to be negative. The value of risk would then be lower — possibly much lower — than the expected value of risk which would represent a risk loving prevention behavior…

These surprising features now being acknowledged, notice that resulting prevention behaviors remain logical. When losses x are not modified:

$$\frac{dV}{V} = (1-a)\frac{dp}{p} \qquad (1)$$

---

[1] In general, V(x,p) = f(px). Imposing V(x,1) = x, which is fairly natural, implies f(u) = V(u,1) ≡ u
[2] More sophisticated versions include an « amber zone » between the green and red zones.

Since a < 1 the value of risk always varies in the same direction as p; a reduction in probability always results in a reduction in risk.

If a is fixed, the percentage reduction of risk is a fixed proportion of the percentage reduction in probability which is not unsatisfactory. And finally, although p^{-a} can become large, reductions in risk are by no means negligible.

> Numerical example: x = 100b, p = 10^{-6}, expected value of risk is 100k/year. With a = 0.33, the risk value is 100 times larger at 10m/year. Dividing probabilities of occurrence by 10 reduces p to 10^{-7}; the value of risk is then reduced by a factor of 4.67 to reach 10k.214 = 2.1m/year.

There is no particular reason why the FRA should be constant over the entire range of risks. In general, it could vary according to x and p. We then have the following variation of risk according to variations in the level of p:

$$\frac{dV}{V} = \underbrace{\frac{d(px)}{px}}_{\text{Risk neutral Variation}} + \underbrace{a(p,x)\frac{dp}{p}}_{\text{Effect of Risk}} \qquad (2)$$

In general, one can always set V(x,p) = px μ(x,p) — which is simply a definition of the multiplier μ. One then immediately derives:

$$\frac{dV}{V} = \frac{d(px)}{px} + a_x(x,p)\frac{dx}{x} - a_p(x,p)\frac{dp}{p} \qquad (3)$$

With $\quad a_x(x,p) = \frac{x\mu'_x}{\mu} \quad a_p(x,p) = \frac{p\mu'_p}{\mu}$

When losses x cannot be mitigated, (3) boils down to (2). When both a_x and a_p are nil, we obtain expected value; when $a_p = 0$ and $a_x \neq 0$ we have EU. The general case features both a_x and a_p non-nil.

Therefore, testing for $a_p = 0$ is a way to test the expected value theory (in addition, testing $a_x = 0$ would test zero risk aversion, i.e. risk neutrality).

3. ELICITATION OF THE FRA

I hope to have shown the above framework to have theoretical value. In practical applications, it would be particularly useful to address questions raised by the prevention of disasters where potential losses x are high and occurrence probabilities p are low. For ordinary risks with "limited" losses and "non- negligible" probabilities, values of a_x and especially a_p could be low or even negligible.

It does, however, remain theoretical as long as values of the FRA cannot be justified. Hence an attempt to find out whether my colleagues at IRSN did exhibit FRAs different from 0 and whether these values might be influenced by the severity of the accident.

Procedures designed to quantify behaviors and preferences have been developed in environmental economics. It has been found that asking direct questions such as: "How much would you be ready to pay for, or think the nation should invest in, the reduction of such risk?" do not provide satisfactory answers. More indirect questions are needed.

3.1. The questionnaire and its validation

Equation (1) directly suggests questions such as: "In order to divide risk by a factor k, by how much would it be necessary to divide its probability of occurrence?" Notice that risk is not defined in such a question and thus refers to "a feeling" of professionals. The answer of respondents should not result from a computation or formula but correspond to (deep) values and (authentic) choice behaviors.

The following questionnaire was proposed:

In order to divide risk by	2	probabilities should be divided by	?
In order to divide risk by	3	probabilities should be divided by	?
In order to divide risk by	5	probabilities should be divided by	?
In order to divide risk by	10	probabilities should be divided by	?

These questions were asked for the severe accident used by IRSN in its estimates of nuclear accident costs in France [2] and then for the major accident referred to in these studies.

In order to test reactions to such an unusual query, the questionnaire was first proposed to 4 colleagues well acquainted with IRSN accident cost estimates. This was performed during face to face interviews which lasted up to half an hour. Interviewees were allowed to ask any question and offer any comments they wished. Results were encouraging. For the severe accident, elicited FRAs ranged from 0.1 to 0.7 with an average of 0.35. For the major accident (with radioactive releases comparable to those of the Fukushima disaster), they ranged from 0.3 to 0.7 with an average of 0.41. FRAs were significantly higher for the major accident than for the severe accident, except for one person.

Although interviewees thought the questionnaire was difficult to answer, the procedure encountered no major difficulty and largely comforted Farmer's intuition.

3.2. The procedure and its difficulties

The questionnaire was then proposed to two groups of professionals who attended a 45 mn presentation of IRSN accident cost estimates. They were not aware of the "survey" and simply attended to get acquainted with accident costs. They were all simultaneously confronted with a printed questionnaire, between the presentation itself and the ensuing Q&A session. After a short general briefing, the questionnaires were handed out and respondents had 5-10 mn to write down their answers.

Results differed from the preparatory face to face interviews. Respondents felt it was extremely difficult to answer; a majority of persons felt they were passing some sort of test, as in school, and were supposed to find "the right answer" although it had been clearly stated that this was not the case. I call the corresponding answers SF — for "deriving from a scholarly formula". The most frequent such routine is to answer with the very same factor as in the question, in which case a = 0, i.e. risk neutrality. Another is to square the factor resulting in a = 0.5; another is to multiply the factor by 10; and so forth.

Overall, it is difficult to judge whether risk neutral answers really mean a fully-conscious risk neutrality or simply result from a difficulty with the notions involved and the search for a simple formula to answer. This could be examined directly with the persons involved ex-post (if they did fill up their name on the questionnaire).

The global feedback — a great difficulty to answer the questionnaire —, suggests that such elicitations are better not conducted through such paper questionnaires but should involve face to face interviews. Several French scholars declare this is a general phenomenon when complex risk questions are involved [3]. For example, they conducted studies of risk perception around a French NPP and say global paper questionnaires give very poor answers. In their experience, it is necessary to acculturate people to the said risk through one-hour interviews. They report that perceptions change in the course of the interview. And also change with time, which was observed by conducting interviews with the same people once a year over several years.

Proceeding through interviews is clearly much more costly in terms of time and resources. The need to ensure comparable standards and interviewing procedures may also make the results easier to criticize. In any case, the results will tend to be protocol dependent…

3.3. Results

In total, 37 questionnaires were received. Details are given in the appendix.

Full risk neutrality ranges from 0% to 30% depending on whether answers which duplicate factors included in the question are interpreted as a routine without deep significance or as fully-conscious expressions of risk neutrality. This means that risk aversion is dominant among IRSN professionals. Risk neutrality appears significantly more plausible for controlled radioactive releases than for the massive releases implied by a major accident: +20% above the previous 30% resulting in 50% risk neutral answers for controlled releases, in the hypothesis most favorable to risk neutrality.

Another general conclusion is that risk aversions elicited in this exploratory attempt are quite high. When leaving aside answers pointing to risk neutrality, they tend to be above the Farmer figure of 0.33, even for the severe accident (between 0.4 and 0.45 in average). There are no answers with positive but low FRAs such as, for instance, a = 0.05; thus, there is a gap between risk neutrality and risk aversion; this reinforces the interpretation that risk neutral answers do not fully correspond to risk neutrality *stricto sensu*.

And a third conclusion is that risk aversion does increase with the severity of the accident as would be expected. It is globally confirmed with FRAs above 0.6 for the major accident.

4. CONCLUSION

This paper suggests that risk aversion with respect to probabilities is a subject worth further investigation. A fairly general formulation is suggested which could perhaps help put the problem on the agenda. The opposing view would argue that EU is well grounded in a solid axiomatic, namely that of Savage [4]. This has been the consensus for many years because the axioms of Savage appear "natural" when formulated. However, many experiments have shown that they do not correctly describe actual behaviors with respect to risk. The intuition put forth by Farmer, not an economist but an engineer, goes back to 1967 when challenging EU was basically unheard of (see however Allais [5]).

The heuristic experiment conducted among IRSN professionals seems to support Framer's intuition and thus constitutes a further argument for considering risk aversion with respect to probabilities.

References

[1] Farmer, F. R. *Siting criteria, a new approach.* Atom, Vol. 128, pp 152-170 (1967). Also presented at the IAEA Symposium on Containment and Siting, Vienna, 1967

[2] Pascucci-Cahen L. and Momal P. *Massive radiological releases profoundly differ from controlled releases.* Eurosafe. 2012

[3] Coanus T., Duchene F., Martinais E. *Prendre en compte la perception des non spécialistes.* Problèmes politiques et sociaux, n° 882, p. 98-101. 2002. Other papers of these authors can also be of interest.

[4] Savage L. *Foundations of Statistics.* New York, John Wiley and Sons, 1954

[5] Allais M. *Le comportement de l'homme rationnel devant le risque : critique des postulats et axiomes de l'école américaine.* Econometrica, vol. 21, p. 503-546. 1953

Appendix

In a first session 19 answers were collected among which 7 were SF with a = 0. Some of these clearly stated that: "Risk is probability". Their answer was thus the same for the two different accidents of largely different severity and simply consisted in copying the factors in the question.

Another 3 respondents copied the factor in the same way for the severe accident (a = 0), but multiplied these factors by 10 for the major accident thus coming up with a ≈ 0.6. There is an obvious suspicion that such answers result from a "formula" rather than a "pure feeling" but this should ideally be tested directly with the respondents.

4 respondents offered identical answers for the severe and the major accident. Their FRAs were thus not influenced by severity of consequences. However, they were fairly high at 0.6, 0.55, 0.4 and 0.2.

The other 5 answers resulted in average FRAs of 0.45 and 0.6 for the two accidents.

In a second session, 18 questionnaires were filled up with 4 risk neutral answers. 5 respondents were neutral for the severe accident (a = 0), but used another routine for the major accident, mainly the "squares" routine; their FRAs were quite high for the major accident, ranging from 0.5 to 0.7. Two questionnaires exhibited equal non-zero FRAs for the two accidents, with very high aversions of 0.75 and 0.85.

The other 7 respondents showed an average FRA of 0.4 for the severe accident and 0.75 for the major accident.

Development of a Methodological Approach to Strategic Fire Service Planning Combining Concepts of Risk, Hazard and Scenario-based Design

Adrian Ridder[*], Uli Barth
University of Wuppertal, Wuppertal, Germany

Abstract: Strategic Fire Service Planning is a quite new field of research because of which there is a need for fundamental research and methodological work. Existing approaches like risk management, the hazard-concept and scenario-based design have been found to not be fully applicable on its respective own to the research question of "how much fire service is necessary in a city". Based on analytical work and the analysis of incident data it is shown that a combined approach of risk and scenario-based methods is a good starting point for further research.

Keywords: risk, hazard, scenario, fire service, data analysis

1. INTRODUCTION

Several scientific disciplines – from economics [1], performance [2] and risk management [3], [4] to operations research [5], systems analysis [6] and probabilistic studies [7], [8] - have already developed methodological approaches to answer parts of the question "how much fire service" is needed to provide for adequate safety against fires and other life-threatening hazards. However, not one single approach of those is able to serve as a holistic blue-print for a comprehensive method for fire service planning on a national scale.

A current work-in-progress research project in Germany called "TIBRO" (German acronym for tactical-strategic innovative fire service and risk-based optimizations) was started in February 2012 and is scheduled for its final report in the first quarter of 2015. The aim of the project is to outline the fundamentals of a comprehensive methodology to derive necessary fire service resources in accordance with the legal tasks to be fulfilled by the fire service.

2. METHODS

2.1 Analytical review of hazard-based, risk-based and scenario approach

In simple terms, the risk-based approach can be described as the composition of a hazard with the probability of its incidence. The hazard-based approach therefore cannot be used to quantify necessary fire service resources for a municipality, because this would result in an overly conservative estimation as only the possible extent of an incident would be considered without paying attention to the probability of it. This can be strikingly depicted with the example of a nuclear power plant: If one would exist in the boundaries of a municipality, a huge hazard of nuclear and radiological incidents exists. However, considering the probability of different accident-scenarios somewhat relativizes the necessary fire service resources needed for the protection of that city. Therefore it would be unsustainable to have all the firefighters, trucks and special equipment on duty every day that would be necessary in a severe accident within the plant.

However, on the other hand fire service planners cannot simply bring small probabilities into play in order to use the small number of large accidents to postulate that very small resources would be adequate as the remaining number of incidents excessing the small scenario could be considered a residual risk. Turn-out data over six years of eight German municipalities ranging in population from ca. 4.000 to 170.000 shows that under a pure probability perspective a fire department in a town with

---

[*]corresponding author: ridder@uni-wuppertal.de

less than 18.750 population would not have to be prepared for a residential fire with persons endangered in an upper floor (a popular design scenario in Germany) as the frequency of those incidents was zero [9]. Therefore it would be enough to purchase equipment and crew the department in a fashion that it could respond to false alarms, incidents resulting from torrential rains and storm winds as well as removing small oil spills (stains) from streets, which together make up more than half of the total number of incidents. For those operations, obviously only very limited equipment, training and personnel would be needed, compared to more complex and life-threatening incidents like house fires or car accidents. However, common sense suggests that it would be irresponsible to regard the more complex incidents as residual risk, disregarding their actual frequency and probability. But it is very difficult to draw a line between types of incidents that have to be regarded and ones that can be disregarded. Using the frequency of incidents in the past alone obviously is not the suitable method to decide that question.

Contributing to those problems with a risk-based approach to fire service planning is the problem of large uncertainties, both aleatory and epistemic, in terms of what type and extent of incidents a fire service will have to deal with one day and what that will require of the fire service. The different interpretations of black swan concepts [10] all apply here. It is important that existing uncertainties are quantified and transparently communicated to the deciders and stakeholders [11], i.e. the locally responsible political body and the citizens.

What is more, the traditional approach of multiplying the probability of an incident with its severity has been found to be not fully applicable to the problem at hand as there exists not one scale on which to measure damage and benefits and also the conversion into monetary dimensions is viewed as controversial. Especially the public often does not understand or accept that kind of engineering calculation when there is a large hazard in place [12]. It can be argued that the prospect of a fire service not being able to confidently deal with a house fire or car accidents seems to be a large hazard to the citizens of the community served by the fire service. At the same time, those citizens and their politicians are the public to which the fire service has to answer and has to go to in order to receive funding, which adds to the complexity of the problem. It is important to bear in mind that though the risk perception of laymen and so-called experts has been found to closely correspond, they are not identical as laymen also include qualitative aspects into the risk assessment, especially dread and put more emphasis on worst-case outcomes [13].

From a scientific point of view, a "safe" risk level cannot be determined as such matters involve not only scientific facts, but also societal and especially ethical questions. Therefore and because the decisions made in that area come with a high demand for political accountability, the decisions rest with the political authorities having jurisdiction, and not with the local fire service, which is only the executer of political will, and not with the scientists, as it is not upon them to decide what should be deemed safe [14].

However, risk science has an important role to play in that process, in that it can offer the deciders valid background information on the decisions at hand, therefore enabling informed decisions. It is especially important that the consequences and outcomes of the different decision options must be made transparent and understandable to the deciders, who are more often than not laymen in this specific area. This so-called analytic-deliberative approach to complex safety and risk questions [11] should be understood by all parties involved, therefore clarifying what to expect from the different parties and where boundaries of competencies lie. Especially the remaining uncertainties of scientific analysis must be made clear to the decision-makers.

To counter the methodological drawbacks of the risk approach outlined above, a scenario approach seems suitable. Scenarios in this case represent independent, representative sets of events and parameters that define the hazards of and the persons exposed to an incident [15] requesting fire service attendance. Using scenarios, the virtually infinite number of tasks and incidents that a fire service possibly has to address are reduced to a manageable number which can directly be linked to operational experience of firefighters. Questions that need to be addressed with this approach are the number of scenarios necessary to achieve a truly representative set of scenarios; currently, methods mainly used in Germany utilize only one scenario of a residential fire with persons trapped in an upper floor. Even a small increase in the number of scenarios from such a limited number obviously offer large advantages in covering the different incidents addressed to the fire service in real life. In addition, also the level of complexity of the covered scenarios must be addressed, as it could range from worst-case scenarios with low-probability and high-impact events (as described above) to simplest-case scenarios [16], both of which are on their own not representative. It is suggested to develop a set of scenarios based on the main tasks the fire service usually has to address (fires, technical rescues, hazmat incidents, water rescue) with each category involving three to four scenarios with increasing levels of scope and complexity. Those 12 to 16 scenarios would cover quite a large spectrum of fire service activities and offer the opportunity to directly link them to different operational readiness levels that could be assigned to different fire stations, as usually not every station has the capability to deal with all possible incidents.

2.2 Analysis of data from fire service turn-outs

As outlined above, further fundamental research is necessary to implement a real risk-based approach. In existing German literature, a number of hypotheses have been found elaborating on risks of different types and heights of buildings, their occupation and the number of units in a building ("risk factors"). The hypotheses in general state that with increasing height, age and size of buildings the fire risk and the demand on the fire service would increase. Similarly, fire risk and demand would be higher in industrial, commerce and special buildings (schools, hospitals, malls etc.) than in private dwellings.

To validate the assumptions of those hypotheses, two studies were conducted with the objective of verifying (or falsifying) them based on turnout data of two German metropolitan fire services (city 1: pop. ca. 350.000, 5 years of data, 1.022 data sets (structure fires only); city 2: pop. ca. 690.000, 4 years of data, 17.093 data sets).

As the work was of a more methodological type, two different methodologies have been specifically developed custom-made to the two available data sets. Knowing that the results of both studies cannot be directly compared, it was considered worthwhile developing two independent ways of data analysis as a fundament for further sophistication, which is still ongoing.

One methodology created absolute risk numbers for the different risk factors, based on a number of items representing the risks (monetary loss and damage compromising the further habitability of the premises, frequency of incidents) and the demand on the fire service operations, covering items like duration of operations on site, number of trucks, personnel and breathing apparatus used, number of rescued civilians and used amounts of extinguishing agents [17].

The second study used a qualitative comparison of the statistical distributions of basically the same items as listed above plus injuries and fatalities of civilians (for which no data was available in the other study), permitting for relative statements like one category of buildings possesses a greater fire risk than others, without giving absolute numbers [18]. Cumulative curves, Q-Q-plots and Box-plots were used in that process.

3. RESULTS

Considering the analytical investigation of the applicability of different methodologies based on hazards, risks and scenarios, it became apparent that a combined approach is necessary, containing both aspects of the risk-based and of the scenario approach. Various problems still remain to be solved, including how to deal with black swans and large uncertainties in general.

To reduce some of the epistemic uncertainty, the two studies of fire service turn-out data tried to substitute assumptions with facts. It has to be admitted, that this objective could not wholly be reached as the results of both studies are inconclusive: Some hypotheses could be proved, some falsified, but not one hypothesis showed the same results in both studies. Some hypotheses could not be conclusively be categorized as the statistical population seems to have been too small to make general statements. Also varying procedures in creating the statistics by the fire services and incomplete data sets (e.g. the building age was not available for all buildings which had an incident in it) contributed to the inconclusive outcome.

However, the outcome of the methodological research attempted here for the first time seems encouraging enough to continue the data analysis with additional cities and counties and with cross-checking the developed methodologies on respectively other municipalities, with which it is hoped to gain some further understanding of the underlying factors. Both studies delivered the proof of principle that it is possible to aggregate and analyse fire service data to a level of detail sufficient for answering questions on risk as posed here and probably also other risk-related problems.

4. CONCLUSION

A risk-based approach seems appropriate to reach individually good fitting fire service protection levels for municipalities of different sizes, populations and risks. The goal should be to determine a level of risk that is adequate, without over-spending because the assumptions are too conservative but at the same time guaranteeing an adequate level of safety.

However, more fundamental research is necessary to create more reliable quantified data on correlations of risk with characteristics of cities, dependencies on socio-demographic factors and the capabilities of the individual fire services. This is hindered by the lack of a national fire service statistic in Germany, for which reason it is necessary to approach single cities and put substantial effort in making their statistics accessible for further analysis. The studies conducted based on existing data are proofs of principle that this approach is worthwhile.

Historic demand on the fire service, based on calls per area and per stations adequately categorized, should be included in further research. Therefore, both a priori (generalized categorization of areas based on risk characteristics drawn from past incidents) and a posteriori (actual historic demand in the area under scrutiny) perspectives could be included in the analysis.

Acknowledgements

The research project on which this report is based has been funded from the German Federal Ministry of Education and Research (grant ID 13N12174). The authors are responsible for the content of this paper.

References

[1] R. Ahlbrandt, Efficiency in the provision of fire services, Public Choice 16 (1973) 1–15. http://link.springer.com/content/pdf/10.1007%2FBF01718802.

[2] G. Bouckaert, Productivity Analysis in the Public Sector: the Case of the Fire Service, International Review of Administrative Sciences 58 (1992) 175–200. http://ras.sagepub.com/content/58/2/175.full.pdf+html.

[3] IFE USA Branch (Ed.), Second International Conference On Fire Service Deployment Analysis, 2002.

[4] Industry and Environment Programme Activity Centre, Hazard identification and evaluation in a local community, 1st ed., United Nations Environment Programme Industry and Environment Programme Activity Centre, Paris, 1992.

[5] G. Carter, J. Chaiken, E. Ignall, Simulation Model of Fire Department Operations: Executive Summary. R-1188/1-HUD, The RAND Corporation, New York City, 1974.

[6] W.E. Walker, Applying Systems Analysis to the Fire Service. P-5456, New York City, 1975.

[7] D. Matellini, A. Wall, I. Jenkinson, J. Wang, R. Pritchard, Bayesian network modelling for fire safety assessment: Part I - a study of human reaction during initial stages of a dwelling fire, in: Advances in safety, reliability and risk management. Proceedings of the European safety and reliability conference, ESREL 2011, Troyes, France, 18-22 September 2011, CRC Press, Boca Raton, FL, 2011, pp. 626–633.

[8] E. Ignall, P. Kolesar, W.E. Walker, Using Simulation to Develop and Validate Analytical Emergency Services Deployment Models. P-5463 The RAND Paper Series, The RAND Corporation, Santa Monica/California, 1975.

[9] T. Hildebrand, Statistische Einsatzdatenanalyse zur Abschätzung der Relevanz des kritischen Wohnungsbrandes für unterschiedlich strukturierte Gemeinden. Bachelor-Thesis, Wuppertal, 6.2.13.

[10] T. Bjerga, R. Flage, On black swans in relation to some common uncertainty classification system, in: Safety, Reliability and Risk Management: Beyond the Horizon. ESREL 2013, Taylor & Francis Group Ltd, London, 2013, pp. 3197–3202.

[11] Committee on Risk Characterization, Understanding Risk: Informing Decisions in a Democratic Society. Informing Decisions in a Democratic Society, National Academies Press, Washington D.C., 1996.

[12] G. Bechmann (Ed.), Risiko und Gesellschaft. Grundlagen und Ergebnisse interdisziplinärer Risikoforschung, Westdt. Verl, Opladen, 1993.

[13] P. Slovic, The perception of risk, Earthscan Publications, London, 2002.

[14] Kommission der Europäischen Gemeinschaften, Mitteilungen der Kommission. Die Anwendbarkeit des Vorsorgeprinzips, Brüssel, 2000.

[15] H.A. Merz, T. Schneider, H. Bohnenblust, Bewertung von technischen Risiken. Beiträge zur Strukturierung und zum Stand der Kenntnisse ; Modelle zur Bewertung von Todesfallrisiken, VDF Hochsch.-Verl. an der ETH, Zürich, 1995.

[16] Referat 5 des Technisch-Wissenschaftlichen Beirates der vfdb, Elemente zur risikoangepassten Bemessung Elemente zur risikoangepassten Bemessung von Personal für die Brandbekämpfung bei öffentlichen Feuerwehren. Technischer Bericht, Vereinigung zur Förderung des deutschen Brandschutzes e.V., Altenberge, 2007.

[17] D. Edner, Analyse von in der Feuerwehr-Bedarfsplanung verwendeten Hypothesen anhand realer Einsatzdaten einer großstädtischen Feuerwehr. Master-Thesis, Wuppertal, 2013.

[18] S. Langer, Analyse von in der Feuerwehr-Bedarfsplanung verwendeten Risikofaktoren für Wohngebäude anhand realer Einsatzdaten. Bachelor-Thesis, Wuppertal, 2013.

Ambiguity in Risk Assessment

Inger Lise Johansen[a] **and Marvin Rausand**[a]

[a] *Norwegian University of Science and Technology, Trondheim, Norway*

Abstract: This paper aims to shed light on the concept of ambiguity in engineering risk assessment. The objectives are to 1) Clarify the meaning of ambiguity in risk assessment; 2) Describe sources and manifestations of ambiguity in preassessment, risk analysis, and risk evaluation/decision-making; and 3) Outline a procedure for approaching ambiguity in practice. To address these objectives, we first review existing definitions of ambiguity, which are argued to be of limited relevance to engineering risk assessment. We then propose a new overall definition of ambiguity as a challenge in risk-informed decision-making, and define linguistic, contextual, and normative ambiguity as distinct categories of ambiguity that have different implications for risk assessment. Based on this, we list concrete sources and manifestations of ambiguity in risk assessment in a set of tables that can be used as a checklist for identifying ambiguity in the assessment process. We finally outline a stepwise procedure for approaching ambiguity in risk assessment, in order to provide practical guidance and stimulate further research on ambiguity in risk-informed decision-making.

Keywords: Ambiguity, Risk Assessment, Risk-Informed Decision-Making

---

1. INTRODUCTION

Ambiguity permeates many strategic decisions that involve major accident risk. There might be ambiguity concerning the information needed to inform the decision-process, the basis for providing it, and its meaning and implication for decision-making. This can be traced back to ambiguity in the interpretation of *risk* [33], which is commonly defined as the answer to three questions: 1) What can happen? 2) How probable is it and/or how uncertain are we? and 3) If it does happen, what are the consequences? [2, 22]. *Risk assessment* can be defined as the process of defining, answering, and evaluating these questions in the phases of preassessment, risk analysis, and risk evaluation/decision-making. There are many sources and manifestations of ambiguity in this process, for example, concerning scope and boundary conditions, critical assumptions, and formulation of decision criteria [34]. Klinke and Renn [24] consider ambiguity as a defining challenge to risk assessment that calls for participatory approaches to risk-informed decision-making, such as the analytic-deliberative approach of NRC [27]. There is, however, limited guidance on how to identify and approach ambiguity in risk assessment by such approaches.

There are several alternative interpretations of ambiguity in the risk and decision-making literature. Some associate it with conflicting values and beliefs about *consequences* in the third question of risk [19, 34], others with incomplete knowledge in the assessment of *probabilities* and *uncertain events* in the first and second question of risk [6, 13], yet others with imperfections in human judgments [9, 25]. Many of the interpretations are problematic in light of foundational research on engineering risk assessment [2] and little has, to our knowledge, been done to scrutinize the concept of ambiguity in this context. In order to improve the role and value of risk assessment in risk-informed decision-making, there is a need for clarifying the meaning, sources, and manifestations of ambiguity in the different phases of risk assessment and to provide guidelines for approaching it in practice.

The purpose of this paper is to: 1) Clarify the meaning of ambiguity in risk assessment, 2) Describe sources and manifestations of ambiguity in the assessment process, and 3) Outline a procedure for approaching ambiguity in practice. The paper is delimited to engineering risk assessment for strategic decisions involving major accident risk. The paper is structured as follows: first, alternative conceptions of ambiguity in the literature are discussed in Section 2, before a new conceptualization is proposed in Section 3. Section 4 describes sources and manifestations of ambiguity in risk assessment, before a procedure for approaching ambiguity is presented in Section 5. Concluding remarks are given in Section 6.

2. LITERATURE ON AMBIGUITY

In this section we present and discuss common conceptions of the term *ambiguity* in everyday English language and in the risk literature. We argue that they are not sufficient for approaching ambiguity in risk assessment, before we present a new definition of ambiguity in the following section.

A dictionary definition of ambiguity is "a word or expression that can be understood in two or more possible ways" [26]. This conception is seen in some risk assessment applications, for example, Colyvan [10] explains ambiguity as "uncertainty arising from the fact that a word can be used in more than one way, and in a given context, it is not clear which way is being used." The distinction between ambiguity and *vagueness* is stressed, where vagueness is something that is "stated in a way that is general and not specific" [26] and hence permits *borderline cases*. Vagueness is about the difficulty of making any interpretation, whereas ambiguity is about making several interpretations. Ambiguity is by this conception a linguistic property of statements that *can* be given multiple meanings depending on the context in which they are interpreted. This explanation is clear, but confined on the grounds that it concerns isolated statements and not how they appear in risk assessment.

A second conception of ambiguity is found in the literature on risk governance [3, 19, 24, 31]. Key to this interpretation is the existence of multiple values and perspectives on the severity, tolerability, and wider meanings of risk; manifested as disputes about framing, ethics, and trust. Ambiguity here corresponds to ambivalence and controversy and refers to the social situation around risk. The International Risk Governance Council [19] coins two types of ambiguity:

1. *Interpretative ambiguity* refers to different interpretations of identical assessment results and "factual" states of the world (e.g., whether an outcome is adverse or not), which is a result of people processing risk information according to their own risk constructs and images.

2. *Normative ambiguity* refers to different perspectives regarding the tolerability of the risk, which comes from differences in applying normative rules for evaluating the states of the world (e.g., fairness and distribution of risk and benefits).

Both are restricted to *consequences* that have an impact on something humans *value* [30]. Stirling [34] elaborates that "under conditions of ambiguity, it is not the probabilities but the possible outcomes themselves that are problematic," and that ambiguity concerns "contradictory certainties" that cannot be objectively described by a single risk picture. While the inference is legitimate, the characterization is conceptually challenging as it implies that situations of ambiguity can be distinguished from situations where risk can indeed be objectively and uniquely described. This is in contrast to a broad line of risk research that acknowledge that risk is not an ontological property that can be objectively and unequivocally assessed [1, 2]. A second limitation is that the categories of interpretative and normative ambiguity are vague and intimately interrelated; one can be a source of the other and it is difficult to draw the line where one stops and the other begins. The prefix *interpretative* is in our view superfluous, and normative ambiguity extends beyond tolerability issues as a distinct source of (interpretative) ambiguity. Tolerability of risk furthermore depends on the other objectives in a decision problem and boils down to not knowing what alternative to choose, which is in our view something wider than ambiguity [15]. The greatest limitation is, however, that ambiguity is restricted to conflicting interpretations of risk assessment results, thus failing to explain how it enters risk assessment in the first place.

A third conception is found within decision theory, statistics, and economics. This is a group of definitions that attribute ambiguity to the assessment of *uncertainties* and/or *probabilities*. Camerer and Weber [8] summarize five such conceptions, concerning *second order probability*, *quality of information*, *weight of evidence*, *source credibility*, or *unknown, but knowable information*. A common feature is that ambiguity is defined in relation to whether the true probability of an exhaustive set of outcomes can be known. Such interpretations are found in some risk assessment applications [5, 12], but are emptied of meaning if probability, like risk, is considered as a subjective construct in line with the arguments above. Another point of critique is that it is unclear how ambiguity differs from mere *uncertainty*, such as in the much cited definition of Ellsberg [13]: "a quality depending on the amount, type, reliability, and unanimity of information, given rise to one's degree of confidence in an estimate of relative likelihoods of future events." A special group of interpretations relate ambiguity to *impreciseness* in subjective expressions of uncertainty. This may either concern the informative basis for expressing uncertainty (e.g., we do not have sufficient information to specify whether our probability is 0.1 or 0.9) and/or vagueness in the description of uncertain events that permit borderline cases concerning their occurrence [6, 17]. Special approaches have developed to deal with such imprecision by non-probabilistic descriptions of uncertainty, for example, fuzzy logic or possibility theory [4, 10]. Some consider imprecision and vagueness as synonymous to ambiguity [21], whereas others stress that they are distinct concepts [10]. This group of interpretations is less controversial from an ontological point of view, but is in our view too limited because it ignores the consequence element in the definition of risk and fails to explain ambiguity as a defining challenge in risk assessment.

A final conception is found within behavioral research on individual and organizational decision-making. March [25] describes "ambiguities of choice" that go beyond the assessment of consequences and probabilities to human judgment and information processing in decision-making. March defines four types of ambiguities: Ambiguities of *preferences* (individual preferences may be vague, inconsistent, or unstable); ambiguities of *relevance* (the usefulness of information for decision-making may be unclear); ambiguities of *intelligence* (there may be several norms for what constitutes rational action); and ambiguities of *meaning* (lack of clarity regarding how one talks about the world and how meaning evolves from information). A study of ambiguity in safety cultures [32] describes the latter as lack of mutual understanding of words, symbols, and cultural manifestations in the process of creating and recreating meaning. Variants of March's [25] conception is also found in risk assessment, for example, as "internal uncertainty" in decision-making that reflects imprecision in human judgments concerning preferences, values, and risk attitudes and may stem from insufficient understanding of problems, modeling assumptions, and so on [9]. Our main objection against these conceptualizations is that attributing ambiguity to limitations in human judgment instead of *what is being judged* circumvents the contextualities of risk and makes ambiguity a constraint rather than a defining challenge to risk assessment.

3. NEW DEFINITION AND CATEGORIZATION

The many conceptions of ambiguity are as such a source of ambiguity. Many of the interpretations are limited to one or two questions in the definition of risk, and are problematic by some risk perspectives. Few of the interpretations are broad, yet specific enough to guide the identification and treatment of ambiguity in risk assessment. A message of this paper is that there are several types of ambiguity that are distinct, yet closely related in the process of defining, answering, and evaluating the three questions of risk. In the following, we propose a new overall definition and clarify three categories of ambiguity that have different sources and manifestations in risk assessment.

3.1 An overall definition

We propose a new overall definition of ambiguity related to risk assessment:

Ambiguity: The existence of multiple interpretations concerning the basis, content, and implications of risk information.

By *risk information* we mean descriptions of risk in a broad sense; covering scenarios, probability distributions, risk metrics, uncertainty factors, sensitivities, and so on [2]. By *content* we refer to what these descriptions express, while we by *basis* refer to the information and judgments that underlie them. By *implications*, we mean how the information fits into and is evaluated in a decision context, but not what solution should be chosen. The ambiguity can be *intra-* and/or *inter-individual* in the sense that it may refer to multiple interpretations by a single individual (e.g., a decision-maker makes two alternative interpretations) or single interpretations by multiple individuals (e.g., a decision-maker makes one interpretation and a stakeholder makes another interpretation). The two are conceptually the same for a decision-maker who makes a decision on behalf of herself and relevant stakeholders. Inter-individual ambiguity may be evident, which may promote conflict and reduce confidence in risk assessment, but be sought resolved in the assessment. It may also go unnoticed until the assessment is finished, which may lead stakeholders to discard the risk analysis and oppose the decision after it is implemented.

The purpose of the overall definition is to clarify ambiguity as a defining challenge to risk assessment. In contrast to Klinke and Renn [24], who consider ambiguity as something that cannot be objectively assessed in distinction from something that is difficult, but possible to assess (i.e., complexity) or for which we lack sufficient scientific knowledge (i.e., uncertainty); we define it in multiple interpretations of risk information *conditioned* on limitations in the understanding of the system (i.e., complexity) and beliefs about what will happen in the future (i.e., uncertainty). Ambiguity is here a distinct challenge that hampers the use of risk assessment in risk-informed decision-making, and may amplify both uncertainty and complexity and influence the decision-maker's tolerability of both.

A key wording in the definition is "the *existence* of multiple interpretations", which implies that ambiguity is not a potentiality as in the dictionary definitions, but a realization in the sense that multiple interpretations do exist (whether evident or not). If it was simply a potentiality, it would neither be defining nor of practical interest because it would be an omnipresent, conceptual challenge: it is first when multiple interpretations exist that the decision-maker has a problem. However, in order to approach ambiguity in risk assessment, it is necessary to also define it by its potential causes. We therefore define three categories of ambiguity in the following, which are potentialities that may or may not be realized as ambiguity according to the overall definition. The relation between the three categories and the overall definition in risk assessment is illustrated in a conceptual map in Figure 1. All categories may influence all the three phases of risk assessment, but have different implications and possibilities for resolution. The outputs of the three phases are the definition, answer, and evaluation of the three questions of risk, which may all be subject to (overall) ambiguity with respect to the basis, content, and implications of risk information.

3.2 Linguistic ambiguity

The first category is linguistic ambiguity, which mirrors the dictionary definition in Section 2:

Linguistic ambiguity: A statement that can be interpreted in two or more possible ways.

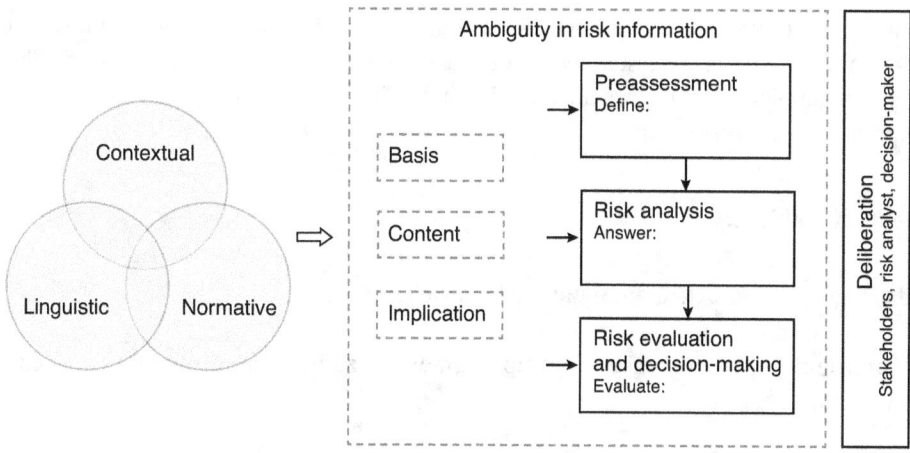

Figure 1: Conceptual map

The statement[1] may be written or oral, and can be a word or phrase, sentence, symbol, model, formula, parameter value, and so on. The statement, A, is formulated by a person (*addresser*), and interpreted by another person (*receiver*). The receiver assigns the statement an interpretation, B, in a process that is written $A \rightsquigarrow B$. A statement is *ambiguous* if it is, by reasonable logical deduction, possible to assign it two or more interpretations, i.e., both $A \rightsquigarrow B_1$ and $A \rightsquigarrow B_2$ (for $B_1 \neq B_2$). Linguistic ambiguity is different from *vagueness*, which concerns *difficulty* in interpretation: $A \rightsquigarrow ?$. However, the practical distinction might be blurred as both concern unclarity in the communication of an intended interpretation (e.g., $A \rightsquigarrow B_1$) and are conditioned on the background knowledge of the receiver.

Linguistic ambiguity concerns the possibility that multiple interpretations can be deduced from a statement. It can therefore be seen as a property of statements that could, in principle, be measured by text analysis or intuitive evaluation. This could say something about the potentiality of a statement for giving rise to multiple interpretations, but not whether multiple interpretations are indeed made. Overall ambiguity as defined in Section 3.1 is a realization of this potentiality, which resides in the conjunction between the statement and its addressers and receivers. This concerns both what is being said and not said ("exformation") and the different associations of the addressers and receivers [28], for example, because they have different professional or cultural background and experience.

The problem of linguistic ambiguity is captured in two "theorems" by Kaplan [21]:

Theorem 1: 50% of the problems in the world result from people using the same words with different meanings.

Theorem 2: The other 50% comes from people using different words with the same meaning.

Linguistic ambiguity implies that an interpretation may be right or wrong compared with the addresser's intention, and thus promote inconsistencies and "wrong" judgments in risk assessment. Linguistic ambiguity can rarely be eliminated because of the inherent ambiguities of language and the impossibility of complete specification [28]. It is, however, possible to reduce it by being vigilant in both the formulation and interpretation of statements in risk assessment. Addressers should delimit the potentiality of multiple interpretations and receivers should be aware of this potentiality. A common strategy for reducing linguistic ambiguity is to standardize definitions of terms and variables by using standards such as [20]. Such standards may, however, be conflicting or of poor quality, and increase linguistic ambiguity in a

---

[1]Ambiguity concerns something that conveys *meaning* in the sense of being perceived, processed, or described by humans. It can therefore be attributed to representations or judgments about the objects under analysis, but not the objects themselves.

specific context. According to Rasmussen [29], any attempt of objective definition must be circular on the grounds that the properties of a statement depends on the context in which it is used. This relates to a different type of ambiguity that is not attributed to individual statements, but the context in which they are used to provide risk information.

3.3 Contextual ambiguity

The second category is contextual ambiguity, which we define as:

Contextual ambiguity: The existence of multiple contextualizations, premises, and knowledge relations in risk information.

This type of ambiguity does not concern multiple interpretations of a statement, but of the context in which it is used and derived. We are broadening the scope from a statement A with its alternative interpretations B_i ($A \rightsquigarrow B_i$), for $i = 1, 2, \ldots$, to alternative statements A_j with their respective interpretations ($A_j \rightsquigarrow B_k$). The issue is not whether A_j or B_k are right or wrong; but that there can be multiple legitimate statements and interpretations. This is a more profound category of ambiguity that cannot be deduced from isolated statements, but the premises for which they are used in creating risk information. Contextual ambiguity is essentially about multiple knowledge relations and descriptions of risk [7].

Contextual ambiguity reflects the recognition that risk assessment produces contextualized knowledge that cannot be judged in relation to traditional scientific notions of objectivity and truth. According to Funtowicz and Ravetz [16], quality is a contextual, rather than universal property of risk assessment, and there is no single privileged point of view for measurement, analysis, and evaluation of risk information. Contextual ambiguity prevails when different framings and approaches to risk assessment provide conflicting risk information [35]. According to Slovic [33], this boils down to ambiguity in the definition of risk, since "there are no right or wrong definitions of risk, just different ones." Ambiguities in the *social* definition of risk (i.e., by decision-makers and stakeholders) imply multiple perspectives on the purpose, scope, and boundaries of the risk analysis, whereas ambiguities in the *technical* definition (i.e., by analysts and researchers) imply multiple perspectices on its validity and approach.

Contextual ambiguity is distinct from, but may be amplified by *context dependence* [10], which is uncertainty arising from the failure to specify the context in which a word or statement is used, recognizing that any statement can assume a multitude of meanings in different contexts [18]. It is also related to IRGC's interpretative ambiguity [19], but goes further in considering multiplicity in the basis and content of risk information and not only in the interpretation of identical risk assessment results.

Unlike linguistic ambiguity, contextual ambiguity cannot be implied from isolated statements, and it is not obvious whether it can be measured. Contextual ambiguity is also a potential in the sense that all risk assessments can be framed in alternative ways, but it is primarily a challenge if recognized in the assessment. A great distinction is that contextual ambiguity should not necessarily be reduced; it may even be desirable as it provides multiple perspectives on the decision problem [35]. Constructive utilization of contextual ambiguity presupposes, however, that both the sources and manifestations of contextual ambiguity are clarified in the assessment, and that discrepancies between the premises and approaches to risk assessment are identified throughout the process. What should be reduced is thus not necessarily the contextual ambiguity, but discrepancy in the creation of meaning within each context.

Social and scientific views are interwoven in the framing and creation of meaning in risk assessment [18], and contextual ambiguity is hence a compound issue that is determined by both scientific/analytical and social factors. To emphasize this connection, but make it possible to untangle it in risk assessment, we distinguish between contextual ambiguity, which concerns multiple knowledge claims, and normative ambiguity as defined below, which concerns multiple value claims.

3.4 Normative ambiguity

The third category is normative ambiguity, which we define as:

Normative ambiguity: The existence of multiple, conflicting, and/or inconsistent values and norms in risk assessment.

This definition is related to, but different from IRGC's [19] conception of normative ambiguity. Our focus is on values and norms that govern the *entire* risk assessment process from preassessment to risk evaluation, but *not* the tolerability of risk as a balance between risk and other values (objectives) in decision-making, which is outside the scope of this paper. Normative ambiguity concerns the normative frame of reference for creating and interpreting risk information. It resides in the convictions of the addressers and receivers that formulate and interpret a statement, rather than the statement as such. It can therefore not be meaningfully explained by the statements A_n and B_n. Normative ambiguity can be seen as a special variant of contextual ambiguity related to *value judgments* in risk assessment.

Risk assessment is influenced by both scientific (e.g., validity, consistency, and explanatory power) and social values (e.g., moral intuitions, ethical principles, responsibilities, or political motives concerning safety, ecology, economy, and so on) [35, 36]. The social values are the objectives that drive decision-making and risk assessment [23], and are evident in the definition and evaluation of consequences in terms of values to be protected and trade-offs to be made. However, our conception of normative ambiguity permeates the entire risk assessment process through value judgments at each phase; from problem structuring and choice of risk metrics to analysis and treatment of uncertainties and sensitivities [11, 33]. It may also be entangled with and amplified by linguistic and contextual ambiguity, such that value conflicts are disguised as conflicts in the interpretation and quality of "factual" statements (i.e., as in IRGC's conception of interpretative ambiguity).

Normative ambiguity is challenging from a risk assessment perspective on the grounds that it implies difficult value judgments and trade-offs. Because of the "inherent ambiguity and value-loaded nature of risk" [18] and because different stakeholders (e.g., operating companies and authorities) inevitably will have different objectives and perspectives, it can rarely be eliminated [35]. It is, however, possible to reduce the potential for normative ambiguity by making value judgments explicit and consistent in the integration between analysis and deliberation, and in particular, to distinguish normative from linguistic and contextual ambiguity.

4. SOURCES AND MANIFESTATIONS OF AMBIGUITY IN RISK ASSESSMENT

In this section we point to concrete *sources* and *manifestations* of the different types of ambiguity in risk assessment. The sources and manifestations are derived from a wide literature (e.g., [14, 27, 30, 34]) and are presented in Tables 1-3, which correspond to the three phases of risk assessment. The main purpose of the tables is to create awareness of ambiguity in risk assessment, but they can also provide practical guidance as checklists for identifying and resolving ambiguity in the integration of analysis and deliberation as outlined in the following section.

The terms "sources" and "manifestations" are blended on the grounds that it may be difficult to tell whether an ambiguity represents a fundamental point of discern or results from earlier ambiguities. This invites challenging questions about the origin and propagation of ambiguity in risk assessment. Is it, for example, possible to say that a statement C that is derived from an ambiguous statement A ($A \rightsquigarrow B_1 \rightsquigarrow$

Table 1: Preassessment

Task	Source/manifestation of ambiguity concerning:	Type
Definition of purpose	Mandate of risk assessment for informing decision-making. Objectives of decision problem with respect to risk (qualitative and quantitative).	C, N
Definition of scope and boundary conditions	Temporal, spatial, and organizational boundaries. External influences and environmental conditions. Desired resolution and precision.	L, C
Definition of central terms	Hazards, hazardous events, consequences, barriers, populations, etc.	L, C
Specification of information needs	Choice of risk metrics and presentation format. Qualitative information needs, e.g., for emergency preparedness and design.	C, N
Collection of background information	Interpretation and relevance of system drawings, references systems, experience data, expert judgments, plans and developments, stakeholder knowledge, etc.	L, C
Identification of stakeholder concerns	Objectives and concerns related to hazards, scenarios, etc.	L, C, N
Identification of constraints and challenges	Understanding of uncertainty and complexity as premises and/or limitations, identification of uncertainty factors and critical assumptions.	L, C
Choice of approach to risk analysis	Relevance, criticality, and fit between risk analysis methods and challenges. Understanding and interpretation of causality, uncertainty, complexity, etc.	C
Choice of approach to risk evaluation/decision-making	Relevance of decision-principles in light of identified constraints and challenges. Principles for integrating analysis and deliberation in decision-making.	C, N
Planning of risk analysis	Required time and resources, e.g., inclusion of disciplinary expertise.	C
Definition of the three questions of risk	*Ambiguity in the basis of risk information*	
What can happen?	Scope for identification and analysis of hazardous events, scenarios, and causal influences	
How probable is it/how uncertain are we?	Understanding of probability and uncertainty as a premise and/or limitation, and required methods of treatment	
What are the consequences?	Scope for specification of endpoints, consequence dimensions, outcomes, and affected populations	

C) is ambiguous even if it is (linguistically) unambiguous? Our answer is that it must necessarily be so because of the web of potential interpretations that are discarded at an earlier stage (e.g., $B_2 \leadsto C_2$). The sources and manifestations may propagate throughout the risk assessment as a compound mix of linguistic (L), contextual (C), and normative (N) ambiguity, and considering statements in isolation is thus somewhat meaningless. A related challenge is whether it is possible to measure the relative criticality of the different sources and manifestations. Our answer is that on their own, they remain potentialities: what we are ultimately interested in is how the sources and manifestations lead to overall ambiguity in the basis, content, and implications of risk information through how the three questions of risk are defined, answered, and evaluated. We have therefore summarized the main manifestations of ambiguity in relation to the three questions of risk at the end of each table.

5. A STEPWISE PROCEDURE

In this section we outline a stepwise procedure for approaching ambiguity in risk assessment. The procedure builds on the previous sections and is extended to consider the use of risk assessment in risk-informed decision-making. The procedure is very coarse and is intended to serve as a point of departure for further research on how to approach ambiguity in risk assessment.

1. Identify sources and/or manifestations of ambiguity during each phase:

    – Use table as a checklist

Table 2: Risk analysis

Task	Source/manifestation of ambiguity concerning:	Type
Detailed planning of risk analysis	Translation of specifications from preassessment into detailed definition of scope, boundary conditions, methodology, etc.	L, C
Acquiring and combining information	System observations (direct, written, or oral), system drawings, mock-up models, operational plans, reliability data, expert judgments.	L, C
Hazard identification	Guidewords and process parameters (e.g., in HAZOP), system drawings, mock-up models, etc. Screening criteria for hazards and scenarios.	L, C
Causal analysis	Accident models and principles for searching for causes and events. Definition of basic and TOP events and system states. Applicability of generic frequency and reliabilty data and need for adjustment, e.g., by organizational factors.	L, C
Consequence analysis	Definition of initiating events, barrier functions, and endpoints (cut-off limits). Definition of credible and worst case conditions and scenarios. Definition of exposures, affected populations, vulnerability and susceptibility models and thresholds.	L, C, N
Calculation of risk metrics	Averaging/aggregation over affected populations, implicit aversion factors and relative weights, discounting future consequences, translation of consequences into monetary value.	C, N
Documentation of critical assumptions	Operational and environmental conditions, e.g., response and duration times, weather conditions, etc.	L, C
Identification of limitations	Documentation of delimitations, simplifications, and approximations.	L, C
Uncertainty analysis	Treatment and presentation by point estimates, distribution, imprecise probabilities, etc. Description of qualitative uncertainty factors and possibility of surprise.	L, C
Sensitivity analysis	Impact and criticality of sensitivities with respect to uncertainty and/or confidence in the results, and their relation to basic premises and assumptions in the analysis.	C
Presentation of results	Multiple and conflicting presentation formats (e.g., numerical ranges and evaluative labels). Tacit assumptions between intermediate and final results.	C, N
Identification of risk reducing measures	Reduction in consequences and/or probabilities within and across hazards and scenarios.	L, N
Third party review and quality assurance	Reference standards for judging the quality of risk analysis.	C, N
Answer to the three questions of risk	*Ambiguity in the basis and content of risk information*	
What can happen?	Specification of hazardous events and accident scenarios. Scenarios that are knowingly or unknowingly left out of the analysis ("Black swans").	
How probable is it/how uncertain are we?	Quantitative, qualitative, and graphical descriptions of probability, frequency, and uncertainty	
What are the consequences?	Categories and descriptions of consequences for affected populations; risk metrics for intermediate and final endpoints	

2. Distinguish types of ambiguity

    – Trace sources

    – Identify and untangle compound manifestations

3. Resolve different types of ambiguities

    – Linguistic: Specify. Seek reduction by showing vigilance in the formulation and interpretation of statements

    – Contextual: Clarify. Seek congruence by eliminating discrepancies between the premises and approaches to risk assessment

    – Normative: Make explicit. Seek consistency by exploring value judgments and preferences

4. In each phase, describe overall ambiguity in relation to the three questions of risk.

Table 3: Risk evaluation and decision-making

Task	Source/manifestation of ambiguity concerning:	Type
Managerial review of risk analysis report	Relevance of premises and assumptions. Criticality of related sensitivities and uncertainty factors.	L, C
Evaluation of confidence in risk analysis	Understanding and tolerability of uncertainty, complexity, and/or ambiguity in risk analysis.	C, N
Comparison of different types of analyses	Discrepancies between terms and contextual premises across analyses.	L, C, N
Balancing preferences and trade-offs	Inconsistency with objectives and concerns defined in preassessment (e.g., labile preferences). Double-counting of confidence-related aspects (e.g., uncertainty).	N
Evaluation of risk reducing measures	Definition, valuation, and distribution of costs and benefits, e.g., in time and across stakeholders.	N
Formulation of decision criteria	Basis and rationale for decision criteria, e.g., values of risk acceptance criteria, disproportionality factors, and discount rates in ALARP. Conditions for invoking the precautionary principle.	L, C, N
Comparison with decision criteria	Adjustment to account for uncertainty (e.g., safety margins); multiple presentation formats and/or criteria.	C, N
Evaluation of the three questions of risk	*Ambiguity in the implications of risk information*	
What can happen?	Generalizability and completeness of scenarios. Relevance with respect to design and operation.	
How probable is it/how uncertain are we?	Interpretation of probability and uncertainty as a premise and/or limitation with respect to confidence in the results and comparison with evaluation criteria.	
What are the consequences?	Applicability of risk metrics as attributes in decision-making. Relevance of consequence categories to specific persons or objects.	

- Describe how the sources and manifestations of ambiguity lead to multiple interpretations of the basis, content, and implication of risk information.
- Give a coarse evaluation of the resulting ambiguity as *high/medium/low* and indicate whether it is mainly linguistic, contextual, or normative

5. Reconsider the approach to risk-informed decision-making in light of unresolved ambiguity, e.g:

- If ambiguity is high and mainly linguistic: the risk assessment should be iterated from the critical phase and be rewritten in deliberation with decision-makers and stakeholders, but can then provide the basis for managerial review and decision by a responsible decision-maker.
- If ambiguity is high and mainly contextual: managerial review and decision should be made in deliberation with relevant stakeholders and experts, comparing alternative risk descriptions with alternative evaluation criteria in light of the decision context.
- If ambiguity is high and mainly normative: consider deliberative principles for decision-making, using risk analysis as vehicle for discussions on preferences and tradeoffs.

6. CONCLUSIONS

The aim of this paper is to shed light on the concept of ambiguity in risk assessment. Ambiguity is in our view an important challenge that has not been fully recognized and explored in risk assessment; neither from a theoretical nor practical point of view. We have reviewed existing definitions of ambiguity and argued that they are too limited or conceptually problematic to be of use in engineering risk assessment. The theoretical contribution of the paper is a new overall definition of ambiguity as a distinct challenge that compromises the role and value of risk assessment in risk-informed decision-making. We have also defined linguistic, contextual, and normative ambiguity as three categories of ambiguity that must be differently approached in risk assessment. The practical contribution of the paper is a set of tables that point to specific sources and manifestations of these categories in preassessment, risk analysis, and risk evaluation/decision-making. The tables also clarify the main implications of ambiguity in relation to how the three questions of risk are defined, answered, and evaluated in the respective phases. The main purpose of the tables is to create awareness of the many sources and manifestations of ambiguity in risk assessment, but they can also be used as a checklist for identifying and discussing ambiguity in an analytic-deliberative process. To assist in such a process, we have finally outlined a stepwise procedure for approaching ambiguity in risk assessment. The procedure is admittedly coarse and is intended to serve as a point of departure for further research on ambiguity in risk-informed decision-making. We have touched upon several challenging issues and have raised questions that have only been partly answered, for example, regarding the ontology, measurability, and propagation of ambiguity in risk assessment. By doing so, we aim to stimulate further work and discussion on ambiguity, and create awareness of an intriguing challenge that has yet to be fully recognized in risk assessment.

References

[1] G. Apostolakis. How useful is quantitative risk assessment? *Risk Analysis*, 24:515–520, 2004.

[2] T. Aven. *Foundations of Risk Analysis. Second edition.* Wiley, Hoboken, NJ., 2012.

[3] T. Aven and O. Renn. *Risk Management and Governance. Concepts, Guidelines and Applications.* Springer Verlag, Berlin Heidelberg, 2010.

[4] T. Aven and E. Zio. Some considerations on the treatment of uncertainties in risk assessment for practial decision making. *Reliability Engineering and System Safety*, 96:64–74, 2011.

[5] M. Basili. A rational decision rule with extreme events. *Risk Analysis*, 26:1721–1728, 2006.

[6] T. Bedford and R. Cooke. *Probabilistic Risk Analysis. Foundations and Methods.* Cambridge University Press, Cambridge, UK, 2001.

[7] M. Brugnach, C. Pahl-Wostl, K.E. Lindenschmidt, J.A.E.B. Janssen, T. Filatova, A. Mouton, G. Hotlz, P. Van Keur, and N. Gaber. *Environmental Modeling, Software and Decision Support*, chapter Complexity and Uncertainty: Rethinking the modeling activity, pages 49–68. Elsevier, 2008.

[8] C. Camerer and M. Weber. Recent developments in modeling preferences: uncertainty and ambiguity. *Journal of Risk and Uncertainty*, 5:325–370, 1992.

[9] M.D. Catrinu and D.E. Nordgård. Integrating risk analysis and multi-criteria decision support under uncertainty in electricity distribution system asset management. *Reliability Engineering and System Safety*, 96:663–670, 2011.

[10] M. Colyvan. Is probability the only coherent approach to uncertainty? *Risk Analysis*, 28:645–652, 2008.

[11] H.E. Douglas. *Science, Policy, and the Value-free Ideal.* University of Pittsburgh Press, Pittsburgh, 2009.

[12] D. Dubois. Representation, propagation, and decision issues in risk analysis under incomplete probabilistic information. *Risk Analysis*, 30:361–368, 2010.

[13] D. Ellsberg. Risk, ambiguity, and the Savage axioms. *The Quarterly Journal of Economics*, 75: 643–669, 1961.

[14] L. Fabbri and S. Contini. Benchmarking on the evaluation of major accident-related risk assessment. *Journal of Hazardous Materials*, 162:1465–1476, 2009.

[15] B. Fischhoff, S. Lichtenstein, P. Slovic, S.L. Derby, and R.L. Keeney. *Acceptable Risk.* Cambridge University Press, New York, 1981.

[16] S. Funtowicz and J. Ravetz. Science fo the post-normal age. *Futures*, 25:739–755, 1993.

[17] A. Hatami-Marbini, M. Tavana, M. Moradi, and F. Kangi. A fuzzy group electre method for safety and health assessment in hazardous waste recycling facilities. *Safety Science*, 51:414–426, 2013.

[18] T. Horlick-Jones. Meaning and contextualization in risk assessment. *Reliability Engineering and System Safety*, 59:79–89, 1998.

[19] IRGC. White paper on risk governance. towards an integrative approach. Technical report, International Risk Governance Council, 2005.

[20] ISO 31000. *Risk management-Principles and guidelines*. International Organization for Standardization, Geneva, 2009.

[21] S. Kaplan. The words of risk analysis. *Risk Analysis*, 17:407–417, 1997.

[22] S. Kaplan and J. Garrick. On the quantitative definition of risk. *Risk Analysis*, 1:11–27, 1981.

[23] R. Keeney. *Value-Focused Thinking. A Path to Creative Decision Making.* Harvard University Press, Cambridge, Massachusetts, 1992.

[24] A. Klinke and O. Renn. A new approach to risk evaluation and management: Risk-based, precaution-based and discourse-based strategies. *Risk Analysis*, 22:1071–1094, 2002.

[25] J. G. March. Ambiguity and accounting: the elusive link between information and decision-making. *Accounting, Organizations and Society*, 12:153–168, 1987.

[26] Merriam-Webster. Online: http://www.merriam-webster.com/.

[27] NRC. *Understanding Risk. Informing Decisions in a Democratic Society.* National Academy Press, Washington, DC, 1996.

[28] T. Nørretranders. *Merk Verden. En beretning om bevissthet (in Norwegian)*. Cappelen, 1992.

[29] J. Rasmussen. Human error and the problem of causality in analysis of accidents. *Phil.Trans.R.Soc.Lond.*, 327:449–462, 1990.

[30] O. Renn. *Risk Governance. Coping with Uncertainty in a Complex World.* Earthscan, London, 2008.

[31] O. Renn, A. Klinke, and M. A. van Asselt. Coping with complexity, uncertianty, and ambiguity in risk governance: A synthesis. *AMBIO*, 40:231–246, 2011.

[32] A. Richter and C. Koch. Integration, differentiation, and ambiguity in safety cultures. *Safety Science*, 42:703–722, 2004.

[33] P. Slovic. The risk game. *Journal of Hazardous Materials*, 86:17–24, 2001.

[34] A. Stirling. Risk, precaution and science: Towards a more constructive policy debate. *EMBO reports*, 8:309–315, 2007.

[35] A. Stirling. Science, precaution, and the politics of technological risk. *Annual New York Academy of Sciences*, 1128:95–110, 2008.

[36] N. Vareman and J. Persson. Why separate risk assessors and risk managers? Further external values affecting the risk assessor *qua* risk assessor. *Journal of Risk Research*, 13:687–700, 2010.

How is capability assessment related to risk assessment? Evaluating existing research and current application from a design science perspective

Hanna Palmqvist[*a], Henrik Tehler[a], and Waleed Shoaib[a]

[a] Division of Risk Management and Societal Safety, Lund University
Lund University Centre for Risk Assessment and Management
Centre for Societal Resilience, Lund University
Lund, Sweden

Abstract: Several countries use capability assessments as a part of their efforts to manage risk. However, it is unclear how such assessments are connected to other risk management activities (e.g. risk assessment). Therefore, the aim of the present paper is to present a study of how capability assessment is related to risk assessment. Capability assessment methods were identified through a scoping study and the Swedish capability assessment method was investigated through interviews with Swedish public actors and analysis of legislative documents. The data was analysed using a design science perspective. The results of the analysis show that the purposes presented for some capability assessment methods are the same or similar to purposes common to risk assessment methods, and the actual form of some of the methods is similar to existing risk assessment methods. Nevertheless, the relationship between capability assessment and risk assessment is unclear. We conclude that if capability assessments are going to continue to be an important part of risk management activities more research is needed to better establish the relationship between risk assessment and capability assessment.

Keywords: capability assessment, risk assessment, method, design science, scoping study

1. INTRODUCTION

Societies and the risks they face are becoming more complex [1,2] and with this increasing complexity the traditional risk management approaches change. This reflects in the introduction of concepts such as societal safety [3], resilience [4,5] and the establishment of national security strategies with a whole-of-government approach [6]. Also, in order to prepare for a large variety of threats and risks, instead of preparing for specific scenarios [7], capabilities-based planning gains more ground in several countries. For example, the Netherlands, Sweden, the United Kingdom and the USA have established methods for assessing capability as part of their emergency preparedness work. Thus, the interest for capabilities-based planning seems to increase and since several countries already have implemented the concept in policy and practice its importance in future efforts to manage risk remains.

Despite the increased interest for capabilities-based planning it is unclear how capability assessment is related to other risk management activities such as risk assessment and vulnerability assessment. Questions like whether risk assessment is part of capability assessment, or vice versa are important to answer both from a scientific as well as a practical perspective. Unless the relationships between capability assessment and other risk management activities are clarified there is a risk of conceptual as well as practical confusion. The present paper attempts to investigate these relationships. We argue

---

[*]*hanna.palmqvist@risk.lth.se*

that (1) the *concept* of capability and how it relates to other important concepts such as risk, vulnerability, and resilience, needs to be investigated. Moreover, (2) we need to study how *methods* for capability assessment relate to methods for risk assessment (and other risk management activities). The present paper deals specifically with this second issue and it is worth noting that the first issue is explored elsewhere [8]. Thus, the present paper aims at investigating how capability assessment is related to risk assessment.

To clarify the relationships between methods for capability assessment and risk assessment we first need to identify capability assessment methods and their use in practice. To do this we performed a scoping study [9,10] focusing on scientific publications in international journals and conference proceedings. To gain a deeper understanding of a method used in practice we chose Sweden as a study case. Swedish authorities have used both risk assessments and capability assessments for some time now, making Sweden a suitable case for the scope of this paper. Identifying methods for capability assessment in literature and practice can be used to produce an overview of the area. However, a theoretical framework is needed to guide the comparative analysis. To that end, we chose to use design science. We consider it suitable since capability assessment methods and risk assessment methods are artefacts designed to achieve some kind of purpose(s).

Design science differs from "traditional" sciences such as natural science and social science mainly because of their different goals, according to several authors from different scientific fields (see e.g. [11-13]). While design science aims at developing "...knowledge for the design and realization of artefacts, i.e. to solve *construction problems*, or to be used in the improvement of the performance of existing entities, i.e. *to solve improvement problems*" [14, p. 224], the sciences that van Aken calls explanatory sciences, including "...the natural sciences and major sections of the social sciences", has the goal "...to describe, explain and possibly predict observable phenomena within its field" [14, p. 224], (see also [11,15]). In other words, the key difference between design science and many other sciences is that the former aims at contributing to changing the world whereas the latter aims at understanding it. You might study an artefact and try to understand it without knowing its purpose, but it is difficult to construct or improve one without a purpose and therefore the purpose(s) of artefacts is a key aspect of design science. Even though design science focuses on constructing and improving artefacts, one can also use other ideas from it to study how existing artefacts work and evaluate them. Recent research has, for example, successfully used ideas from design science to evaluate risk and vulnerability assessments and accident investigations [16-18].

Methods for capability assessment (and risk assessment) are artefacts, i.e. created by humans for some purpose(s). Therefore, we argue that a design science perspective is suitable to use in the present context. However, design science is not a homogenous research area but made up of several disciplines. For this study we have chosen an approach presented by Brehmer [19] who suggests using Rasmussen's [20] abstraction hierarchy to describe, design and evaluate artefacts (in this case capability and risk assessment methods). The hierarchy in Brehmer's version has three perspectives: Purpose, Function and Form. A description of an artefact from the respective perspective corresponds to answering the questions "*Why* does the artefact exist?" (Purpose), "*What* does the artefact do to fulfil its purpose?" (Function), "*How* does it do it?" (Form). A number of researchers [16,18] provide more detail on the use of the abstraction hierarchy in the present context. Thus, we have analysed a number of papers dealing with capability assessment methods to determine if they describe *why* the suggested method should be used in practice (Purpose), *what* it does to achieve the purpose(s) (Functions), and *how* it does it (Form). Using the same approach, we analysed the empirical data from Sweden. These analyses then formed the basis for a comparison with risk assessment methods using the same set of questions.

Following this introduction section, we first present both the approach and results from the scoping study, followed by the study of the Swedish capability assessment method. Finally, we discuss and present conclusions regarding capability assessment's relation to risk assessment.

2. SCOPING STUDY

In order to identify scientific papers presenting methods for capability assessment, we performed a scoping study [9,10]. Thus, below, we first present the method used in the scoping study. The section concludes with the findings which include the analysis of the identified papers using the abstraction hierarchy.

2.1. Method

We identified papers presenting capability assessment methods through searches in the databases Scopus and Web of knowledge. The term *capability assessment* and the related concepts *capacity assessment* and *ability assessment* via Boolean operators (see Table 1) were used to construct the search queries. In order to further identify relevant papers we manually hand searched reference lists and relevant journals[†]. Papers were included if they addressed capability assessment related to emergency, crisis, disaster, or catastrophe response management. Papers were not included if they focused on assessments of an individual's capability. Instead, the focus was on different kinds of organisations' capability. All searches and methods of data handling were recorded. Hence, the searches, performed between April and December 2013, resulted in the screening of 4544 unique titles. Papers that did not seem relevant based on the title were excluded, resulting in 62 possibly relevant papers based on the title. The references that were manually searched for, added another 11 possibly relevant papers based on the title. The abstracts of these 73 papers were read and 54 were still found relevant. We were not able to download the full-length version of 10[‡] of these, resulting in that 44 papers were read in full length.

[†] Disaster management and response 2003-2007, Disaster prevention and management 1992-2013, Disasters 1977-2013, International journal of critical infrastructure 2004-2013, International journal of disaster risk reduction 2012-2013, International journal of disaster risk science 2010-2013, International journal of emergency management 2001-2013, Jamba 2006-2013, Journal of contingencies and crisis management 1993-2013, Journal of risk and uncertainty 1988-2013, Journal of risk research 1998-2013, Natural hazards 1988-2013, Reliability engineering and system safety 1988-2013, Risk analysis 1981-2013, Safety 1991-2013.

[‡] **J. Von Kanel, E. W. Cope, L. A. Deleris, N. Nayak, and R. G. Torok**, "Three key enablers to successful enterprise risk management," *IBM J. Res. Dev.*, vol. 54, no. 3, 2010.; **Unknown**, "Strengthening our emergency response systems," *Public Work.*, vol. 134, no. 6, 2003.; **J. H. Gu, X. Y. Wu, and H. Y. Wu**, "Capability assessment for earthquake emergency rescue based on the reported death toll rate," *J. Harbin Inst. Technol. (New Ser.*, vol. 16, no. 2, pp. 145–149, 2009.; **S. Y. Wang and G. J. Tang**, "Research on evaluation of urban disaster emergency capability based on unascertained measure," *J. Harbin Inst. Technol. (New Ser.*, vol. 16, no. 2, pp. 109–113, 2009.; **S. Y. Wang and J. Liu**, "Checking of city disaster emergency capacity evaluation," *J. Harbin Inst. Technol. (New Ser.*, vol. 16, no. 2, pp. 119–123, 2009.; **S. J. Cannon, T. Kontuly, and H. J. Miller**, "GIS-based emergency response planning in a Mexico-US border community," *Appl. Geogr. Stud.*, vol. 23, no. 2, pp. 227–246, 1998.; **X. Jianguang and X. Ruhe**, "Research on model and method of emergency capability assessment," in *Proceedings of the 2nd International Conference on Modelling and Simulation, ICMS 2009, 7*, 2009, pp. 484–489.; **X. Sun**, "SEM-based Capability Assessment of Emergency Management Agency," in *Proceedings of ISCRAM China 2010: Fourth International Conference on Information Systems for Crisis Response and Management*, 2010, pp. 456–460.; **Z. Guo and M. Qi**, "Comprehensive Assessment Method of Urban emergency Response Capability based on FAHP," in *International Conference on Management Science and Engineering Location: Wuhan, Peoples R China, Oct 17-18, 2010*, 2010, pp. 273–276.; **S. Wang and Y. Sun**, "The Function of the Disaster Background on Urban Disaster Emergency Capability Assessment," in *International Disaster and Risk Conference (IDRC) Location: Chengdu, Peoples R China, Jul 13-15, 2009*, 2009, pp. 453–457.

Table 1: Search strategies

Search term	Database: Scopus	Database: Web of knowledge
Capability assessment	1. (ALL({capability assessment}) AND LANGUAGE(english))	1. Topic=("capability assessment") OR Title=("capability assessment")
	2. (ALL("capability assessment") AND LANGUAGE(english))	2. Topic=("capabilit* assessment") OR Title=("capabilit* assessment")
Capacity assessment	3. (ALL({capacity assessment}) AND LANGUAGE(english))	3. Topic=("capacity assessment") OR Title=("capacity assessment")
	4. (ALL("capacity assessment") AND LANGUAGE(english))	4. Topic=("capacit* assessment") OR Title=("capacit* assessment")
Ability assessment	5. (ALL({ability assessment}) AND LANGUAGE(english))	5. Topic=("ability assessment") OR Title=("ability assessment")
	6. (ALL("ability assessment") AND LANGUAGE(english))	6. Topic=("abilit* assessment") OR Title=("abilit* assessment")

Note: All searches in Web of knowledge were refined by: Languages=(ENGLISH); Timespan=All years; Search language=Auto. The searches in Web of knowledge covered the years 1864-2013, in Scopus 1960-2013.

2.2. Findings

The information found in the papers was coded related to general information and the three perspectives discussed earlier (Purpose, Functions and Form).

2.2.1. General information

The majority of the papers originate from China (N=33), followed by the United States (4), Sweden (3), Australia (2) and the Netherlands (1). Researchers from both Australia and Finland authored one (1) paper. 32 papers were found in conference proceedings, 9 were articles and 3 review papers in international scientific journals. The majority (N=14) was published in 2011, followed by 8 in 2010, 6 in 2012, 6 in 2009, 3 in 2013, 2 each year in 2005, 2006 and 2007, and 1 in 2008. The databases categorised the papers and the majority of the relevant papers were found in the category "decision sciences", followed by "business, management and accounting" and "engineering". 36 of the papers suggest methods for capability assessment. The remaining papers evaluate or compare methods for capability assessment, or evaluate the capability of an organisation after an incident.

The methods identified are intended to be used on various administrative levels in society, ranging from the emergency department of a city (e.g. [21,22]) and the city as a whole (e.g. [23-28]), to regions [29] and nations [30]. Other papers present methods to be used by companies (e.g. [31-34]) or for certain infrastructures such as subway systems [35] or power systems [36]. Thus, the contexts in which the proposed methods are intended to be used cover a broad spectrum.

2.2.2. Purpose

Studying an artefact based on its purpose means trying to answer the question "*Why* does the artefact exist?" A purpose of the capability assessment method is presented in 19 of the 36 papers. The most common purpose is to provide support to decision-makers in some way [26,28,37-39], which is also an important purpose of risk assessments, (see e.g. [40]). Other relatively frequent purposes of the identified capability assessment methods are to either increase capability or to identify weaknesses in capability [41-47]. Thus, in the cases [26,28,37-39] there seems to be a connection between capability assessment and risk assessment; they aim to provide support to decision-makers. However, a majority of the identified methods does not explicitly describe the purpose(s) of conducting capability assessments. Therefore, it is difficult to say if the conclusion is valid for only a small part of the capability assessment methods or for a majority of them. Moreover, possible connections to risk assessment are not discussed in the studied papers. Therefore, we cannot say if it was the intention of the designer to use a purpose similar to that of a risk assessment or if it was just a coincidence.

An indirect way of getting insights about the purpose of a capability assessment method is to look for evaluations of it. If a method is evaluated it seems reasonable that it should be evaluated with respect to its intended purpose. Therefore, even though the purpose might not be explicitly stated in a description of a capability assessment method it might be possible to use an evaluation of the method to derive the purpose. Testing the method and then analysing how it performed in the test is the most likely approach to use when evaluating a method. In 25 of the 36 studied papers a real or fictional case was used to demonstrate and test the method. However, the most common way to present the cases is to describe the result of the capability estimate, for example the authors might conclude that the capability in the case used is 23.4, but not provide any further comment or evaluation of the result. We have not found an evaluation of the suggested method in relation to the stated purpose, for example by reporting if the produced capability assessment improves the support for decision-makers, in any of the papers.

To summarise, 17 of the 36 papers do not present a purpose for the suggested method at all. In the remainder of the papers (19) common purposes are to provide support to decision makers, and to improve capability or identify weaknesses in it. Moreover, none of the papers provide an evaluation of the method in question to show that it actually fulfils the purpose in question.

2.2.3. Functions
Trying to identify the functions of a method involves asking the question "*What* does the method do to fulfil its purpose?" (see [18] for more details on analysing functions). For example, common functions of risk assessment methods are risk identification, risk analysis, and risk evaluation [40]. These functions help fulfil one common purpose of risk assessments and that is to provide support for decision-making as is discussed in the next few lines. In producing an assessment of risk one first needs to identify the risk, which means finding and describing it in terms of the events that might lead to something undesirable. Moreover, one also needs to analyse the risk, which means determining the level of the risk in some way. Finally, one needs to evaluate the risk, which means comparing the level of the risk to some criteria. All these help decision-makers determine if the risk is acceptable/tolerable or if actions to reduce the risk are necessary [40].

Thus, when analysing the methods for capability assessment found in the literature we looked for descriptions of what should be done in order to arrive at the end result to see if there are similarities to the functions found for risk assessment. However, none of the papers identified in the scoping study discuss what needs to be done for the purpose of the method to be fulfilled. Instead, one needs to investigate the form of the methods to derive the needed functions. For example, most of the methods for capability assessment found in the literature are so called index methods. This means that they specify various indicators that should be assessed. Each indicator is then weighed together with other indicators and the result is the overall capability index. Studying the structure of the capability index provides information of the aspects (the different indicators) that are necessary to assess in order to arrive at the final result. Thus, it provides information regarding *what* needs to be done to assess capability and therefore it can be seen as a representation of the functions that need to be fulfilled. However, the indicators used in the studied capability assessment methods (see examples of indicators under "Form" below) are very different from the key functions of risk assessment (identification, analysis, evaluation). Even some specific "products" of risk assessment (e.g. likelihood and consequence estimates) are lacking from the capability assessment methods (see the discussion below). Therefore, we conclude that capability assessment methods appear to be quite different from risk assessment methods when studying them from a function's perspective.

2.2.4. Form
Our analysis on the form perspective focuses on *how* capability assessment methods are constructed and *how* they produce the output of the functions. As noted above the output of every function performed by a capability assessment method, or a risk assessment method, might in practice be produced in many ways. For example, a method for risk assessment needs to identify risks in some way (a function) but *how* it does it varies much between different methods. One might, for example, use table-top exercises involving different stakeholders with relevant knowledge of the context of interest. The exercises might be of ad-hoc or "brainstorming" character or very systematic using fir

example guidewords to help identifying critical events for example such as the HAZOP procedure [40]. Thus, the form perspective deals with the concrete design of the method.

In total, we identified 36 papers suggesting capability assessment methods (see above). 5 of them suggest methods that are based on: analysing emergency response processes and their failures [39,46,48], using table-top exercises for assessing capability [47] or using a military planning framework to assess the capability to respond to a natural disaster [49]. The remaining 31 contain suggestions that can be classified as index methods. Such a method makes use of indicators of capability, and often assigns each indicator a numerical value and derives a final score that reflects the capability. Half of these 31 papers report that they improve previously established methods (e.g. by suggesting other indicators or another weighting system). Although there are risk assessment methods that are index methods, for example the Disaster Risk Index by UNDP [50], they seem less common[§]. Nevertheless, even though the capability assessment methods do not make use of common components in risk assessments (e.g. hazards, scenarios, vulnerability and consequences) they might still resemble the risk assessment methods if the indicators reflect such key components. For example, the Disaster Risk Index contains the indicators: hazard, population living in the exposed area, and vulnerability [50, p. 100]. If index methods for capability assessment contained such components we could conclude that some index methods for risk assessment resemble capability assessment methods seen from the form perspective. Examples of indicators in the identified capability assessment methods are resources, management and plan (see [33]). Such broad indicators are then usually broken down in more specific indicators like technological capabilities, staff quality, organisational structure and flexibility. Although the indicators are probably important for how a scenario develops and what the consequences will be, they are difficult to relate to concepts commonly used in risk assessment and therefore it is difficult to compare them to risk assessment methods.

2.2.5. Conclusion

The empirical data from the scoping study provide limited opportunities to compare capability assessment methods with risk assessment methods. The primary reason for this is that the descriptions of the capability assessment methods are seldom detailed enough to allow for an in-depth comparison. However, we can conclude that the purposes presented for some of the capability assessment methods are the same or similar to purposes common to risk assessment methods. Moreover, we can conclude that the actual form of some of the methods resembles the form of some existing risk assessment methods such as the index methods. Nevertheless, from a functional perspective, when studying what the methods actually do, they appear to be very different. Moreover, the material found in the scoping study does not allow us to draw any conclusions regarding whether the capability assessment methods actually fulfil the stated purposes. In other words, we found no evaluations of the suggested methods.

3. THE SWEDISH METHOD FOR CAPABILITY ASSESSMENT

Swedish legislation requires 290 local municipalities, 21 regional counties, and 23 governmental agencies to perform risk and vulnerability assessments as part of their emergency preparedness work [51-54]. According to the statutory instructions [53,54], the risk and vulnerability assessment must include a capability assessment and the actors shall assess capability according to an ordinal four-level scale: (1) Good capability, (2) Good capability in general, (3) Some, but inadequate, capability, and (4) No or very inadequate capability. To facilitate the assessment, the instructions provide indicators that the actors shall take into account (e.g. cooperation, information security, material resources, personnel resources, and practical experience). In addition, the Swedish Civil Contingencies Agency (MSB) performs a national capability assessment based on a request from the government through the

---

[§] We have not investigated this in detail, i.e. we have not performed the same type of scoping study for risk assessment methods as we have for capability assessment methods. However, the ISO standard for risk management provides a summary of 31 tools for risk assessment and it is clear that a majority of them are built on other principles than an index method. For example, they use events, scenarios, causes, consequences and likelihood as building blocks rather than different contextual factors that are weighted together.

yearly letter of appropriation. In order to receive input to this assessment, MSB constructs scenarios, sends them to selected regional and national actors and asks them to perform a special capability assessment based on the indicators and assessment scale mentioned above.

3.1. Data collection

Semi-structured interviews with representatives from 15 municipalities, 5 counties and 5 governmental agencies were conducted in a previous research project in 2011 [55]. The counties were selected to obtain a geographical spread: two in the southern part, two in the central part and one in the northern part of Sweden. Three municipalities were chosen in each county based on the size of the municipality: one small with less than 15,000 inhabitants, one medium with between 15,000 and 90,000 inhabitants and one large with more than 90,000 inhabitants. Five governmental agencies were chosen to represent five different areas of expertise. On the national level, MSB representatives responsible for performing the national capability assessment were interviewed in 2013. The interviews covered capability assessment, critical infrastructure and interdependencies. An interview guide consisting of some sixty questions (twenty related to capability assessment) was used to guide the interviews. In total, representatives from 26 actors were interviewed and 25 of the 26 interviews were recorded. Only the recorded interviews are included in this study: 15 municipalities, 4 counties, 5 governmental agencies, and MSB.

3.2. Method for analysis

All the recorded interviews were transcribed into text and saved as Microsoft Word documents using intelligent verbatim transcription with the purpose to facilitate the analysis of the material. The transcribed interviews were searched in order to find sentences or text segments that could be classified according to the three perspectives Purpose, Function and Form. First, we searched the text using the Swedish word for capability. This also allowed us to find concepts such as "emergency management capability" and "capability assessment". We thereafter read the text in close proximity to the search results and marked the text segments that were related to capability. Following this stage, we coded the marked segments according to the perspectives. We finally skimmed the entire transcribed interview with the aim of finding text related to capability that had not been found through the search. This resulted in a document for each interview with highlighted text sections according to the three perspectives. A highlight meant that the text section was judged to contain information related to a perspective and one section could be categorised as several perspectives. Only text sections relevant to the different perspectives were coded, not the entire document.

In addition to the interviews we have studied the legislator's intention with the Swedish capability assessment method through legislation, guidelines, and reports. We did this using the same perspectives (Purpose, Function, Form) as when analysing the research papers and the interviews. Below we present the results from the study of the legislation, guidelines and reports, and the interviews.

3.3. Findings

3.3.1. Purpose

The purpose answers the question "*Why* does the artefact exist?". From the legislator's perspective, the four purposes of the risk and vulnerability assessment, including the capability assessment, are to "provide a basis for decisions to decision makers and those in charge of operations; provide the public with an information basis of society's risks; provide basic data for community planning; and contribute to providing a risk profile for all of society" [56, p. 16]. Thus there seems to be a connection between capability assessment and risk assessment from the legislator's perspective especially since they have the exact same purpose.

The interviews did not explicitly address the practitioners' views of the purpose of capability assessments. Still, some related answers were given to for example the asked question if they have had any difficulties in assessing capability, and in their general descriptions of how they work with capability assessments. A couple of municipalities and a county administrative board wonder how they can use the result of the assessments in their own organisations and they find the process of performing the assessment more useful than the final report. All respondents were also asked about their view on how risk assessment and capability assessment are related. The majority says that they are related but they have difficulties in explaining what the relationship is.

3.3.2. Functions

The functions answer the question *"What* does the method do to fulfil its purpose?". The process for developing the Swedish method (indicators and assessment scale) is presented in a report [57] published by the Swedish Emergency Management Agency, later MSB. The report presents no functions explaining how the method helps fulfil some purpose. In the non-statutory guidelines for risk and vulnerability assessments [56], the legislator presents parts that could be included in the risk and vulnerability assessment (e.g. risk identification, risk analysis, probability and consequence assessment). These could be seen as functions of risk and vulnerability assessment, similar to the functions identify, analyse and evaluate [40]. However, according to the non-statutory guidelines capability assessment is also part (or a function) of the risk and vulnerability assessment.

The interviews did not reveal any conclusive information concerning the functions of capability assessment methods. One possible reason is that the respondents had difficulties expressing the purpose of the capability assessment method they employed. Therefore, it also became difficult for them to explain *what* the methods do to fulfil the purpose (functions). The interviews did therefore not contribute to our analysis of the connection between risk assessment and capability assessment when using the function perspective.

3.3.3. Form

When studying an artefact from the form perspective we are interested in *how* the actual method fulfils the functions and the purpose(s) and *how* the method is constructed. The legislator describes the Swedish method, consisting of a list of indicators and an assessment scale, in the statutory instructions [53-54]. Contrary to the index methods found in the scoping study discussed above, the Swedish method says nothing about how to translate the indicators into the assessment scale (e.g. through a weighting system). Even though the instructions present detailed information about indicators to use in the capability assessment, there are no statutory instructions on how to assess risk. However, in the non-statutory guidelines [56] several specific methods and tools for performing risk assessment are presented, for example seminar-based scenario methods, and traditional risk assessment methods. But in the descriptions of these risk assessment methods we find no similarity to the indicators and assessment scale used for assessing capability.

During the interviews, respondents from the local level expressed that the indicators and the assessment scale helped them in their work. But still the majority express that they find it difficult to assess capability and some say that they do not see how the indicators relate to capability and that they do not know how to interpret the indicators. Moreover, how the capability assessment relates to risk assessment also seems to be unclear, as mentioned above.

3.3.4. Conclusion

The empirical data from the study of the Swedish capability assessment method depicted that from the legislators' perspective the overarching purposes of both risk assessment and the capability assessment are the same. Moreover, they see the capability assessment as a function of the risk assessment (i.e. a capability assessment is needed to produce a risk assessment). On the form level, we only found differences between the stipulated capability assessment method and the suggested risk assessment methods. The respondents say that there is some connection between risk and capability but they have difficulties in explaining the connection. Apart from this, we cannot say more about how risk assessment relates to capability assessment based on the empirical data studied.

4. DISCUSSION

This paper aims at investigating how capability assessment is related to risk assessment. However, we found that we cannot say much more about the relationship than that the purposes presented for some of the capability assessment methods are the same or similar to purposes common for risk assessment methods, and that the actual form of some of the methods is similar to some existing risk assessment methods (i.e. index methods). One important reason for us not finding out more about the relationship is that possible connections between capability assessment and risk assessment are not discussed extensively in the identified papers. However, if the strategy of capabilities-based planning is to prepare for a large variety of threats and risks instead of a single risk [7] it seems reasonable to expect a clear relationship between the two, especially when the two assessments are used together, similar to the Swedish case where the capability assessment is part of the risk assessment. An explicit relationship would probably lead to less confusion, better use of resources when it comes to performing the assessments, and a clearer goal when designing the methods.

The scoping study was limited to only include scientific papers. Our study could have also included grey literature and comparisons between other countries' methods for risk assessment and capability assessment. However, we chose to limit our study to Sweden since the Swedish capability assessment method is part of the risk assessment method, and has been so for some years, and this could allow us to study the relationship in detail. But even in this particular case we did not find a clear relationship. Another limitation of the scoping study is that we did not perform similar searches in order to identify risk assessment methods for which we could identify purposes, functions and forms. This means that we only compared the identified capability assessment methods with the corresponding perspectives presented in the ISO standard [40]. However, since we had difficulties in finding information regarding the perspectives for the studied capability assessment methods, the results from such an extended comparison would most likely be similar to those presented here. Furthermore, the case study is limited to 25 Swedish public actors of some total 300. Despite the limited number of respondents, we covered different parts of Sweden when it came to geography, sectors and size of municipalities. It is worth noting that views other than the ones we reported regarding the Swedish capability assessment method are likely to exist.

5. CONCLUSION

We studied how capability assessment is related to risk assessment, using a design science approach to compare purposes, functions and forms of the two types of methods. The results show that in general it is unclear how capability assessment and risk assessment are related to each other. Nevertheless, we can conclude that the purposes presented for some of the capability assessment methods are the same or similar to purposes common to risk assessment methods. Moreover, the actual form of some of the capability assessment methods is similar to existing risk assessment methods (i.e. index methods). However, from a functional perspective (i.e. when studying what the different methods do or produce), they appear to be very different. The fact that it is difficult to establish a clear comparison, and that the available empirical data points in different directions depending on which perspective one assumes in the analysis we believe that the relationships between the methods for capability assessment and risk assessment deserve further attention, especially since capabilities-based planning gains more ground within the risk management work all over the world.

Acknowledgements

We thank the Swedish Civil Contingencies Agency for funding the research on which the present paper is based. We would also like to thank Marcus Abrahamsson, Kerstin Eriksson, Henrik Hassel and Kurt Petersen for help with conducting the interviews.

References

[1] C. N. Calvano and P. John. *"Systems engineering in an age of complexity"*, Systems Engineering, 7(1), pp. 25–34, (2004).

[2] OECD. "*Emerging Risks in the 21st Century: An agenda for action*", Organisation for Economic Co-Operation and Development, OECD, 2003, Paris.

[3] O. E. Olsen, B. I. Kruke, and J. Hovden. *"Societal Safety: Concept, Borders and Dilemmas"*, Journal of Contingencies and Crisis Management, 15(2), pp. 69–79, (2007).

[4] A. Boin and A. McConnell. *"Preparing for Critical Infrastructure Breakdowns: The Limits of Crisis Management and the Need for Resilience"*, Journal of Contingencies and Crisis Management, 15(1), pp. 50–59, (2007).

[5] S. Somers. *"Measuring Resilience Potential: An Adaptive Strategy for Organizational Crisis Planning"*, Journal of Contingencies and Crisis Management, 17(1), pp. 12–23, (2009).

[6] S. L. Caudle and S. De Spiegeleire. *"A New Generation of National Security Strategies: Early Findings from the Netherlands and the United Kingdom"*, Journal of Homeland Secturity and Emergency Management, 7(1), article 35, (2010).

[7] Programme National Security. *"National Security, Strategy and Work programme 2007-2008"*, Programme National Security, Ministry of the Interior and Kingdom Relations, 2007, Breda.

[8] H. Palmqvist, H. Tehler, K. Eriksson, and T. Aven. *"Untitled work: Definition of capability"*, Manuscript in progress.

[9] H. Arksey and L. O'Malley. *"Scoping studies: towards a methodological framework"*, International Journal of Social Research Methodology, 8(1), pp. 19–32, (2005).

[10] K. Davis, N. Drey, and D. Gould. *"What are scoping studies? A review of the nursing literature"*, International Journal of Nursing Studies, 46(10), pp. 1386–400, (2009).

[11] A. G. L. Romme. *"Making a Difference: Organization as Design"*, Organization Science, 14(5), pp. 558–573, (2003).

[12] K. Peffers, T. Tuunanen, M. A. Rothenberger, and S. Chatterjee. *"A Design Science Research Methodology for Information Systems Research"*, Journal of Management Information Systems, 24(3), pp. 45–77, (2007).

[13] I. Horva. *"A treatise on order in engineering design research"*, Research in Engineering Design, 15(3), pp. 155–181, (2004).

[14] J. E. van Aken. *"Management Research Based on the Paradigm of the Design Sciences : The Quest for Field-Tested and Grounded Technological Rules"*, Journal of Management Studies, 41(2), pp. 219–246, (2004).

[15] J. E. van Aken. *"Management Research as a Design Science: Articulating the Research Products of Mode 2 Knowledge Production in Management"*, British Journal of Management, 16(1), pp. 19–36, (2005).

[16] M. Abrahamsson and H. Tehler. *"Evaluating risk and vulnerability assessments : a study of the regional level in Sweden"*, International Journal of Emergency Management, 9(1), pp. 76–92, (2013).

[17] H. Tehler, B. Brehmer, and E. Jensen. *"Designing societal safety: A study of the Swedish crisis management system"*, 11th International Probabilistic Safety Assessment and Management Conference And The Annual European Safety and Reliability Conference 2012, PSAM 11 ESREL 2012 Volume: 5, 25-29 June 2012, pp. 4239-4248 (2012).

[18] A. Cedergren and H. Tehler. *"Studying risk governance using a design perspective"*, submitted to Safety Science.

[19] B. Brehmer. *"Understanding the Functions of C2 Is the Key to Progress"*, The International C2 Journal, 1(1), pp. 211–232, (2007).

[20] J. Rasmussen. *"The role of hierarchical knowledge representation in decisionmaking and system management"*, IEEE Transactions on Systems, Man, & Cybernetics, 15(2), pp. 234–243, (1985).

[21] X. Liu, Y. Ju, and A. Wang. *"A Dynamic Vague Multiple Attribute Decision-Making Method for Emergency Capability Assessment"*, Journal of Convergence Information Technology, 6(11), pp. 86–94, (2011).

[22] Y. Ju, A. Wang, and X. Liu. *"Evaluating emergency response capacity by fuzzy AHP and 2-tuple fuzzy linguistic approach"*, Expert Systems with Applications, 39(8), pp. 6972–6981, (2012).

[23] Y. Song, and J. Han. *"A Comprehensive Evaluation of City Emergency Management Capacity based on Analytic Hierarchy Process and Fuzzy Mathematics Method"*, 2011 International Conference on Management and Service Science (MASS), 12-14 August 2011, (2011).

[24] J. Liu, J. Y. Su, W. Wang, and D. H. Ma. *"Information Entropy Method for Evaluating Regional Earthquake Relative Disaster-Carrying Capability"*, Applied Mechanics and Materials, 166–169, pp. 2070–2073, (2012).

[25] G. Xiong and X. Yan. *"Research on the Synthetic Evaluation Indicators System of City Public Department Emergency Management Capability"*, International Conference on Management Science & Engineering (ICMSE), 20-22 August 2007, pp. 2499–2504, (2007).

[26] C. Juan. *"Disaster Emergency Capability Evaluation for Readiness of Urban Community - Based on Multi-level Fuzzy Comprehensive Evaluation Model"*, 2011 2nd IEEE International Conference on Emergency Management and Management Sciences (ICEMMS), 8-10 August 2011, pp. 400–403, (2011).

[27] L. Zhao, M. Tang, and M. Chen. *"The Fuzzy Synthetic Assessment on the City Disaster Emergency Capability Based on Fuzzy Pattern Recognition"*, International Conference on Management and Service Science (MASS '09), 20-22 September 2009, pp. 1-4, (2009).

[28] D. M. Simpson. *"Disaster preparedness measures: a test case development and application"*, Disaster Prevention and Management, 17(5), pp. 645–661, (2008).

[29] W. Cao, H. Xiao, and Q. Zhao. *"The comprehensive evaluation system for meteorological disasters emergency management capability based on the entropy-weighting TOPSIS method"*, International Conference on Information Systems for Crisis Response and Management (ISCRAM), 25-27 November 2011, pp. 434–439, (2011).

[30] Z. Jiang and L. Yu. *"Application of TOPSIS in Evaluating Public Crisis"*, International Conference on Management Science & Engineering (ICMSE), 20-22 September 2012, pp. 1953–1957, (2012).

[31] Y. Yi-lin and L. Guang-li. *"Study on evaluation of the city gas emergency response capability"*, International Conference on Logistics Systems and Intelligent Management (Volume 3), 9-10 January 2010, pp. 1683–1686, (2010).

[32] H. Shang, X. Wang, and X. Liu. *"Application of Fuzzy Assessment Method to Emergency Response Capability in Hazardous Materials Transportation"*, Second International Conference on Innovative Computing, Information and Control (ICICIC '07), 5-7 September 2007, pp. 413–413, (2007).

[33] X. Tan and Y. Ren. *"Comprehensive Evaluation of Enterprise Emergency Response Capability Based on Grey-AHP Method"*, International Conference on Management and Service Science (MASS), 24-26 August 2010, pp. 1–4, (2010).

[34] D. Lin, M. Kong, L. Zhou, and Y. Changsheng. *"Research on the evaluation of the enterprise emergency capability with the butterfly catastrophe theory"*, 2011 2nd IEEE International Conference on Emergency Management and Management Sciences, 8-10 August 2011, pp. 29–32, (2011).

[35] G. Yang and X. Xu. *"Assessment of emergency capacity to major public incidents on urban subway"*, 2011 International Conference on Information Systems for Crisis Response and Management (ISCRAM), 25-27 November 2011, pp. 34–39, (2011).

[36] M. Fan, S. Liu, and Z. Zhang. *"The capability assessment of emergency power supply in urban power network"*, 20th International Conference on Exhibition on Electricity Distribution - Part 1, 8-11 June 2009, pp. 1–4, (2009).

[37] L. Zhang and W. Chen. *"Studies on the emergency response ability evaluation of gas pipeline leakage accident based on GI"*, 16th International Conference on Industrial Engineering and Engineering Management (IE&EM '09), 21-23 October 2009, pp. 1294–1298, (2009).

[38] Y. Wenan and Z. Qingzhu. *"Evaluation of Emergency Maintenance Capability for Expressway Based on Triangle Whitening Weight Function"*, International Journal of Advancements in Computing Technology, 5(4), pp. 914–921, (2013).

[39] H. Zhengqiang and J. Deng. *"Emergency process Capability Assessment based on Stochastic Petri nets"*, 2010 IEEE International Conference on Emergency Management and Management Sciences (ICEMMS), 8-10 August 2010, pp. 367–370, (2010).

[40] *"Risk management - Principles and guidelines"*, ISO Standard 31000:2009.

[41] S. Hao. *"Research on capability evaluation indicators of government emergency management in the public emergency"*, 2011 International Conference on E-Business and E-Government (ICEE), 6-8 May 2011, pp. 1–5, (2011).

[42] H. Zhang, Y. Xiao, B. Yu, and X. Jang. *"Comprehensive Evaluation of Maritime Emergency Capability"*, 2010 Second International Conference on Computer and Network Technology (ICCNT), 23-25 April 2010, pp. 452–456, (2010).

[43] Y. Yi-dan. *"Study of the evaluation index system of the emergency capability of mass incident based on management through overall process"*, 2011 2nd IEEE International Conference on Emergency Management and Management Sciences (ICEMSS), 8-10 August 2011, pp. 224-227, (2011).

[44] Y. Han and Y. Yu. *"Study on Emergency Response Capability Assessment of Urban Disaster"*, 2009 International Conference on Public Administration (ICPA 5th), pp. 399–404, (2009).

[45] M. Hu and Y. Lu. *"Priliminary research on emergency response capacity evaluation system"*, 2011 2nd IEEE International Conference on Emergency Management and Management Sciences (ICEMSS), 8-10 August 2011, pp. 439-442, (2011).

[46] Z. Han and J. Deng. *"Process oriented emergency capability assessment"*, 2010 IEEE International Conference on Emergency Management and Management Sciences (ICEMMS), 8-10 August 2010, pp. 496–499, (2010).

[47] J. Nilsson. *"What's the Problem? Local Officials' Conceptions of Weaknesses in their Municipalities' Crisis Management Capabilities"*, Journal of Contingencies and Crisis Management, 18(2), pp. 83–95, (2010).

[48] B. A. Jackson, K. Sullivan Faith, and H. H. Willis. *"Are We Prepared? Using Reliability Analysis to Evaluate Emergency Response Systems"*, Journal of Contingencies and Crisis Management, 19(3), pp. 147–157, (2011).

[49] P. Tatham, R. Oloruntoba, and K. Spens. *"Cyclone preparedness and response: an analysis of lessons identified using an adapted military planning framework"*, Disasters, 36(1), pp. 54–82, (2012).

[50] UNDP. *"Reducing disaster risk - A challenge for development"*, United Nations Development Programme, 2004, New York.

[51] SFS 2006:942. *"Emergency Management and Heightened Alert Ordinance"*, Sweden.

[52] SFS 2006:544. *"Act on municipal and county council measures prior to and during extra-ordinary events in peacetime and during periods of heightened alert"*, Sweden.

[53] MSBFS 2010:7. *"The Swedish Civil Contingencies Agency's instructions on governmental authorities' risk and vulnerability analyses"*, Sweden.

[54] MSBFS 2010:6. *"The Swedish Civil Contingencies Agency's instructions on municipalities' and county councils' risk and vulnerability analyses"*, Sweden.

[55] M. Abrahamsson, K. Eriksson, H. Hassel, K. Petersen, and H. Tehler. *"Kritiska beroenden, förmågebedömning och identifiering av samhällsviktig verksamhet [Critical interdependencies, capability assessment and identifying critical infrastructure]"*, Lund University Centre for Risk Assessment and Management (LUCRAM), 2011, Lund.

[56] Swedish Civil Contingencies Agency (MSB). *"Guide to risk and vulnerability analyses"*. Swedish Civil Contingencies Agency (MSB), 2012, Karlstad.

[57] Swedish Emergency Management Agency (SEMA). *"Indikatorer på krisberedskapsförmåga [Indicators of emergency management capability]"*, Swedish Emergency Management Agency (SEMA), 2007, Stockholm.

ANALYSES OF AP1000® EXPANDED EVENT TREE SEQUENCES BASED ON BEST-ESTIMATE CALCULATIONS

J.Montero-Mayorga*, C.Queral, J.Gonzalez-Cadelo and G. Jimenez

Universidad Politecnica de Madrid, Calle Alenza 4, 28003, Madrid, Spain
fj.montero@alumnos.upm.es

Abstract: The Westinghouse **AP1000®** reactor is an advanced design whose safety systems are based on natural mechanisms such as gravity or natural circulation, namely, they are passive safety systems. Because of the passive nature of the safety related systems and its dependency on small changes on certain variables (e.g. pressure), it is necessary to confirm that when core cooling is achieved, uncertainties are bounded. The thermal-hydraulic (T/H) uncertainty evaluation process performed by Westinghouse Electric Company (WEC) identified a set of low T/H margin by expanding probabilistic risk assessment (PRA) event trees. Expanded event trees contain more branches than classic event trees, including all possibilities for system actuation. Then detailed conservative computer codes were applied in order to analyze the bounding sequences that were significant to the core damage frequency and demonstrating that the T/H uncertainty was bounded. The UPM group has analyzed the low-margin sequences obtained by WEC with the best estimate computer code TRACE in order to verify the previous results and also to study the phenomenology of such sequences through a best estimate code. This paper presents the results obtained for the DVI line break case confirming that it does not exist damage in the bounding sequence selected for that case.

Keywords: AP1000, DVI line break, Expanded event tress, Focused PRA, Passive safety systems

1. INTRODUCTION

An exhaustive range of activities as part of the **AP1000/AP600** certification process were performed in order to provide confidence on design capabilities and especially on the performing and reliability of the passive safety system. Due to the limited operational experience of the passive safety systems the inherent uncertainties related with the use of such systems must be considered since small changes in any of the physical parameters involved in a system performance (pressure, temperature, etc) could lead to different conclusions on the success core cooling.

The PRA provides insights into any plant vulnerability, so that, as in a standard one, in **AP1000** PRA [1] an event tree is constructed for each initiating event category, in order to model the accident sequences that may result. In the same way, it is necessary to determine the minimum number of systems and components that are necessary to provide adequate core cooling, namely, to define safety systems success criteria. This non-negligible work requires of extensive thermal-hydraulic analyses in order to justify the success criteria used in each event tree.

In the **AP1000** such justification was addressed, in part, through analysis performed in the Deterministic Safety Analysis (DSA) as part of the Design Control Document (DCD) [2] but in sequences which involved Automatic Depressurization System (ADS) actuation, (e.g. small break LOCA) the DSA T/H analysis was not applicable due to the assumption of single failure in such analyses. This fact leads to the performance of a large number of simulations due to the multiple combinations between events and failure combinations. Such analysis is only acceptable if fast computer codes are used. Therefore, MAAP4 code was used for this purpose. This issue was a licensing problem since the NRC required a more detailed analysis in order to bound the potential T/H uncertainties associated to the use of best estimate assumptions in MAAP4 and the limited details in such code [3].

In order to cope with this issue, Westinghouse developed an approach to bound the T/H uncertainty in the **AP1000** PRA success criteria analysis, see references [3], [4], [5] and [6]. This approach must demonstrate that the sequences which have been considered as success sequence in PRA are bounded by T/H uncertainties, namely, the success criteria have been defined for enclosing a range of accident conditions with enough margin to core damage.

The main task of this process is the expansion of the event trees in order to take into account more possibilities for system performance than those which are considered in success criteria and thereafter to identify which sequences are worthy to perform exhaustive analysis. In the following a description of the process as well as the analysis performed by the Universidad Politecnica de Madrid (UPM) will be described.

2. THERMAL-HYDRAULIC UNCERTAINTY EVALUATION PROCESS

This process must demonstrate that the sequences which have been considered as success sequence in PRA are bounded by T/H uncertainties. For this purpose, low-margin risk-significant sequences must be determined and the main way for finding such sequences is to expand the event trees of Focused PRA [3]. The Focused PRA is a sensitivity study to the **AP1000** PRA which does not include active systems for accident mitigation. The event-tree expansion of the Focused PRA contributes to distinguish the failed equipment from the functioning equipment since by expanding the trees, all the possibilities and not only success criteria are considered. Figure 1 shows an example of event tree expansion. For instance, the Core Makeup Tank (CMT) actuation possibilities are 1 out of 2 (success criteria) or zero while in the expanded tree also the possibility of 2 CMT actuation is taken into account.

Once the expansion of the event trees has been completed, the success sequences are grouped into two categories.

- OK category which contains success sequences in which the core remains covered during the whole transient
- UC category which contains success sequences in which some core uncovery is detected (low margin).

The success sequences with UC end states are conservatively considered to lead to core damage in order to allow quantification of their risk importance and are collected and ranked by their contributions to core damage frequency CDF and large release frequency (LRF), as shown in Table 1 (only the first 25 sequences have been collected for this paper. After that, only the sequences which contribute to the CDF or LRF with more than 1% of the base CDF or LRF are considered risk important. Being the base **AP1000** CDF and LRF frequencies 2.41E-07/year and 1.95E-08/year, respectively [1]. So that Westinghouse identified the 13 sequences which are gathered in Table 2.

Subsequently, the 13 risk important sequences are bounded by 5 short-term and 2 long-term sequences, see Tables 2 and 3. The final step of the process is to analyze such sequences by using DSA T/H computer codes (e.g. NOTRUMP) and methods to show if adequate core cooling is achieved and therefore T/H uncertainties are bounded. A schematic view of the whole T/H uncertainty evaluation process is shown in Figure 2.

Figure 1: Event tree expansion process

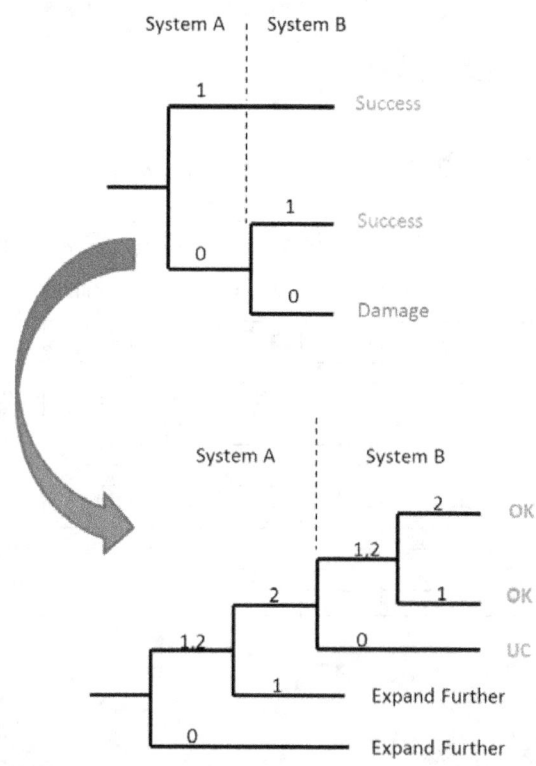

Table 1: PRA sequences sorted by CFD and LRF

Number	Initiating Event	Sequence CDF	Sequence LRF	Percentage CDF	Percentage LRF	CI	IRWST & RECIRC	CMT	ACCUM	ADS-4	ADS 2,3	BOUNDED By
1	SILB	8.96E-07	5.37E-08	371.66	275.6	Yes	Yes	1	0	4	2-4	C
2	SADS	4.58E-07	2.75E-08	190.05	140.93	Yes	Yes	2	1	4	2-4	E
3	SILB	3.05E-07	1.83E-08	126.76	94	Yes	Yes	0	1	4	2-4	A
4	MLOCA	2.89E-07	1.73E-08	119.85	88.88	Yes	Yes	0	2	4	2-4	AB
5	CMT	1.34E-07	8.05E-09	55.67	41.28	Yes	Yes	0	2	4	2-4	AB
6	SADS	9.12E-08	5.47E-09	37.82	28.05	No	Yes	2	2	4	2-4	E
7	MLOCA	3.01E-08	1.81E-09	12.48	9.26	Yes	Yes	2	0	4	2-4	C
8	LLOCA	8.51E-09	8.5lE-09	3.53	43.63	No	Yes	2	2	4	2-4	D
9	CMT	6.42E-09	3.85E-10	2.67	1.98	Yes	Yes	1	0	4	2-4	C
10	MLOCA	2.44E-09	1.47E-10	1.01	0.75	Yes	Yes	0	1	4	2-4	A
11	SILB	2.09E-09	1.25E-10	0.87	0.64	Yes	Yes	1	0	3	2-4	C
12	SILB	1.64E-09	9.83E-10	0.68	0.5	Yes	Yes	1	0	4	0-1	C
13	SILB	1.52E-09	1.52E-09	0.63	7.77	No	Yes	1	0	4	2-4	C
14	CMT	1.14E-09	6.85E-11	0.47	0.35	Yes	Yes	0	1	4	2-4	A
15	SADS	1.07E-09	6.42E-11	0.44	0.33	Yes	Yes	2	1	3	2-4	
16	SADS	8.40E-10	5.04E-11	0.35	0.26	Yes	Yes	2	1	4	0-1	E
17	SADS	7.77E-10	4.66E-11	0.32	0.24	No	Yes	2	1	4	2-4	E
18	SILB	7.21E-10	4.32E-11	0.3	0.22	Yes	Yes	0	1	3	2-4	
19	MLOCA	6.92E-10	4.15E-11	0.29	0.21	Yes	Yes	0	2	3	2-4	
20	SADS	6.76E-10	4,05E-11	0.28	0.21	Yes	Yes	1	1	4	2-4	E
21	MLOCA	6.44E-10	3.86E-11	0.27	0.2	Yes	Yes	0	2	4	0-1	AB
22	SILB	6.15E-10	3.69E-11	0.26	0.19	Yes	Yes	0	1	4	0-1	A
23	SILB	5.16E-10	5.16E-10	0.21	2.65	No	Yes	0	1	4	2-4	A
24	MLOCA	4.88E-10	4.88E-10	0.2	2.5	No	Yes	0	2	4	2-4	A
25	CMT	3.17E-10	1.90E-11	0.13	0.1	Yes	Yes	0	2	3	2-4	

Table 2: PRA risk important sequence

AP1000 Thermal-Hydraulic Uncertainty Low Margin/Risk Important sequences

Case/LM	Initiating event	CI	IRWST & RECIRC	CMT	ACC	ADS-4	ADS 2/3	PRHR	CDF	LRF	%CDF	%LRF	Bounding sequence
1	SILB	Yes	Yes	1	0	4	02-abr	N/A	8.96E-07	5.37E-08	317.7	275.6	C
2	SADS	Yes	Yes	2	1	4	2-4	N/A	4.58E-07	2.75E-08	190.1	140.9	E
3	SILB	Yes	Yes	0	1	4	2-4	Yes	3.05E-07	1.83E-08	126.8	94	A
4	MLOCA	Yes	Yes	0	2	4	2-4	Yes	2.89E-07	1.73E-08	119.9	88.9	B
5	CMT	Yes	Yes	0	2	4	2-4	Yes	1.34E-07	8.05E-09	55.7	41.3	B
6	SADS	No	Yes	2	2	4	2-4	N/A	9.12E-08	5.47E-09	37.8	28	E
7	MLOCA	Yes	Yes	2	0	4	2-4	N/A	3.01E-08	1.81E-09	12.5	9.3	C
8	LLOCA	No	Yes	2	2	4	2-4	N/A	8.51E-09	8.51E-09	3.5	43.6	D
9	CMT	Yes	Yes	1	0	4	2-4	N/A	6.42E-09	3.85E-10	2.7	2	C
10	MLOCA	Yes	Yes	0	1	4	2-4	Yes	2.44E-09	1.47E-10	1.0	0.8	A
11	SJLB	No	Yes	1	0	4	2-4	N/A	1.52E-09	1.52E-09	0.6	7.8	C
12	SILB	No	Yes	0	1	4	2-4	Yes	5.16E-10	5.16E-10	0.2	2.6	A
13	MLOCA	No	Yes	0	2	4	2-4	Yes	4.88E-10	4.88E-10	0.2	2.5	A
							Totals =		2.22E-06	1.44E-07			

Table 3: Low margin bounding sequences (WEC results)

Bounding Sequences Analyzed for T/H Uncertainty

Analysis case	Initiating event	Cont. Isol	IRWST & RECIRC	CMT	ACC	ADS-4	PRHR HX	Bounds Risk Important Case	Core Peak Clad Temp
Short-term Cooling									
A	Reactor coolant system hot leg (3.0")	No	Yes	0	1	4	Yes	3,10,12,13	No uncovery
B	Double-ended CMT balance line (6.8")	Yes	Yes	0	2	4	Yes	4,5	No uncovery
C	Double-ended DVI line (4")	No	Yes	1	0	3	No	1,7,9,11	1127 K
D	Double-ended cold-leg LLOCA	No	Yes	2	2	4	Yes	8	1288 K
E	Spurious ADS-4	No	Yes	1	1	4	Yes	2,6	844 K
Long-term Cooling									
F	Double-ended DVI line (4")	Yes	1/1 & 1/1	1	0	3	No	1-5,7,9,10	No uncovery
G	Double-ended DVI line (4")	No	1/1 & 1/2	1	0	4	No	6,8,11-13	No uncovery

Figure 2: T/H uncertainty evaluation process

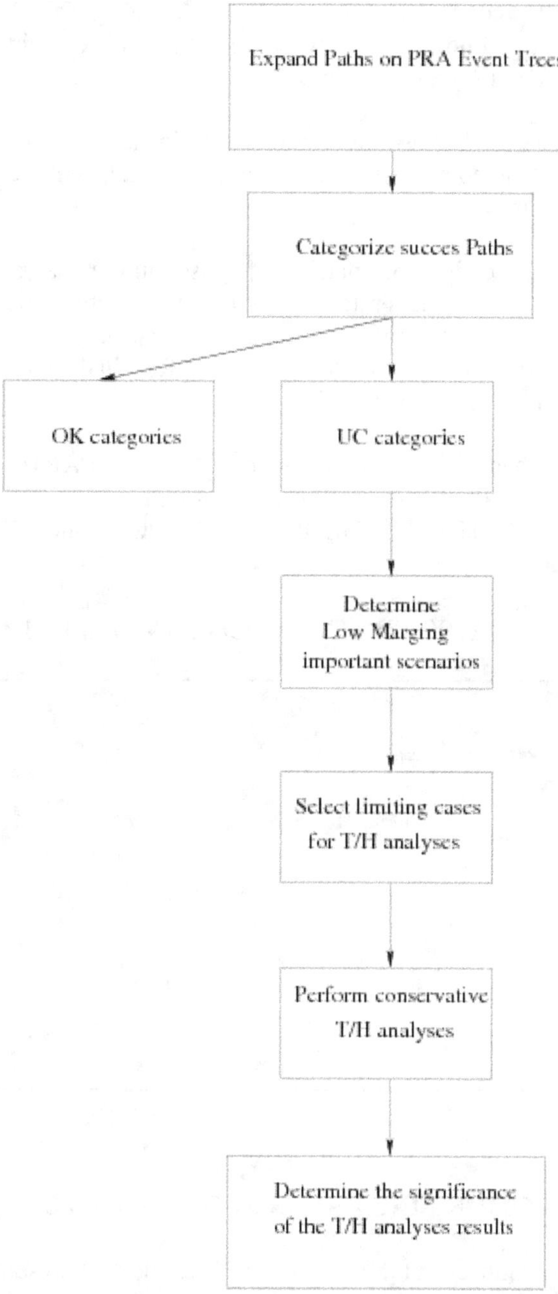

```
┌─────────────────────────────┐
│ Expand Paths on PRA Event Trees │
└──────────────┬──────────────┘
               ▼
┌─────────────────────────────┐
│   Categorize succes Paths   │
└──────┬───────────────┬──────┘
       ▼               ▼
┌──────────────┐  ┌──────────────┐
│ OK categories│  │ UC categories│
└──────────────┘  └──────┬───────┘
                         ▼
              ┌────────────────────┐
              │     Determine      │
              │    Low Marging     │
              │ important scenarios│
              └─────────┬──────────┘
                        ▼
              ┌────────────────────┐
              │ Select limiting cases│
              │  for T/H analyses  │
              └─────────┬──────────┘
                        ▼
              ┌────────────────────┐
              │ Perform conservative│
              │    T/H analyses    │
              └─────────┬──────────┘
                        ▼
              ┌────────────────────┐
              │Determine the significance│
              │ of the T/H analyses results│
              └────────────────────┘
```

3. ANALYSES OF AP1000 LOW MARGIN SEQUENCES BASED ON BEST ESTIMATE CALCULATIONS

As described in the previous section, once the low margin risk important sequences have been identified, detailed DCD computer codes and assumptions are used to evaluate these sequences and to demonstrate that T/H uncertainties are bounded. As shown in Table 3, five short-term bounding sequences and two long-term bounding sequences were determined [4]. The results obtained by Westinghouse (last column of table) shows that only three out of the seven sequences present core uncovery but they do not exceed Peak Cladding Temperature (PCT) limit in any case.

The bounding sequence "C" is especially important since the direct vessel injection (DVI) line break initiating event has been categorized as the even with the largest contribution to CDF in AP1000 PRA accounting for 39.4% of the total [1]. Moreover, this sequence bounds the low margin sequence with largest contribution to CDF and LRF being 317% and 275,6% respectively, see Table 2. The analysis performed with NOTRUMP predicts for this bounding sequence a PCT of 1127K, Table 3. This particular sequence presents the following systems availability for the accident mitigation: 1 CMT, 3 out of 4 ADS stage-4 valves and IRWST injection.

The UPM group has analyzed the DVI short-term low-margin risk-important sequence with the best estimate TRACE code in order to compare the results obtained with a detailed DCD conservative code against a more realistic analysis.

The AP1000 model for TRACE code V 5.0 patch 2, [8] used in the analysis consists of all the main components such as vessel, Steam Generators, Pressurizer, Reactor Coolant Pumps (RCPs) and connecting pipes as well as the passive safety systems, Core Makeup Tanks (CMTs), Accumulators (ACCs), Automatic Depressurization System (ADS), Passive Residual Heat Removal system (PRHR). No active systems are implemented in this model.

The total number of thermal hydraulic components presented in the **AP1000** TRACE model is 368, being 86 PIPEs 191 HTSTRs, 3 POWERs, 44 VALVEs, 4 PUMPs, 24 BREAKs, 10 TEEs, 3 FILLs and 3 VESSELs. In addition 119 TRIPS, 377 CONTROL BLOCKS and 447 SIGNAL VARIABLES complete the model, see Figure 3.

Figure 3: AP1000 TRACE model in SNAP nodalization

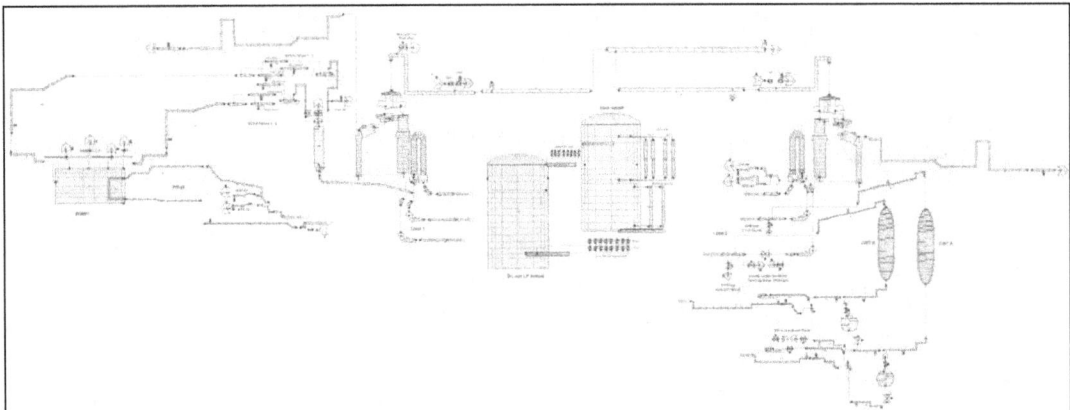

4. DIRECT VESSEL INJECTION LINE BREAK SEQUENCE EVOLUTION

The size of the DVI line is 6.8 inches but the inlet nozzle (vessel side) presents a 4-inch flow restrictor which limits the effective break size and hence the maximum flow that can be depleted through the break. Accordingly, the event is classified as a medium LOCA.

An important difference respect to other MBLOCAs (hot leg and cold leg) is that, in this kind of sequence, 1 CMT, 1 ACC and 1 IRWST injection line became unavailable due to the location of the break. In addition, since the outlet of the normal residual heat removal system (active system) is connected to both DVI, it is also assumed that the water from this system is lost through the broken line [1].

As in a standard LOCA, the event results in a reactor trip and safety injection signal "S" that causes RCP trip and CMT actuation. The ADS-1 actuates after CMT low level signal (67%) and the accumulators should inject when the pressure is lower than 49 bar. When a full depressurization has been achieved, the IRWST begin to drain into the reactor. The main setpoints involved in this kind of sequence are listed in Table 4 [2].

It must be pointed out that the DVI sequence which is simulated in this analysis (bounding sequence "C" of T/H uncertainty evaluation process) is not a typical LOCA sequence but presents some restrictions, which are the following ones: 1 CMT is available, both accumulators are not able to inject, the ADS stages 1, 2 and 3 do not actuate neither and the full depressurization is achieved only with 3 out of 4 valves of ADS stage 4 whose setpoint is reached when the level in the available CMT is 20%. Thereafter the low pressure injection is achieved through the IRWST. Moreover, the PRHR is not credited since it is not a part of success criteria in sequences with automatic ADS leading to IRWST gravity injection [5], see Table 3 case "C".

Table 4: Main setpoints for tipycal DVILB sequence. (*) unavailable system for the analyzed bounding sequence "C".

Function	Setpoints assumed in DVILB analysis
Reactor trip on low pressurizer pressure	124 bar
"S" signal on low-low pressurizer pressure	117 bar
Reactor coolant pumps trip	"S" signal + 6 sec
PRHRS valve starts to open (*)	"S" signal
CMT injection starts (1 out of 2)	"S" signal + 2 sec
ACC injection starts on low RCS pressure (*)	49 bar
ADS-1 (*)	Low CMT level (67.5%)
ADS-2,3 (*)	Delay with respect ADS-1
ADS-4 (3 out of 4)	Low-Low CMT level (20%)
IRWST valve opens	ADS-4 actuation

5. ANALISIS OF LOW MARGIN DVI RISK IMPORTANT SEQUENCE WITH TRACE CODE.

This section presents the simulation performed with TRACE code for DVI line break low-margin sequence. As described before, the system availability of this sequence is as follows: 1 out of 2 CMT (intact loop), 0 out of 2 ACC, stages 1, 2 and 3 of the ADS do not actuate, 3 out of 4 ADS stage 4 valves and IRWST injection. Figure 4, shows the AP1000 RCS with availability of systems in this sequence. The results are plotted comparing with those obtained with NOTRUMP by Westinghouse [4]. The chronology of the sequence is presented in Table 5.

As it is depicted in Figure 5, the pressure decrease (blowdown phase) presents a similar trend in both cases until the pressure stabilizes below 100 bar. After that, TRACE predicts a longer stabilization at the secondary pressure.

In this sequence there is not core cooling through the Passive Residual Heat Removal (PRHR) system and therefore the available CMT is the only way to cool the core until the ADS stage 4 allows the actuation of the IRWST. Although the CMT flow predicted by TRACE is slightly greater Figure 6, the trend is quite similar. The time when the CMT reach the 20% of its level and therefore the ADS stage 4 actuates, is appreciable in both cases since the flow rate is increased at this time. This moment is also appreciable in the pressure plot where a depressurization step between about 25 bar and containment pressure is achieved very quickly.

Later, when the IRWST injection is achieved, the system is again pressurized above the IRWST injection pressure, which produces a short interruption in the flow rate, Figure 7. The minimum core level is reached at about 1900 seconds producing a fast rise in the clad temperature. When the water from the IRWST flows into the system the core level is recovered and the temperatures remain low. It is appreciable that the PCT value obtained in the analysis performed with NOTRUMP for this sequence (1127 K), see Table 3, is higher than the obtained with TRACE code (600 K), Figure 8. It must be taken into account that NOTRUMP analysis includes more conservative models and assumptions than TRACE

Figure 4: DVILB low magin sequence. Availability emergency systems

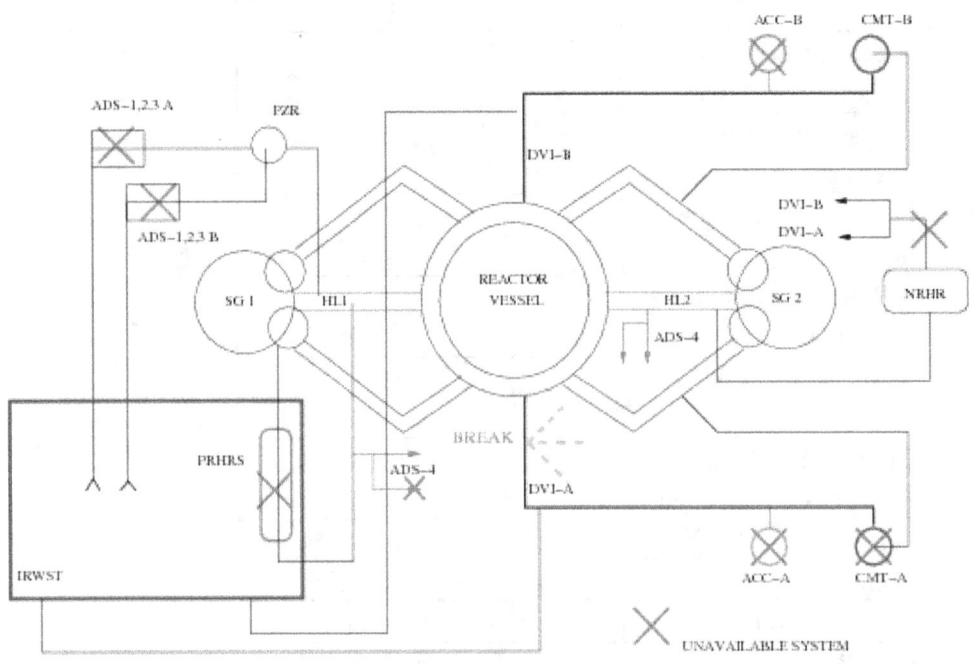

Table 5: DVI line break low margin sequence chronology. TRACE code.

Event	Time (s)
Break opens	0
Reactor trip	24.3
"S" signal	27.2
RCP trip	33.2
CMT injection starts	39.2
Auto ADS-4	1355
IRWST injection	1870.3
Maximum PCT	1995.1

Figure 5: DVILB low magin sequence. Pressure transient. TRACE vs NOTRUMP

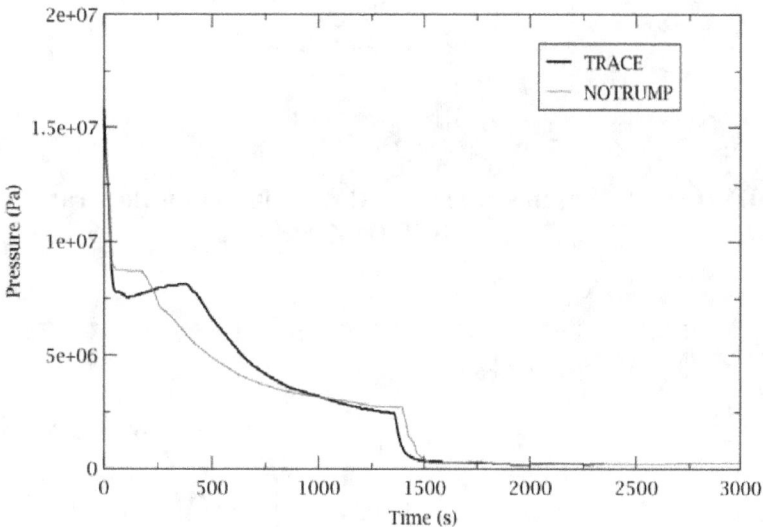

Figure 6: DVILB low magin sequence. CMT injection flow rate. TRACE vs NOTRUMP

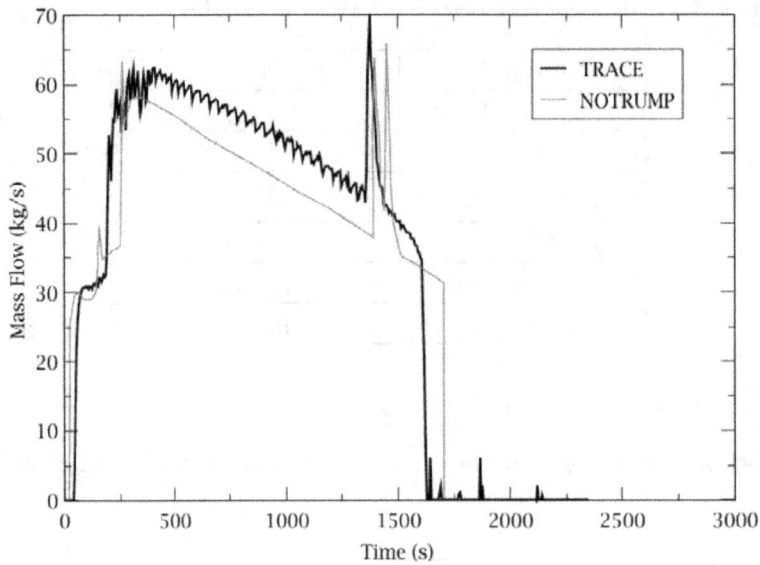

Figure 7: DVILB low magin sequence. IRWST injection flow rate. TRACE vs NOTRUMP

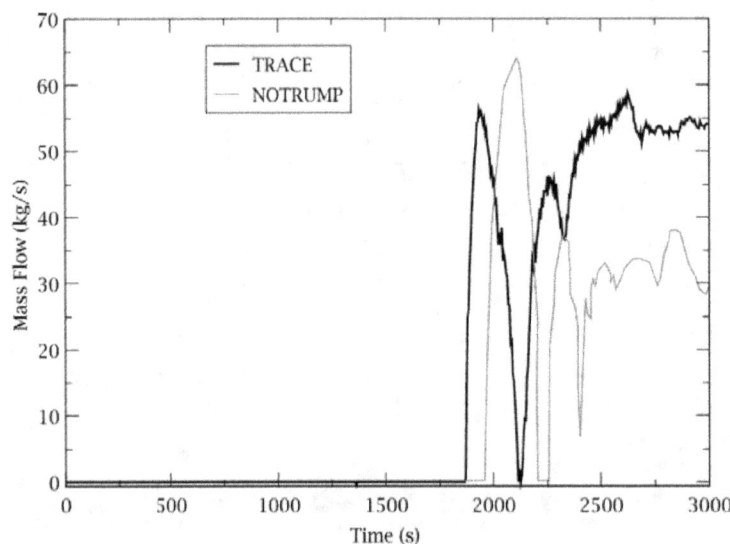

Figure 8: DVILB low magin sequence. Peak cladding temperature. TRACE

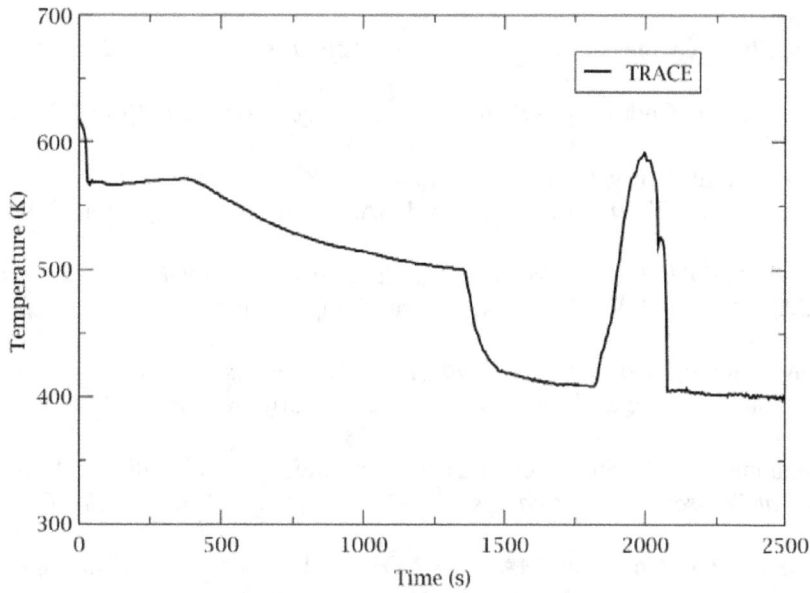

6. CONCLUSIONS

This paper sumarizes the thermal hydraulic uncertainty evaluation process developed by Westinghouse in order to bound the possible uncertainties that can affect to the PRA succes criteria due to the use of passive safety systems in **AP1000** reactor.

The UPM group has analyzed the DVILB low margin sequence with the best estimate TRACE computer code in order to verify how the behaviour of the plant would be in a more realistic case. The results show a very similar trend between TRACE and NOTRUMP simulations, but the PCT value obtained with TRACE remain well below as the predicted by NOTRUMP. The result is the expected since, the NOTRUMP code is the DSA computer code used by WEC for SBLOCA analysis.

This result has allowed to verify and confirm the **AP1000** thermal-hydraulic uncertaintiy evaluation process peformed by WEC for the most risk-important low-margin sequence

AKNOWLEDGEMENTS

This work has been funded by the Spanish Ministry of Competitiveness and Economy within ISAMAR project: ENE2011-28256. Its support is gratefully acknowledged. The authors would also like to thanks to Westinghouse Electric Company and to USNRC for the technical support.

REFERENCES

[1] **Westinghouse Electric Company.** *AP1000 Probabilistic Risk Assessment Report.* 2003

[2] **Westinghouse Electric Company.** *AP1000 Design Control Document.* Rev19. 2008

[3] **Selim Sancaktar and Terry Schulz.** *Risk Informing PRA Success Criteria;* Application to the AP1000 Proceedings of ICAPP '04 Pittsburgh, PA, USA, June 13-17, 2004 Paper 4239

[4] **Ohkawa et al.** *"AP600 PRA Thermal/Hydraulic Uncertainty Evaluation for Passive Systems Reliability".* WCAP-14800-NP. Westinghouse electric Company. June 1997

[5] **Westinghouse Electric Company.** *AP1000 Probabilistic Risk Assessment Report.* Chaper 59 Appendix A. Thermal Hydraulic analysis to Support Success Criteria.

[6] Advisory Committee on Reactor safeguards subcommittees on Reliability and probabilistic Risk Assessment. *Official Transcrip of proceedings ACRST-3228.* NRC. 23 January, 2003

[7] **Westinghouse Electric Company** MAAP4 / NOTRUMP Benchmarking to Support The Use of MAAP4 for AP600 Success Criteria Analysis, WCAP-14869, 1997

[8] **US Nuclear Regulatory Commission.** *TRAC-RELAP Advanced Computational Engine (TRACE) User's Manual Input Specification.* 2007.

DISCLAIMER

This paper is the result of the analysis carried out by the research group of the Technical University of Madrid and therefore Westinghouse Electric Company and the USNRC were not involved in such investigations and they are not responsible about the contents here exposed.

AP1000 is a trademark or registered trademark in the United States of Westinghouse Electric Company LLC, its subsidiaries and/or its affiliates. This mark may also be used and/or registered in other countries throughout the world. All rights reserved. Unauthorized use is strictly prohibited. Other names may be trademarks of their respective owners

Application of Web-based Risk Monitor in Tianwan Nuclear Power Plant

Hao Zheng[a*], Wei Wang[b], Xiaohui Gu[c], Yong Qu[c], Zhenli Bao[c], Xuhong He[b]

[a] Lloyd's Register Consulting, Beijing, China
[b] Lloyd's Register Consulting, Stockholm, Sweden
[c] Jiangsu Nuclear Power Co., Lianyungang, China

Abstract: As one of the specific applications of Living PSA, Risk Monitor, which is a real-time analysis tool used to determine the instantaneous risk based on actual plant configuration, has been widely used in risk-informed decision-making process during plant operation. A web-based Risk Monitor application for Tianwan nuclear power plant (NPP) is currently being used onsite to improve the PSA applications, popularize the concept of risk-informed management, and then enhance the whole risk management level in Tianwan NPP. Compared with the traditional windows based tools, this web-based Risk Monitor application is a natural multi-user program with great advantages. It has good interface with plant's existing information system and can automatically update the risk information upon changes of component status/configurations.

This paper presents the overview of the Risk Monitor application used in Tianwan NPP, including challenges and experience of implementing Web-based Risk Monitor, and major feature improvements which facilitate the application of Risk Monitor. Example PSA applications implemented in Tianwan NPP will also be presented together with future plan and challenges.

Keywords: PSA Application, Risk Monitor, Tianwan Nuclear Power Plant.

1. INTRODUCTION

Since the beginning of 1990s risk monitors based on PSA models have been used at more and more Nuclear Power Plants (NPP) for operation risk management. Benefits of risk monitor application on following aspects have been demonstrated by many successful cases during the past twenty years:

- To increase plant safety via an enhanced ability to improve risk awareness, and assess and manage risk
- To increase operational flexibility resulting from support of risk-informed regulation to achieve more economic operation

China National Nuclear Safety Administration (NNSA) published the technical policy "Application of PSA in nuclear safety" for guidance of PSA applications in February 2010 and initiated several PSA application pilot projects in August 2012. As a part of PSA application project, Tianwan nuclear power plant started the application of the web-based risk monitor tool via agreements with Lloyd's Register Consulting and China Nuclear Power Engineering Co. (CNPE) in 2012.

Tianwan NPP has two operating units of VVER-1000/V320 type reactor with some advanced safety features including the double containment, N+3 redundant safety systems, a core catcher, fully digital control system (DCS) etc. The risk monitor tool currently being used in Tianwan NPP is the Web-based RiskSpectrum® RiskWatcher (RWWeb) developed by Lloyd's Register Consulting. It has distinct advantages in comparison with the traditional desktop based tools, when it comes to PSA application in nuclear power plants.

This paper presents an overview of the Risk Monitor application used in Tianwan NPP, including challenges and experiences of implementing Web-based RiskWatcher, and major feature

---

[*] email address: hao.zheng@lr.org

improvements which facilitate the application of Risk Monitor. Some PSA applications implemented in Tianwan NPP based on the Risk Monitor will also be presented together with future plan and challenges.

2. RISK MONITOR OVERVIEW

The Tianwan Risk Monitor RWWeb is designed to be used by both PSA knowledgeable and non-PSA knowledgeable personnel. Most of the Risk Monitor users are assumed to not be familiar with PSA model or PSA jargon. The application therefore uses the normal plant equipment IDs and descriptions and a minimum level of PSA related terms.

RWWeb enables plant operators and schedulers to evaluate the plant risks associated with scheduling and approving online and outage maintenance activities, and will help plant personnel better understand the risks in any plant configuration.

The RWWeb supports the following primary functions:

- Managing plant operational safety
- Logging historical records of actual plant configurations
- Online maintenance planning
- Long term scheduling
- Defence-in-depth analysis
- Providing information on the risk importance of the components that are in service as well as out of service

One of the key features in the Tianwan RWWeb is that all model related data is edited in the baseline PSA model and no changes need to be introduced "afterwards" in the separate risk monitor model. The information about changes in plant configuration are stored separate from the PSA model data and then dynamically combined into the PSA model as boundary conditions when doing model re-quantification. This principle simplifies the process of going from a living PSA baseline model to a functional risk monitor model and will greatly simplify continuous update work - i.e. maintaining a true living PSA model.

The Tianwan RWWeb supports the blended approach of risk-informed decision making by providing qualitative and quantitative evaluations. The instantaneous risk can be quantified according to the changes of plant configuration. Following risk information can be presented in Tianwan RWWeb:

- A risk profile showing the risk level over time
- Comparison of different risk curves
- Cumulative risk during any user defined period of time
- Indication of current risk level at a given time point in the form of a number (relative or absolute risk), and in the form of colour indication e.g. green, yellow, orange and red
- Qualitative "defence-in-depth" status, which shows whether safety functions, systems, sub-systems and components are available, degraded or unavailable e.g. green, yellow, orange and red
- Importance measures showing how important components, systems are in terms of contributing to current risk, or in terms of possible reduction of current risk

3. ADVANCED FEATURES

The Tianwan RWWeb is designed to be an advanced risk assessment and management platform to facilitate the risk-informed decision-making process, and it has great advantages compared to other traditional desktop-based risk monitor tools. Several important features are introduced during the development to facilitate its application on site, some are the characteristics of the web application, and others are due to the new innovative design.

3.1. Web-based Application

The Tianwan RWWeb is a web-based application. In comparison with traditional desktop-based applications, the web application has some distinct advantages:

- Easy to deploy
- Easy to maintain and update
- More accessible
- More traceable
- Platform independent, better adaptability and compatibility

The web applications avoid the burden in deploying on each client machine. No installation is required for the users, since all the deployment and maintenance work will be concentrated on the server side, and the users only need the standard web browser to access to the application. These characteristics have made it ideal for the Risk Monitor application. The web-based Risk Monitor is easier to use, and has the possibility to be more widely used, which is of great significance for the promotion of Risk Monitor application, and thereby be very helpful to the plant safety management.

Web applications are more traceable since all data are centralized on server side and users may only be able to make changes via pre-designed operation interfaces. It is easy to observe the log information and user browsing patterns as every request is sent to the server and can be logged. This makes the web-based Risk Monitor well consistent with the quality assurance requirements of nuclear power plants.

Risk Monitor runs the PSA model, reads, interprets and displays the results of the quantification, and stores them for future retrieval and review. The users may specify how to display the results, as well as other specific information from the quantification. A Risk Monitor should be designed for different personnel in nuclear power plant. It is preferred if Risk Monitor can provide interface and capabilities for users without PSA experience while providing functions and features that are useful for PSA practitioners. The RWWeb is customized to fulfil these requirements.

The high customization and extensibility of web application also makes it a plant-specific application. The function modules are properly designed to be independent along with clear boundaries. So it can be easily customized to fully support specific needs in functionalities and interfaces.

3.2. Multi-user Operation

The RWWeb is a natural multi-user application, and it is easy to provide collaboration between multiple users, as all data are centralized. Multiple users are able to access the same model data and plant configuration information at the same time. Multiple users can simultaneously perform many standard functions on the same model dataset without affecting other users or their data. These standard functions include but are not limited to, viewing risk profile, tracing operation log, performing individual What-if analysis, printing reports, etc. In the Tianwan RWWeb, logic check has been introduced before editing operation event log to prevent timing errors due to event log edited by more than one user at the same time.

However, risk quantification in Tianwan Risk Monitor is performed asynchronously as the model calculation is not instantaneous. As a result it is possible that quantification might have been expired before it is accomplished due to that the base data of the calculation (plant configuration) have been superseded. The result of these quantifications will be tagged to tell users that they are no longer valid and that there are new results available.

3.3 Integrated Model Management

The PSA model used by the risk monitor is based on, and is consistent with, the Living PSA for the facility, and should be updated with the same frequency as the Living PSA.

The Tianwan RWWeb adopts an integrated model management strategy, which greatly improves traceability of the model and plant configuration. When the Living PSA model has changed, the Risk Monitor model should be updated accordingly, and then it will be asked whether it is a model upgrade indicating the real changes in the actual plant or whether it is a PSA model correction to fix a known error. For model upgrading, the new model will be applied continuously. For model correction, the new model will replace the previous one. Different versions of the model are preserved and seamlessly integrated, the model changes are well kept and easily tracked, therefore activities at any particular time point can be linked to the true real-time plant model.

3.4 Calculation Scheduling

Calculation speed is one of the key issues for a real-time Risk Monitor application system. It becomes even more important for a web-based Risk Monitor since there might be many concurrent calculation requests. In the Tianwan RWWeb, a series of measures have been taken to improve the performance of quantifications.

- Perform calculation asynchronously, risk profile can be refreshed when calculation is finished
- Use distributed calculation architecture, and the calculation capability can be extended with increasing of calculation cores and servers
- Intelligent scheduling algorithm has been adopted to manage the calculation resources more efficiently. Calculation priority is considered in the scheduling algorithm, for instance the online quantification should always be given higher priority than planning
- Avoid duplicate calculation, if the same plant configuration has been quantified, the results can be re-used directly. a new quantification is only started when the current plant configuration has been changed and the new configuration has never been analysed before

Generally, quantification algorithms can be developed based on either fault tree logic or pre-solved minimum cut sets with different accuracy. Quantification based on fault tree logic from PSA model solved for each new plant configuration is applied for Tianwan RWWeb due to requirements of accuracy.

3.5 Automatic Operation Log Import

Tianwan RWWeb has good interfaces with the plant's existing information system. A standalone tool was developed and embedded in RWWeb which can automatically download all relevant status signals of equipments from information system, convert to equipment status based on mapping rules and update in RWWeb upon the changes of status/configuration. Calculation will be automatically initiated after configuration change and the risk profile will be refreshed after calculation completed.

In theory no operator intervention is required during the whole process of operation risk updating by using this function. It can reduce the workload of using risk monitor for operational risk monitoring, and increase the accuracy and real-time characteristics of risk monitor. Feedback from Tianwan NPP indicate that there is almost no time delay caused by automatic input of plant configuration change. The major time delay for on-line operational risk monitoring is the PSA calculation time. For quality assurance of the automatic updating of the plant configurations, the users should review all the imported logs.

4. APPLICATIONS IN TIANWAN NPP

Since December 2013, Tianwan NPP had started the trial use of the RWWeb on site. The scope of PSA applied in risk monitor for Tianwan NPP is Level 1 PSA of internal events for both full power and low power/shutdown modes. Four colour risk bands (Green, Yellow, Orange and Red) have been used for representing risk levels from low to unacceptable. Management measures related to each risk level has been established and integrated into overall risk-informed management system.

4.1 Risk Awareness

It was found safety concept and safety culture change are the most challenging things in the beginning of using risk monitor in Tianwan NPP. Currently most safety management activities in the NPPs are derived from either requirements in regulations/rules or results from deterministic safety analysis. Those prescriptive requirements are familiarized and well accepted by plant staff, and implemented in their daily work as well. One goal of risk monitor application is to increase risk awareness of each working staff and introduce risk insights from PSA point of view into existing safety/risk management framework.

During the period of risk monitor trial use, several external and internal training courses on PSA, risk monitor and risk-informed management were carried out for plant staff to increase knowledge of the whole concept and functionality of the tools. The deployment and configuration of the Tianwan RWWeb makes it accessible for everyone in plant to view risk profile and carry out risk analysis by using "What-if" module. It increases the involvement of each staff on risk management and perceptual experience of risk concept.

Good communication is ensured during the whole process. Questions or suggestions raised in use of risk monitor can be emailed to dedicated PSA engineer onsite for discussion and problem solving.

4.2 Online Risk Monitoring

Online risk monitoring is considered as one of the major applications at the current stage in Tianwan NPP. One dedicated computer is deployed in the main control room for risk monitoring and a dedicated account is assigned to be used in main control room with permission for online module. With automatic operation log import function, this system can continuously and steadily monitor risk status of NPP unit, update and refresh risk profile without operator's intervention. ()

Both quantitative risk measure (CDF) and qualitative risk measures (defence-in-depth status) are monitored and treated separately in the system. The qualitative risk measures provide risk insights at the safety function/system level/sub-system level in addition to the quantitative risk insights including CDF and risk importance measures.

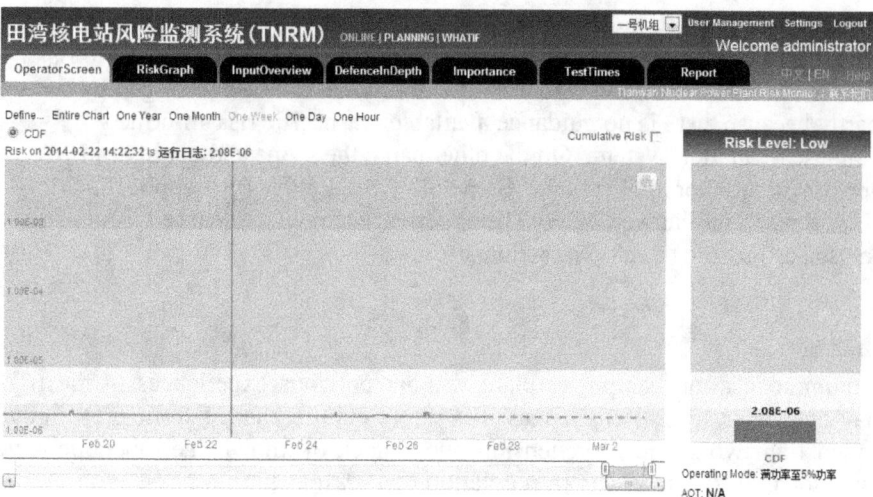

Figure 1 - Online Risk Monitoring with Risk Curves in Upcoming Days

4.3 Risk of Maintenance Planning

Planning module in RWWeb can be used for both day-to-day maintenance and maintenance scheduling purpose. The planning module is mainly used for day-to-day maintenance risk management in Tianwan NPP so far.

As a part of the planning work process every day, draft daily plan (three-day rolling plan) will be imported into risk monitor for risk calculation. A risk report for daily plan will be generated and used by planning staff as input for further adjustment. The finalized daily plan with its risk insights will be published in the risk monitor as part of online risk profile. In other words, operation staff can easily get risk insights of the upcoming activities in addition to current risk level by using RWWeb.

4.4 Safety Supervision

Dedicated PSA engineers onsite use risk monitor for plant routine risk supervision. Risk report for each unit will be prepared each week together with traditional STA safety report. Based on the plant management procedure, detailed risk evaluation report will be published if there is a significant risk increase (e.g. reach the risk level of Orange) and suggestions on risk reduction and control will be prepared based on integrated consideration which includes the risk insights from the RWWeb.

4.5 Future Plans

The RWWeb was put into trial use since the last December and only several applications have been implemented in Tianwan NPP. Tianwan NPP has a long term plan for promoting PSA application and establishing risk-informed management system. Following aspects are considered in the future plan for application of risk monitor.

Enhancement of current practice on application of risk monitor including online risk monitoring, day-to-day maintenance risk management etc. is to be carried out in 2014. Risk management activity based on risk monitor is to be integrated into operation management procedures in a better way. Improvements on RWWeb are also planned to facilitate application. One example is to have alarm in main control room when there is significant change of risk (e.g. risk increase from Green to Yellow) in order to alert operator for attention.

Risk monitor is planned to be used as a support tool for plant engineering and technology modification. A professional local team will be established to implement application of risk monitor with support from external experts.

More applications based on risk monitor for more economic operation is to be planned in Tianwan NPP. It can be found that current applications of the RWWeb in Tianwan NPP are all related to risk monitoring and risk awareness to enhance safety. Operation flexibility has not been changed by the current practices. It is partly because there is no guidance available on specific risk-informed applications in current Chinese regulation system. On the other hand, the scope and quality of PSA model also limit the scope of risk monitor applications. Some pilot projects for PSA application on operation flexibility will be planned in Tianwan NPP. The experience and lesson learned from these pilot projects can also be used as input in developing guidance.

5. CONCLUSION

Application of the risk monitor in a nuclear power plant without previous experiences of the risk-informed application may encounter several challenges like safety culture change from prescriptive framework to risk-informed framework, incomprehension due to lack of knowledge, uncooperative staff due to additional workload, no requirements in regulation etc. A transition process with proper educational sessions and good communication is significant for success of risk monitor application. A well designed and customized web-based risk monitor tool can facilitate the whole process.

The RWWeb developed for Tianwan NPP fulfills the general requirements for risk monitor. In addition to that, the web-based framework brings lots of inherent advantages including the flexibility to deploy and manage the application, better adaptability and compatibility, more accessible, etc. Improvements on calculation scheduling make speed of fault tree solution within an acceptable level. Automatic operation log import significantly reduces the workload of operator in using risk monitor and increases the precision of configuration input.

However, there are some limitations of current risk monitor used in Tianwan.

- Uncertainty is not addressed. Only point value of CDF is provided in current risk monitor. As the RWWeb is to re-quantify the PSA model each time, it is possible to have uncertainty results for each plant configuration. However, the uncertainty results are not required either from Tianwan NPP or authority.
- Unable to apply different risk bands for different operation modes. One risk band definition will be used for all operation modes. This may result in that the baseline risk in some shutdown modes is at Yellow band.
- Effect of different failure modes for one component is not considered. All failure modes for one component are mapped together to the component in current version of risk monitor. This inaccuracy treatment may cause the result more conservative and the risk monitor model does not reflect the plant in a exact way.

Those limitations will be further investigated in further development.

References

[1] OECD NEA, "Risk Monitors: State of the Art in their Development and Use at Nuclear Power Plant", NEA/CSNI/R(2004)20, 2004
[2] NNSA Technical Policy, "Application of PSA in Nuclear Safety", 2010
[3] Lloyd's Register Consulting, "RiskSpectrum® RiskWatcher Web User Manual", 2013

Analyzing system changes with importance measure pairs: Risk increase factor and Fussell-Vesely compared to Birnbaum and failure probability

Janne Laitonen*, Ilkka Niemelä

Radiation and Nuclear Safety Authority (STUK), Helsinki, Finland

---

Abstract: Importance measures are used to rank components of a system according to a selected criterion depending on the decision problem. Sometimes, more than one importance measure may be used. In risk-informed decision making, a component that is critical to safety is usually prioritized higher in allocating activities, e.g. maintenance or inspection. One desired effect of this prioritization is to improve the reliability of the critical components. Changes in the system or component reliability affect their importance measures. If these feedbacks are taken into account, new ranking for the components may be obtained.

This paper examines the properties of risk importance measure pairs in analyzing system changes with fault tree analysis. A common approach is to use risk increase factor (or risk achievement worth) and Fussell-Vesely importance measure. This approach is compared to an alternative method which utilizes Birnbaum importance measure and the failure probability of a basic event. It is shown that the first approach may lead to difficulties in understanding the effect of system changes whereas the latter seems to provide simpler and more robust alternative. The paper includes examples to show and compare the differences between the two methods. The key advantages of the alternative method are that it reflects the absolute instead of relative change, the variables are independent, and that the interpretation of the importance measures is straightforward, reflecting risk in terms of safety margin and failure probability.

Keywords: Importance measure, basic event ranking, PRA, PSA

---

1. INTRODUCTION

Risk-informed decision making is becoming more and more common in many industrial sectors [1]. In nuclear industry, it is used by both licensees and regulatory bodies (see, e.g., [2]). For instance, in Finland the foundation for risk-informed licensing, regulation and safety management of a nuclear power plant is laid in the nuclear safety legislation and the requirements on the use of risk-informed methods are included in about ten nuclear regulatory guides (see, e.g., nuclear regulatory guide YVL A.7 [3]). Typical subjects for decision making are ranking and prioritizing systems, structures and components (SSC) for enhanced maintenance, inspection, testing or safety categorization [1]. Sometimes, more than one importance measure may be used in decision making. For example, NUMARC 93-01 provides the implementation guidelines for maintenance rule: in determining the risk-significant SSCs risk reduction worth (RRW), risk achievement worth (RAW) and the top 90% of the cut sets contributing to core damage frequency are to be applied (for more information see, e.g., [4], [5], [6]).

In addition to RRW and RAW, Fussell-Vesely (FV) importance is commonly used, also in combination with RAW [7] which is also called risk increase factor (RIF, which is used in this paper). But even as appealing as it might be, using several importance measures in risk-informed decision making may be more complex than at first one might assume. This has been recognized also in [7] where the authors advice to use Birnbaum importance instead of RAW in combination with FV. This paper tries to contribute on this discussion by studying the use of importance measure pairs: risk increase factor and Fussell-Vesely are compared to Birnbaum and failure probability in analyzing system changes.

First, the importance measures used are introduced. Then, the dependence between RIF and FV is shown which is also given in [1] although the discussion in [8], [9] and [10] states that there would be no functional relation or direct relation between FV importances and RAWs. The influence of the dependence is briefly discussed and illustrated graphically with FV-RIF -map. After this, an alternative method utilizing independent variables failure probability and Birnbaum is discussed briefly. Finally, the two methods are compared with a simple pump line system which goes through various changes including both structural and component reliability improvements. Since living PRA, system modifications or even safety classification (one goal may be to improve reliability by increasing tests, maintenance and inspections) all introduce feedback and updates to PRA model, the authors consider relevant to study which method should be used in analyzing them.

2. IMPORTACE MEASURES

It is assumed that the reliability model is presented in the form of minimal cut sets. The following discussion does not deal with initiating events since they have different dimension compared to failure probabilities and therefore their Birnbaums cannot be compared (see, e.g., [11], [12]). The total risk, R, is represented as a (multi)linear function $R(X_i) = aX_i + b$ where aX_i represents all the cut sets containing the event i with failure probability X_i and b all other cut sets. Using this notation risk increase factor (or risk achievement worth, RAW) is defined as

$$RIF(X_i) = \frac{a+b}{aX_i+b} \tag{1}$$

Fussell-Vesely is defined as

$$FV(X_i) = \frac{aX_i}{aX_i+b} \tag{2}$$

Birnbaum is defined as

$$B = \frac{\partial R(X_i)}{\partial X_i} = a \tag{3}$$

Risk increase factor is a weak (almost independent) function of X_i and Fussell-Vesely is proportional to X_i when $aX_i \ll b$ (see, e.g., [7] and [13]). The independence of Birnbaum from X_i is evident. In addition, the following relation holds between risk reduction worth and Fussell-Vesely: $RRW = 1/(1 - FV)$ [7]. Hence, although the following discussion deals with RIF, FV, Birnbaum and failure probability, similar treatment for RAW and RRW is possible.

It must be noted that PRA codes may approximate RIF by setting the corresponding failure probability X_i equal to unity without reducing the Boolean minimal cut set expression (i.e., reminimization). Hence, the result may not be consistent with the value that is obtained if the Boolean value of the event i is set to 'true' [11], [13]. For simplification and clarity, it is assumed in this paper that RIF is approximated without reminimization. This means that a and b do not change whether we calculate RIF, FV or Birnbaum.

3. RIF AND FV MAPPING

Using equations (1) and (2) (and assuming that RIF is approximated without reminimization) one obtains $RIF(X_i) \approx FV(X_i)(a+b)/aX_i$ and noticing that $FV(X_i)b/aX_i = 1 - FV(X_i)$ one yields to

$$RIF(X_i) \approx 1 + FV(X_i)\left(\frac{1}{X_i} - 1\right) \tag{4}$$

Thus, for any X_i, possible combinations of RIF and FV are defined by equation (4) above. Similar relation is also presented in [1]. The equation shows that for a constant failure probability, RIF and FV have an (almost) exact relation, i.e., in FV-RIF -map the plotted basic events can only appear on the curves defined by equation (4) regardless of the system. Let us call these curves of equal failure probability equiprobability curves which are shown in figure 1 for different failure probabilities on FV-RIF -map. The failure probabilities X_i for each equiprobability curve are given on the right. The dashed lines illustrate arbitrary criteria by which the basic events can be categorized into high, medium and low risk areas, i.e., upper right corner (FV > 1.00E-2 and RIF > 1.00E+2) is considered high risk area, lower right and upper left corners medium risk areas, and lower left corner (FV < 1.00E-2 and RIF < 1.00E+2) low risk area.

The dependence of RIF and FV may lead to difficulties in understanding the effect of system changes in, e.g., risk-informed decision making. Since RIF, FV and failure probability are tied by equation (4) it is clear that if two of these variables are known, the third can be calculated regardless of the failure logic of the system. Since failure probabilities are needed for calculating any risk importance measure, either FV or RIF can be detemined by the corresponding equiprobability curve. Hence, the third variable can be explained by the other two and therefore does not add any information about the importance of the basic event.

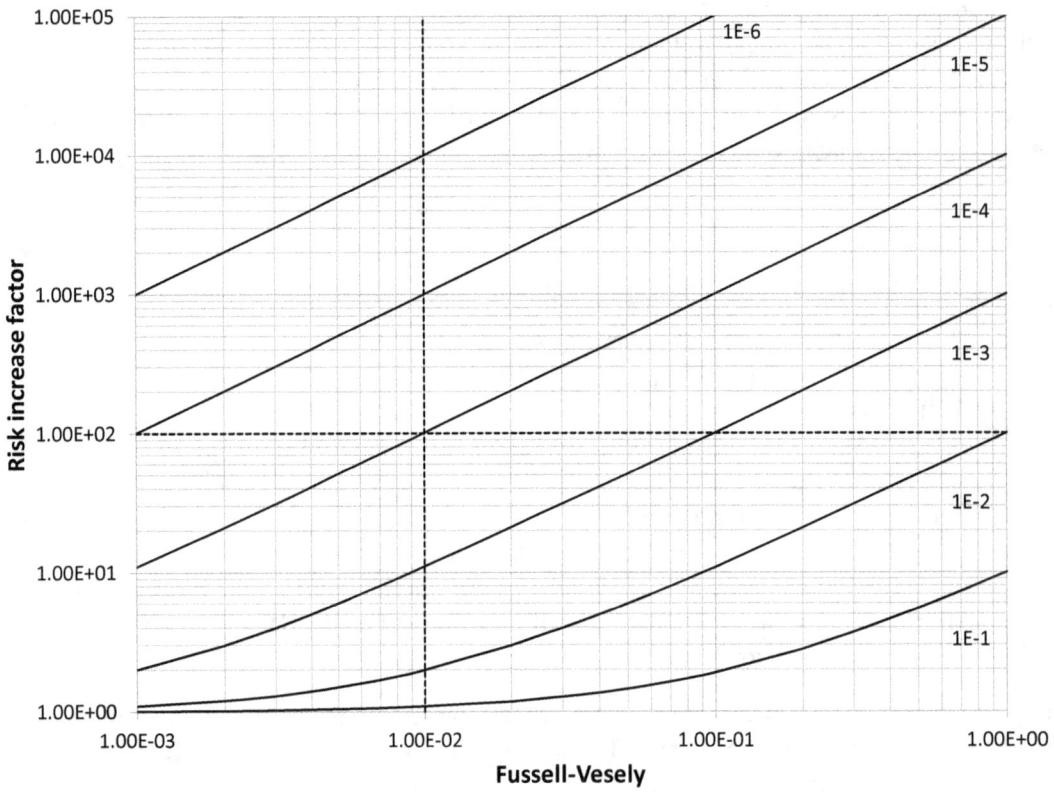

Figure 1: The black curves (called equiprobability curves in this paper) illustrate the dependence of FV and RIF as defined by equation (4). The failure probabilities X_i for each equiprobability curve are given on the right. For example, basic event whose failure probability is 1.00E-4 can only appear on the corresponding curve regardless of the system. The dashed lines illustrate arbitrary criteria by which the basic events can be categorized into high, medium and low risk areas.

4. BIRNBAUM AND FAILURE PROBABILITY MAPPING

Risk inherent to a component can be expressed in terms of failure probability of the component and the remaining safety margin if the component is to fail. This is reflected by choosing the importance measures failure probability X_i and Birnbaum as the two attributes for analyzing system changes or ranking components according to their criticality to safety. Birnbaum, see equation (3) above, is completely dependent on the structure of the system and independent of the corresponding failure probability X_i [7]. Therefore, this choice avoids the problematic issues of the dependence between FV and RIF that were discussed above and shown in equation (4). The interpretation of both importance measures is also quite straightforward or even intuitive since failure probability is a measure of reliability or quality characteristic to the component and Birnbaum measures the safety margin inherent to the remaining structure if the component fails.

These findings motivate to consider the use of failure probability X_i and Birnbaum instead of Fussell-Vesely and risk increase factor for analyzing system changes or ranking components. Figure 2 presents the X_i-B -map. The black lines represent contours of equal risk and are defined as a product of failure probability X_i and Birnbaum. Let us call these lines equirisk lines. The numerical value of risk is given next to the corresponding equirisk line. The lines divide the space into separate risk zones, e.g., high, medium and low risk zones that may be used in classification of the components. The components (or basic events) can move only horizontally or vertically, both of which have a practical explanation: Horizontal movement naturally means change in the component reliability and vertical movement changing the defense-in-depth of the system.

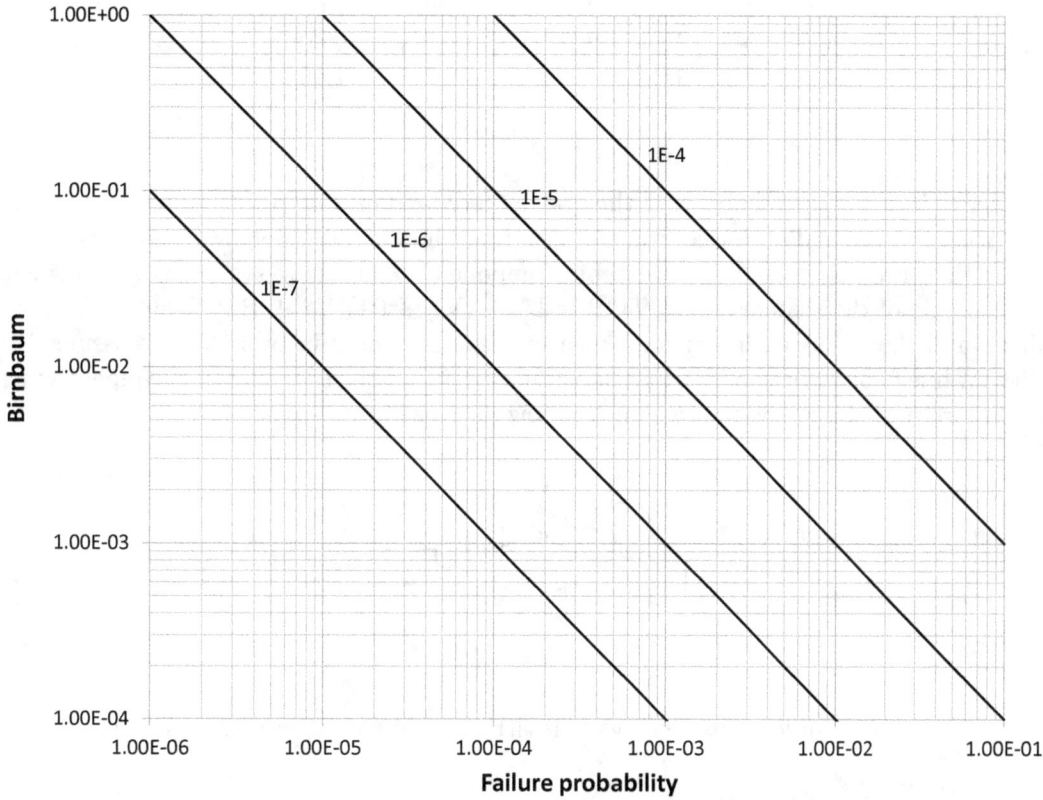

Figure 2: X_i-B -map. The black lines represent contours of equal risk (called equirisk lines in this paper) and the numerical value is given next to the corresponding equirisk line. The lines divide the space into separate risk zones, e.g., high, medium and low risk zones. The components (or basic events) can move only horizontally or vertically since X_i and B are independent on each other.

5. COMPARISON BETWEEN (FV, RIF) AND (X_I, B) IN ANALYZING SYSTEM CHANGES

The following simple example compares the two methods discussed above, i.e., utilizing importance measure pair (FV, RIF) or (X_i, B), and illustrates some potential problems that may arise when FV and RIF are used in analyzing system changes.

Consider a simple pump line system shown in figure 3 containing basic events L1 (line fails), V1 (valve fails) and P1 (pump fails). The Boolean logic of the system is TOP = L1 + V1 + P1. The basic event failure probabilities and the discussed importance measures are presented in table 1. The total failure probability for the system is 1.01E-2.

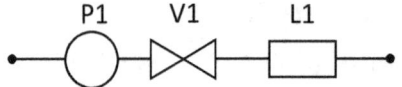

Figure 3: Simple pump line system with one pump and valve.

Table 1: Basic event (BE) failure probabilities X_i and the discussed importance measures Fussell-Vesely (FV), risk increase factor (RIF) and Birnbaum (B) for a single pump line system shown in figure 3.

BE	X_i	FV	RIF	B
L1	1E-5	9.79E-4	9.89E+1	1.00
V1	1E-4	9.79E-3	9.89E+1	1.00
P1	1E-2	9.89E-1	9.89E+1	1.00

Now, in order to increase the reliability of the system, a redundant pump P2 and valve V2 are installed in parallel with the originals. Notice from table 1 that the Fussell-Vesely values are the highest for P1 and V1, they are also the most unreliable components. The new system is shown in figure 4. The Boolean logic of the system is now TOP = L1 + (V1 + P1)(V2 + P2). The basic event failure probabilities and the discussed importance measures are presented in table 2 (it is assumed that the probability for line failure does not increase substantially although the length of the pipeline increases). The total failure probability for the system has now decreased to 2.11E-5.

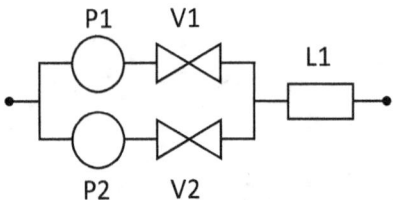

Figure 4: Simple pump line system with two redundant pumps and valves.

Table 2: Basic event (BE) failure probabilities X_i and the discussed importance measures Fussell-Vesely (FV), risk increase factor (RIF) and Birnbaum (B) for a simple pump line system with two redundant pumps and valves shown in figure 4.

BE	X_i	FV	RIF	B
L1	1E-5	4.74E-1	4.74E+4	1.00
V1	1E-4	5.21E-3	5.31E+1	1.10E-3
P1	1E-2	5.21E-1	5.26E+1	1.10E-3
V2	1E-4	4.78E-2	4.79E+2	1.01E-2
P2	1E-3	4.78E-1	4.79E+2	1.01E-2

This system change is illustrated in figures 5(a) and 5(b). In figure 5(a) each basic event is plotted on the FV-RIF -map. The black diamonds represent the original system with only one pipeline and the blue triangles the new system with two redundant lines. The black arrows show how the basic events have moved on the map due to the system change. Notice that all three basic events move along the equiprobability curves since their failure probabilities have not changed. As seen, the FV and RIF importances of basic event L1 (line fails) increase dramatically although the total failure probability for the system has changed from 1.01E-2 to 2.11E-5. The FV and RIF importances of basic events V1 and P1 decrease. In figure 5(b) each basic event is plotted on the X_i-B -map. Again the black diamonds represent the original system and blue triangles the new one. Basic events V1 and P1 move down since their Birnbaum importance decreases due to introducing redundant valve and pump. It can be seen that the change has improved the reliability of the system.

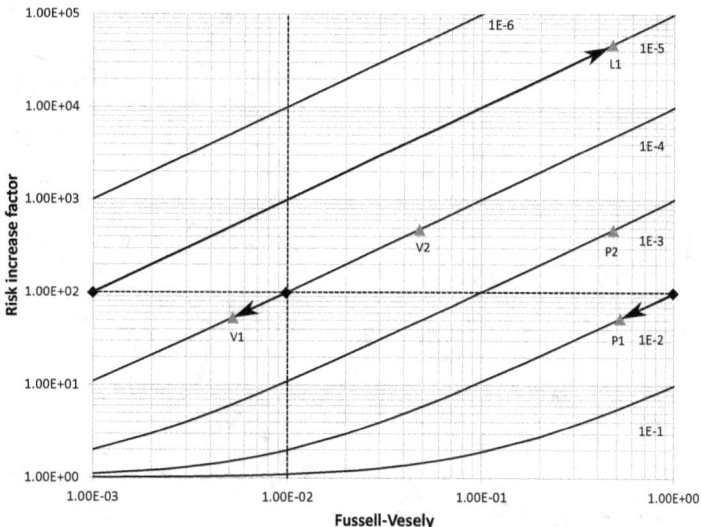

(a) FV-RIF -map. The importance measures of V1 and P1 decrease but the importance measures of L1 increase due to the system improvement. The basic events are located on and move along the equiprobability curves defined by equation (4) since FV and RIF are not independent.

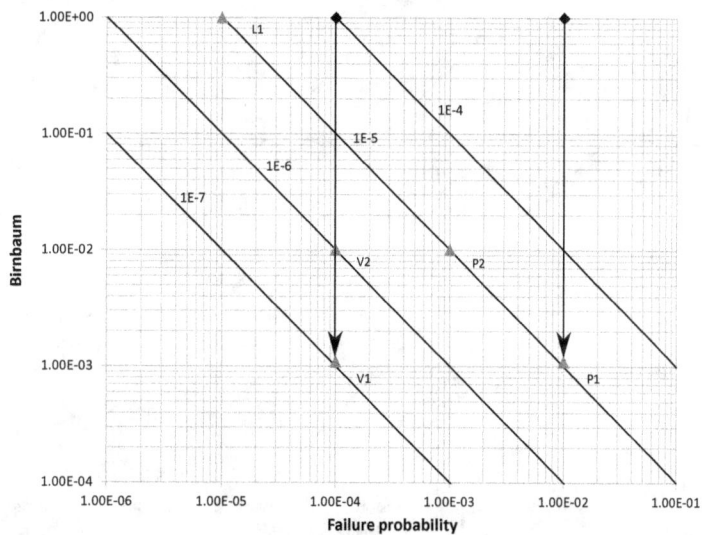

(b) X_i-B -map. The Birnbaum of V1 and P1 decreases due to the system improvement. It can be seen that the system has been improved. X_i and B are independent. After the system improvement L1, P1 and P2 are located on the same equirisk line.

Figure 5: The system is changed and improved by introducing redundant valve and pump. The black diamonds represent the original system with only one pipeline (see fig. 3) and the blue triangles the new system with two redundant lines (see fig. 4).

By using the figure 5(a) one might now consider the basic event L1 the most important but this conclusion can be misleading since L1, P1 and P2 all contain the same risk as seen in figure 5(b) (the blue triagles are located on the same equirisk line). Let us assume that L1 can be made ten times more reliable, i.e., now L1 has a failure probability of 1E-6. The total failure probability for the system decreases from 2.11E-5 to 1.21E-5 and affects the values of the importance measures. This change is illustrated in figures 6(a) and 6(b). The new values for importance measures are shown in red squares. The blue arrows show how the basic events have moved due to the reliability improvement. As seen in figure 6(a), basic events V1, P1, V2 and P2 move along the equipropability curves since their failure probability has not changed whereas L1 changes the curve due to the reliability improvement. On the other hand, it is fairly straightforward to observe the reliability improvement of L1 as well as of the system from figure 6(b).

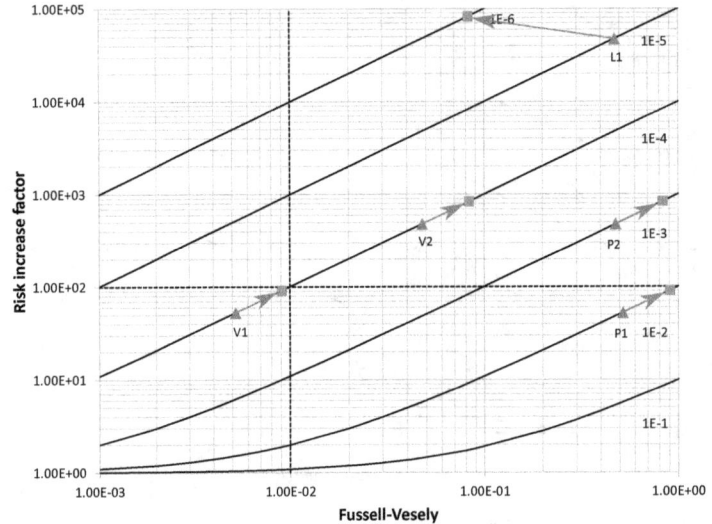

(a) FV-RIF -map. The reliability improvement of L1 increases the FV and RIF importances of the other basic events and, e.g., V1 almost moves from low risk area to high risk area.

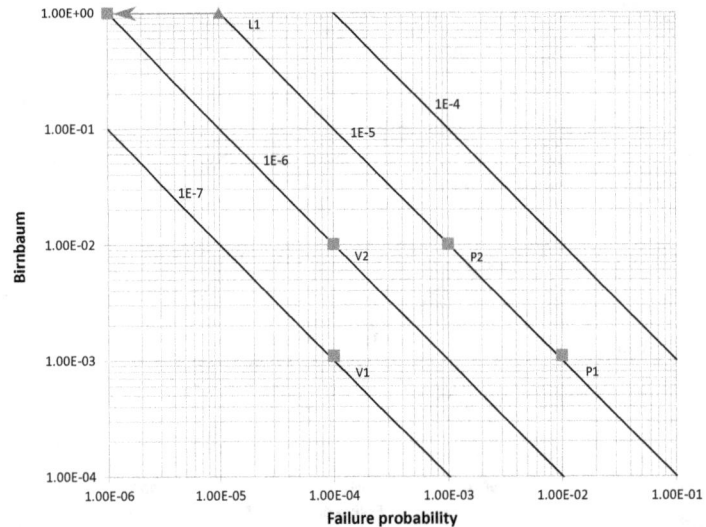

(b) X_i-B -map. The reliability improvement of L1 does not affect the other basic events.

Figure 6: The system is changed and improved by making L1 ten times more reliable. The old values for importance measures are shown in blue triangles and the new values in red squares.

Last, let us study how making P2 ten times more reliable affects the importance measures and their graphical illustration. Thus, the failure probability of P2 is changed from 1E-3 to 1E-4. The total failure probability for the system is further decreased from 1.21E-5 to 3.02E-6. Figures 7(a) and 7(b) illustrate the change. The new values for importance measures are shown in green circles and the red arrows show how the basic events move due to the reliability improvement. Figure 7(a) shows that if the criteria for the high risk area were, e.g., FV > 1.00E-1 and RIF > 1.00E+3 (instead of FV

> 1.00E-2 and RIF > 1.00E+2 as illustrated in the figure 7(a) and discussed before), the reliability improvement of P2 would actually move it from medium risk area to high risk area and V2 from low risk area directly to high risk area. On the other hand, figure 7(b) clearly shows that the decrease in the failure probability of P2 affects the Birnbaums of V1 and P1 via improving their safety margin. The importance of L1 and V2 is not influenced by the system improvement because neither their failure probability nor safety margin is changed.

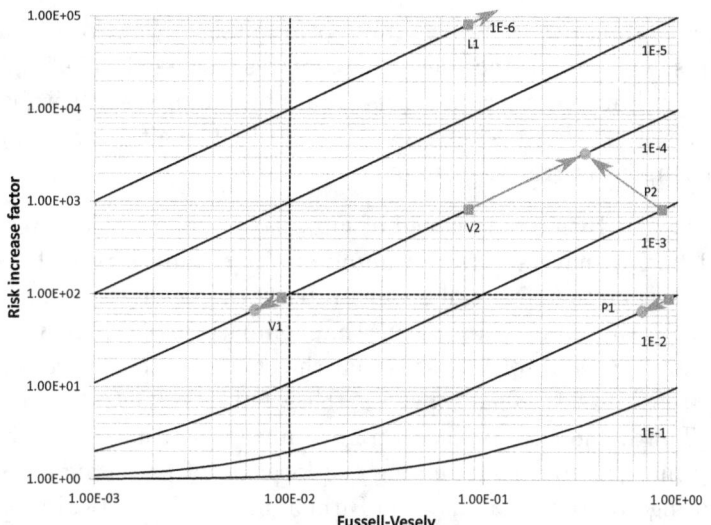

(a) FV-RIF -map. It is rather challenging to see that the system has been improved.

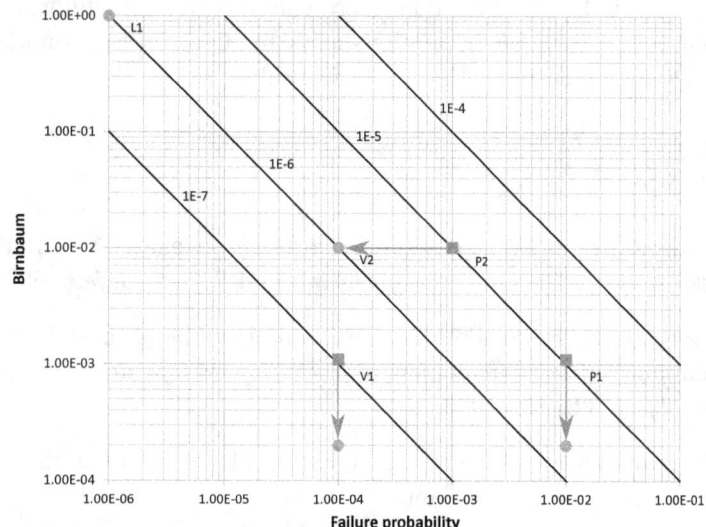

(b) X_i-B -map. Seeing the effects of reliability improvement is fairly straightforward compared to FV-RIF -map.

Figure 7: The system is changed and improved by making P2 ten times more reliable. The old values for importance measures are shown in red squares and the new values in green circles.

The simple example above demonstrated some of the ambiguities of using FV and RIF that the authors have found problematic. Utilizing FV and RIF in analyzing system changes (both component reliability improvement and changing the system failure logic) may lead to complex behavior and difficulties in interpreting the results if the feedback is taken into account in the reliability model or PRA. In particular, the equiprobability curves defined by equation (4), which shows that FV and RIF are not independent, along with the division of FV-RIF -space into four quadrants cause that some basic events may move from low risk area to high risk area even with a relatively small change.

Living PRA, system modifications or even safety classification (one goal may be to improve reliability by increasing tests, maintenance and inspections) all introduce feedback and updates to PRA model. Hence, it is relevant to consider which method should be used in analyzing them.

6. CONCLUSION

In this paper analyzing system changes, including both system failure logic changes and basic event reliability improvements, with importance measure pairs was discussed. The importance measure pairs were Fussell-Vesely and risk increase factor versus failure probability and Birnbaum. The dependence between risk increase factor and Fussell-Vesely was shown. Utilizing the two importance measure pairs was compared with a simple illustrative example.

Using failure probability and Birnbaum instead of Fussell-Vesely and risk increase factor in analyzing system changes was found to have several advantages: First, failure probability and Birnbaum are independent on each other and therefore can be illustrated orthogonally. Second, the failure probability and Birnbaum measures are absolute which means that comparison of different system configurations is evident. Another consequence of absoluteness is that the change on basic event importances expresses real change and only those basic events move on the graphical illustration that are really affected by the modification. Third, the axes of the graphical illustration have clear interpretations, failure probability and safety margin, so understanding the location of the basic events is straightforward. These features make using failure probability and Birnbaum map a practical tool for decision maker to assess wheter the system under consideration should be improved by increasing component reliability or safety margin.

The authors recognize that each importance measure contains different information and has its own use but the findings made in this paper hopefully shed some light on the complexity that lies behind using two or more importance measures in component ranking.

REFERENCES

[1] Jussi K. Vaurio. Ideas and developments in importance measures and fault-tree techniques for reliability and risk analysis. *Reliability Engineering and System Safety*, 95:99–107, 2010.

[2] Risto Himanen, Ari Julin, Kalle Jänkälä, Jan-Erik Holmberg, and Reino Virolainen. Risk-informed regulation and safety management of nuclear power plants - on the prevention of severe accidents. *Risk Analysis*, 32:1978–1993, 2012.

[3] Radiation and Nuclear Safety Authority (STUK). YVL A.7 Probabilistic risk assessment and risk management of a nuclear power plant.

[4] U.S. Nuclear Regulatory Commission. Regulatory guide 1.160: Monitoring the effectiveness of maintenance at nuclear power plants.

[5] Nuclear Energy Institute. NUMARC 93-01, revision 4A: Industry guideline for monitoring the effectiveness of maintenance at nuclear power plants, 2011.

[6] H. Duncan Brewer and Ken S. Canady. Probabilistic safety assessment support for the maintenance rule at duke power company. *Reliability Engineering and System Safety*, 63:243–249, 1999.

[7] M. van der Borst and H. Schoonakker. An overview of PSA importance measures. *Reliability Engineering and System Safety*, 71:241–245, 2001.

[8] Michael C. Cheok, Gareth W. Parry, and Richard R. Sherry. Use of importance measures in risk-informed regulatory applications. *Reliability Engineering and System Safety*, 60:213–226, 1998.

[9] W. E. Vesely. Supplemental viewpoints in the use of importance measures in risk-informed regulatory applications. *Reliability Engineering and System Safety*, 60:257–259, 1998.

[10] Michael C. Cheok, Gareth W. Parry, and Richard R. Sherry. Response to 'Supplemental viewpoints on the use of importance measures in risk-informed regulatory applications'. *Reliability Engineering and System Safety*, 60:261, 1998.

[11] E. Borgonovo and G. E. Apostolakis. A new imporatance measure for risk-informed decision making. *Reliability Engineering and System Safety*, 72:193–212, 2001.

[12] E. Borgonovo. Differential, criticality and Birnbaum importance measures: An application to basic event, groups and SSCs in event trees and binary decision diagrams. *Reliability Engineering and System Safety*, 92:1458–1467, 2007.

[13] Mohammad Modarres. *Risk Analysis in Engineering: Techniques, Tools, and Trends*. Taylor & Francis Group, 2006.

Energy loss optimization in basic T-shaped water supply piping networks for probabilistic demands

KW Mui[a], LT Wong[a], and CT Cheung[a]
[a] Department of Building Service Engineering,
The Hong Kong Polytechnic University,
Hong Kong, China

Abstract: Minimization of energy loss of water supply networks is a major concern of pump power reduction for sustainable water systems in buildings. This paper presents a mathematical model for energy loss optimization in a common basic T-shaped water supply piping network that serves infinite probabilistic demands. Optimized designs based on proper network pipe sizes are analyzed. Optimal pipe radius ratios ($2^{1/7}$ to $2^{3/7}$) and their corresponding energy implications in the network are also discussed. The results show that existing piping designs are not optimized for probabilistic demands and there is potential for energy loss reduction.

Keywords: Buildings, Energy loss, Piping Network, Probabilistic Demands, Water Supply System

List of symbols

C, c	Constant and its matrix element
E, e	Energy and its matrix element (J)
F	Function as defined
f	Friction factor (-)
i, j, l	Dummy variables as defined
L	Length (m)
m	Number of independently operated appliance
n	State of a binary operated appliance
P, p	Pressure and its matrix element (Pa)
ϑ, φ	Demand probability and its matrix element (-)
r	Radius (m)
V, v	Volumetric demand flow rate and its matrix element ($m^3 s^{-1}$)
Δ	Change of
$[\ ], \{\ \}$	Matrix notation and matrix element group
ρ	Density (kg m^{-3})
λ	Dummy parameter group as defined

Superscript
'	Derivatives

Subscript
$0, 1, 2, 3, 4, \ldots$	Of conditions 0, 1, 2, 3, 4, ... as defined
c	Of constant
i, j, l, m, n	Of i-th, j-th, l-th, m-th, n-th conditions as defined
max	Of maximum value
opt	Of optimum
R	Of pipe radius ratio
V, v	Of volumetric flow rate

1. INTRODUCTION

Potable water demands are unsteady, random and intermittent [1]. As survey studies indicated that the end-use water demand probability is generally less than 0.1 even during daily rush hours [2,3], demand overload is legitimate in the design of water supply pipe network inside the building. Statistical methods for estimating appliance usage patterns and associated instantaneous water demands have been developed to size mains water pipes [4]. To ensure supply certainty rather than optimality, the existing design approach to determine fixture units may overestimate the simultaneous water demand and oversize water pipes [5]. Alternatively, Monte Carlo simulations can be used to decide the failure probability density function of the water supply system, which is influenced by the occupant load profile, for meeting the demand and assessing the performance [6].

Distributing a supply of water as uniformly as possible over a territory through piping networks is a classical problem of optimization. An urban water supply study showed that 45% of the total pumping energy needed to deliver water from the treatment plants to households was consumed inside buildings [7]. Apart from building height, energy loss at supply pipes is another concern. Under steady conditions, Bejan's constructal law of the generation of flow configuration is a proven useful tool for optimizing the geometric layout of schemes by minimizing pumping power requirements for distributing water uniformly over an area [8,9]. However, with respect to the optimal operation of water supply networks in buildings where flow rates are unsteady, there are no existing models systematically optimizing energy efficiency and interrelated issues.

This paper dealt with the energy loss optimization problems of a common basic T-shaped water supply piping network that serves infinite probabilistic demands and established a mathematical model for the required minimum water pumping energy under a fixed pipe volume constraint. Energy loss reduction potential through proper pipe size was also investigated. The results were discussed in terms of their implications for theory and practice.

2. ENERGY LOSS

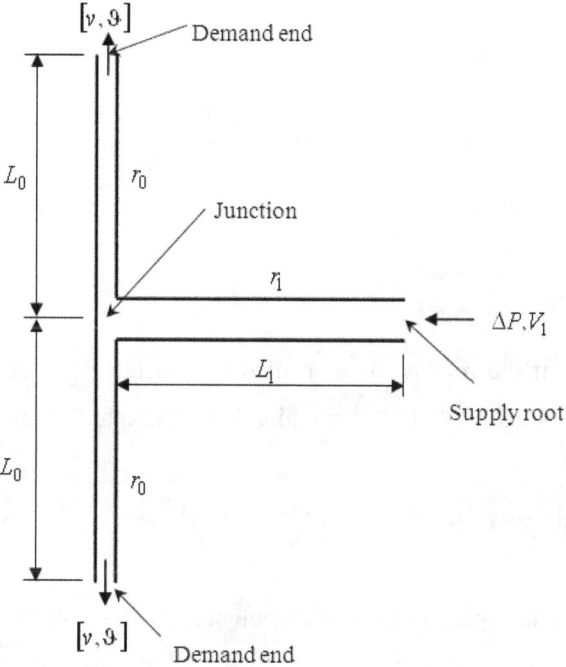

Figure 1. A basic T-shaped water supply piping network

Figure 1 shows a basic T-shape water supply piping network that a branch pipe of length L_0 and of radius r_0, fed by a centre main water pipe of length L_1 and radius r_1. The network consists a supply root and two demand ends as in a T-shape construct [8]. A demand in the network is due to a number of independently operated appliances $n_1, n_2, ... n_m$ at each of the two demand ends and each appliance demand is characterised by a constant flow rate $v = C_v$ and a probability $\varphi = \vartheta$. The flow rate between two consecutive demands is assumed zero with a probability $\varphi = 1 - \vartheta$. As an appliance is either 'in demand' ($n=1$) or 'not in demand' ($n=0$), an appliance demand can be described by,

$$n_i : \{0,1\}; \ v_i : \{0, v_i\}; \ \varphi_i : \{1 - \vartheta_i, \vartheta_i\}; \ i = 1,2,...,m \tag{1}$$

Appliances are arranged according to their 'in demand' flow rates $[v_1, \vartheta_1]$, $[v_2, \vartheta_2]$, ..., $[v_m, \vartheta_m]$ such that $v_1 \leq v_2 \leq ... \leq v_m$ and expressed the flow rates in terms of $v_1 = C_v$ and $v = \left[C_v \left\{ 0, \frac{v_1}{v_1}, \frac{v_2}{v_1}, ..., \frac{v_{m-1}}{v_1}, \frac{v_m}{v_1} \right\} \right]$, then a demand at the demand end is given by,

$$v = C_v \{ 0, 1, C_{V_2}, ..., C_{V_{m-1}}, C_{V_m} \}; \ C_{V_1} = 1 \leq C_{V_2} \leq ... \leq C_{V_{m-1}} \leq C_{V_m} \tag{2}$$

The demands $v_{i,j}$ at the two branches i, j fed by the m binarily operated appliances are probabilistic. There are 2^m combinations of demands in each branch pipe, and $(2^m)^2$ cases at the junction where two branches meet, throughout the centre pipe or at the root. For a pair of demand ends i, j, the demand probability at any instant is $\varphi_i \varphi_j$ (denoted as φ_{ij}) and the corresponding demand probabilities are expressed by,

$$\varphi = \left\{ \varphi_{11}, ..., \varphi_{(1)(2^m)}, ..., \varphi_{(2^m)(1)}, ..., \varphi_{(2^m)(2^m)} \right\}$$
$$\varphi_{ij} = \varphi_i \varphi_j; \ \varphi_i, \varphi_j = \prod_{l=1}^{m} \vartheta_l^{n_l} (1 - \vartheta_l)^{1-n_l} \tag{3}$$

i and j are dependent on the number of end demands m,

$$i, j = 1 + \sum_{l=1}^{m} 2^{l-1} n_l; \ n_l = [0,1] \tag{4}$$

The centre pipe demand at the root is the sum of demands due to branch pipes, and given by an expression below, where a total of $(2^m)(2^m)$ combinations of demand pairs are encountered,

$$V_1 = V_1[v, \varphi]; \ v = \left\{ v_{11}, ..., v_{1(2^m)}, ..., v_{(2^m)(1)}, ..., v_{(2^m)(2^m)} \right\}; \ v_{ij} = v_i + v_j \tag{5}$$

Taking $C_0 = \dfrac{f \rho C_v^2}{\pi^2}$ as the unit pipe friction of turbulent flow, the total pressure loss of the network ΔP is determined by the following equations, where ΔP_0 and ΔP_1 are the pressure losses in the branch and centre pipes respectively,

$$\Delta P[p,\varphi] = \Delta P_0[p,\varphi] + \Delta P_1[p,\varphi] \; ; \; p = \left\{ p_{11},\ldots,p_{1(2^m)},\ldots,p_{(2^m)(1)},\ldots,p_{(2^m)(2^m)} \right\}$$

$$\Delta P_0 : p_{ij} = \begin{cases} 0 & ; ij = 11 \\ \dfrac{C_0 L_0}{C_v^2 r_0^5} v_{i,j}^2 & ; ij \neq 11 \; ; \; v_{i,j} = \max(v_i, v_j) \end{cases} \; ; \; \Delta P_1 ; p_{ij} = \dfrac{C_0}{C_v^2} \dfrac{L_1}{r_1^5} v_{ij}^2 \quad (6)$$

It is noted for the above expression, the pressure required at the junction between i and j is maintained for the higher flow rate $\max(v_i, v_j)$.

As the pressure loss at the network is flow rate dependent and hence transient, the energy required for the pressure loss at the demands is chosen as the optimization parameter. The minimum energy E required for the water supply network is given by,

$$E[e,\varphi] = \Delta PV_1 \; ; \; e = \left\{ e_{11},\ldots,e_{1(2^m)},\ldots,e_{(2^m)(1)},\ldots,e_{(2^m)(2^m)} \right\} \; ; \; e_{ij} = p_{ij} v_{ij} \; ;$$

$$E = e_{11}\varphi_{11} + \ldots e_{1(2^m)}\varphi_{1(2^m)} + \ldots e_{(2^m)1}\varphi_{(2^m)1} + \ldots e_{(2^m)(2^m)}\varphi_{(2^m)(2^m)} = \sum_{ij} e_{ij} \varphi_{ij} \quad (7)$$

For a constant pipe volume C, the energy in Eq. (7) can be minimized by choosing a proper pipe radius ratio $C_R = (r_1 / r_0)$ for pressure terms p_{ij} in Eq. (6) such that,

$$2L_0 r_0^2 + L_1 r_1^2 = C \quad (8)$$

$$r_1 = \left(\dfrac{C}{L_1} - \dfrac{2L_0}{L_1} r_0^2 \right)^{1/2} \quad (9)$$

Let F be the frictional losses in the branch and centre pipes, where λ is an arbitrary constant in the pressure terms in Eq. (6), the general solution for the optimal pipe radius ratios $C_{R,opt}$ can be determined by taking derivative $F' = 0$,

$$F = \lambda_0 L_0 r_0^{-5} + \lambda_1 L_1 r_1^{-5} = \lambda_0 L_0 r_0^{-5} + \lambda_1 L_1 \left(\dfrac{C}{L_1} - \dfrac{2L_0}{L_1} r_0^2 \right)^{-5/2} \quad (10)$$

$$F' = -5\lambda_0 L_0 r_0^{-6} + \lambda_1 L_1 \left(\dfrac{-5}{2} \right) \left(\dfrac{C}{L_1} - \dfrac{2L_0}{L_1} r_0^2 \right)^{1/2(-7)} \left(-\dfrac{4L_0}{L_1} r_0 \right) = 0 \quad (11)$$

$$-5\lambda_0 L_0 r_0^{-7} + 5 \left(\dfrac{2\lambda_1 L_1 L_0}{L_1} \right) r_1^{-7} = 0 \quad (12)$$

$$C_{R,opt} = \dfrac{r_1}{r_0} = \left(\dfrac{\lambda_0}{2\lambda_1} \right)^{-1/7} = \left(\dfrac{2\lambda_1}{\lambda_0} \right)^{1/7} \quad (13)$$

The optimal pipe radius ratios $C_{R,opt}$ for Eq. (6) with probabilities φ_{ij} expressed below can be determined by taking derivative $C'_{R,opt} = 0$ with an optimum demand probability ϑ_{opt},

$$C_R = [c,\varphi] \; ; \; c = \left\{ c_{11},\ldots,c_{1(2^m)},\ldots,c_{(2^m)(1)},\ldots c_{(2^m)(2^m)} \right\} \quad (14)$$

$$C_{R,opt} = \sum_{ij} c_{ij} \varphi_{ij} (\vartheta_{opt}) \quad (15)$$

3. OPTIMAL PIPE RADIUS RATIOS

Three cases for c_{ij} in calculating the pressure loss at pipe r_1 and r_0 are given below,

$$c_{ij} = \left(\frac{2\lambda_1}{\lambda_0}\right)^{1/7}_{ij} = \begin{cases} 2^{1/7} \\ 2^{1/7}\left(\frac{v_i + v_j}{max(v_i, v_j)}\right)^{2/7} \\ 2^{3/7} \end{cases} ; v_{ij} = \begin{cases} v_i, v_j \to 0 \\ max\{v_i, v_j\} \quad ; v_i, v_j \neq 0 \\ v_i + v_j \end{cases} \quad (16)$$

It is noted for all positive demands, the term $1 \leq \left(\frac{v_i + v_j}{max(v_i, v_j)}\right) \leq 2$ for all demand probabilities φ_{ij}, the optimal pipe radius ratio $C_{R,opt}$ exists only in a range between $2^{1/7}$ and $2^{3/7}$, i.e.

$$2^{1/7} \leq C_{R,opt} \leq 2^{3/7} \quad (17)$$

According to an earlier study, the case of demand probability $\vartheta = 1$ is actually a steady flow condition and the optimal pipe radius ratio determined for it is $2^{3/7}$ [8].

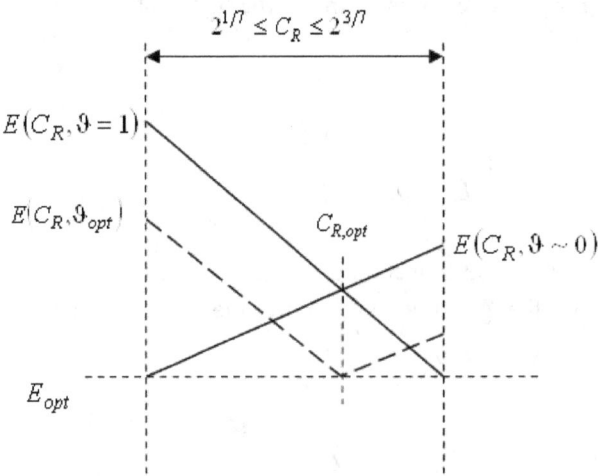

Figure 2. Relative pumping energy

Figure 2 illustrates the schematic relationships between the energy loss E for probabilistic demands and pipe radius ratio $2^{1/7} \leq C_R \leq 2^{3/7}$. Pipe radii at $C_R = 2^{3/7}$ optimized for the maximum demand (at the demand probabilities $\vartheta_{ij} = 1$) are not energy loss optimized for the cases of minimum demands (i.e. $\vartheta_{ij} \sim 0$) and vice versa. An optimized probability ϑ_{opt} exists as the energy loss at the boundaries of C_R.

The minimum energy loss E_{opt} can be derived by the two extreme demand cases, i.e. $\vartheta_{ij} \sim [0,1]$. The optimal probability, given by F below, is determined by taking such that,

$$F = c_{11}\varphi_{11} + c_{(2^m)(2^m)}\varphi_{(2^m)(2^m)} \quad (18)$$

$$F = 2^{3/7}(1 - \vartheta_{ij})^{2m} + 2^{3/7}\vartheta_{ij}^{2m} \quad (19)$$

$$F' = -2^{3/7}(2m)(1-\vartheta_{ij})^{2m-1} + 2^{3/7}(2m)\vartheta_{ij}^{2m-1} = 0 \qquad (20)$$

$$\vartheta_{opt} = 2^{-1} \qquad (21)$$

$C_{R,opt}$ can be determined for the two extreme demand cases (i.e. $C_{R,opt}$) using the general equations of order $m = 1$ and validated via the general equations of order $m = 2$. It should be noted that Eq. (20) approximates the optimum without taking the influences of demand flow rates C_{V_i} in elements c_{ij}, where $ij \neq (1)(1),(2^m)(2^m)$, into account. This point shall be examined in the validation section below.

4. OPTIMAL PIPE RADIUS RATIOS (USING GENERAL EQUATIONS OF ORDER $m = 1$)

When $m = 1$, water is supplied to two identical binary operated appliances $n : \{0,1\}$, one on each side of the T-shaped piping network. Hence, there are $2^m = 2$ demand combinations in each branch pipe and $(2^m)^2 = 4$ demand combinations at the junction, throughout the centre pipe or at the root of the network.

$$n_1 : \{0,1\} \; ; \; v : C_v\{0,1\} \; ; \; \varphi : \{1-\vartheta_1, \vartheta_1\} \qquad (22)$$

Centre pipe demand V_1 is determined by $V_1 = V_1[v, \varphi]$,

$$v = \{v_{11}, v_{12}, v_{21}, v_{22}\} = \{v_1 + v_1, v_1 + v_2, v_2 + v_1, v_2 + v_2\} = C_v\{0,1,1,2\} \qquad (23)$$

For the binary operated appliance n_1,

$$n_1 = \begin{cases} 0 & ; i,j = 1 \\ 1 & ; i,j = 2 \end{cases}; \; \varphi_1 = (1-\vartheta_1) \; ; \varphi_2 = \vartheta_1 \qquad (24)$$

$$\varphi = \{\varphi_{11}, \varphi_{12}, \varphi_{21}, \varphi_{22}\} = \{\varphi_1\varphi_1, \varphi_1\varphi_2, \varphi_2\varphi_2, \varphi_2\varphi_2\}$$
$$= \{(1-\vartheta_1)^2, \vartheta_1(1-\vartheta_1), \vartheta_1(1-\vartheta_1), \vartheta_1^2\} \qquad (25)$$

Taking the total pressure loss of the network $C_0 = \dfrac{f\rho C_v^2}{\pi^2}$, the total pressure losses of the network, branch and centre pipes ΔP, ΔP_0, ΔP_1 are determined by the following equations,

$$\Delta P[p, \varphi] = \Delta P_0[p, \varphi] + \Delta P_1[p, \varphi] \; ; \; p = \{p_{11}, p_{12}, p_{21}, p_{22}\} \; ;$$

$$\Delta P_0 : p = \left\{0, \frac{C_0 L_0}{C_v^2 r_0^5}v_2^2, \frac{C_0 L_0}{C_v^2 r_0^5}v_2^2, \frac{C_0 L_0}{C_v^2 r_0^5}v_2^2\right\} = C_0\left\{0, \frac{L_0}{r_0^5}, \frac{L_0}{r_0^5}, \frac{L_0}{r_0^5}\right\}$$

$$\Delta P_1 : p = \left\{0, \frac{C_0 L_1}{C_v^2 r_1^5}v_2^2, \frac{C_0 L_1}{C_v^2 r_1^5}v_2^2, \frac{C_0 L_1}{C_v^2 r_1^5}(2v_2)^2\right\} = C_0\left\{0, \frac{L_1}{r_1^5}, \frac{L_1}{r_1^5}, \frac{4L_1}{r_1^5}\right\}$$

$$\Delta P : p = C_0\left\{0, \frac{L_0}{r_0^5} + \frac{L_1}{r_1^5}, \frac{L_0}{r_0^5} + \frac{L_1}{r_1^5}, \frac{L_0}{r_0^5} + 4\frac{L_1}{r_1^5}\right\} \qquad (26)$$

The energy loss E at the network is expressed by,

$$E : e = \{e_{11}, e_{12}, e_{21}, e_{22}\} = C_0 C_v \left\{ 0, \left(\frac{L_0}{r_0^5} + \frac{L_1}{r_1^5} \right), \left(\frac{L_0}{r_0^5} + \frac{L_1}{r_1^5} \right), 2\left(\frac{L_0}{r_0^5} + 4\frac{L_1}{r_1^5} \right) \right\}$$

$$E = C_0 C_v \left(0 + 2(1-\vartheta_1)\vartheta_1 \left(\frac{L_0}{r_0^5} + \frac{L_1}{r_1^5} \right) + 2\vartheta_1^2 \left(\frac{L_0}{r_0^5} + 4\frac{L_1}{r_1^5} \right) \right) \qquad (27)$$

Energy loss E is optimized under a given pipe volume constraint C. Taking $\vartheta_{opt} = 2^{-1}$, the optimal pipe radius ratio $\vartheta_{opt} = 2^{-1}$, is given by,

$$C_R : c = \{ 2^{3/7}, 2^{1/7}, 2^{1/7}, 2^{3/7} \} \qquad (28)$$

$$C_R = 2^{3/7}(1-\vartheta_1)^2 + (2^{1/7})2(1-\vartheta_1)\vartheta_1 + 2^{3/7}\vartheta_1^2 \qquad (29)$$

$$C_R' = 2^{3/7}(1-\vartheta_1)^2 + (2^{1/7})2(1-\vartheta_1)\vartheta_1 + 2^{3/7}\vartheta_1^2 = 0 \qquad (30)$$

$$C_{R,opt} = 2^{3/7}(1-\vartheta_{opt})^2 + (2^{1/7})2(1-\vartheta_{opt})\vartheta_{opt} + 2^{3/7}\vartheta_{opt}^2 = 2^{2/7} \qquad (31)$$

Figure 3 illustrates the relative energy loss E/E_{opt} required at a T-shaped water supply piping network when $m = 1$ and Figure 4 is the corresponding graph of optimal pipe radius ratio against demand probability. They confirm the range of optimal pipe radius ratios $C_{R,opt} \in [2^{1/7}, 2^{3/7}]$ at the boundary conditions illustrated in Figure 2, i.e. the minimum and maximum demand probabilities are $\vartheta \sim 0$ and $\vartheta \sim 0$ respectively.

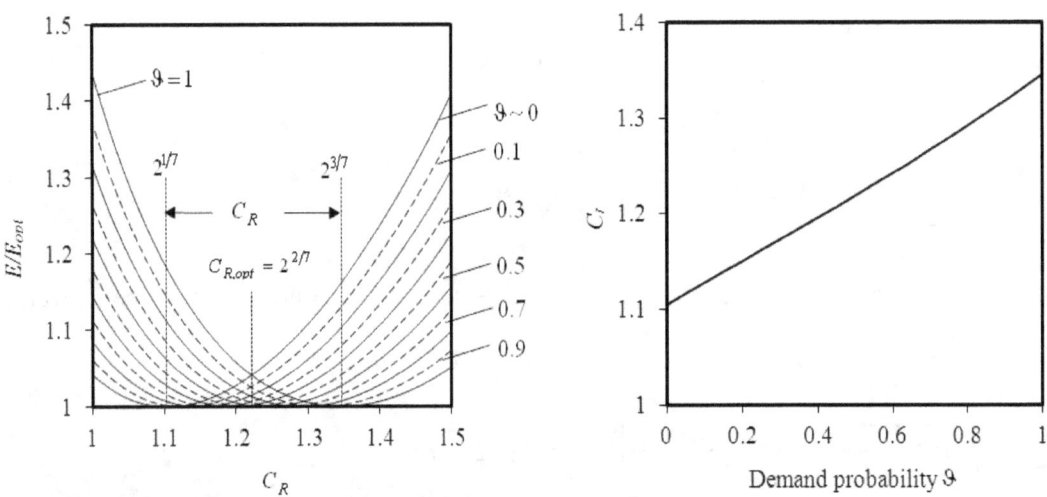

Figure 3. Relative energy loss at a T-shaped water supply piping network when $m = 1$

Figure 4. Optimal pipe radius ratios, $m = 1$

Eq. (30) gives $\vartheta_{opt} = 0.5$, which when substituted into Eq. (29) yields $\vartheta_{opt} = 0.5$, as shown in Figure 3. If $C_{R,opt} = 2^{2/7}$ is applied to all demand probabilities, up to 4.2% more energy loss can be produced as compared with any optimal cases with a single set of demand probabilities. If $C_{R,opt} = 2^{3/7}$ is optimized for the steady flow, the additional energy loss is 13-16% as compared with the optimal cases where demand probabilities are in between 0.001 and 0.1.

5. OPTIMAL PIPE RADIUS RATIO VALIDATION (USING GENERAL EQUATIONS OF ORDER $m = 2$)

Validity of the pipe radius ratio $C_{R,opt} = 2^{2/7}$ is tested in this case. When $m = 2$, two binary operated appliances (n_1, n_2) demand probabilities $(\vartheta_1, \vartheta_2)$ and thus two demand flow rates $(C_v, C_v C_V)$ exist at each demand end, i.e.

$$n_1 : \{0,1\} \; ; \; v : C_v \{0,1\} \; ; \; \varphi : \{1-\vartheta_1, \vartheta_1\} \tag{32}$$

$$n_2 : \{0,1\} \; ; \; v : C_v C_V \{0,1\} \; ; \; \varphi : \{1-\vartheta_2, \vartheta_2\} \tag{33}$$

There are $4 (= 2^m)$ demand combinations in each branch pipe,

$$v = \{v_1, v_2, v_3, v_4\} = \{0, 1, C_V, 1+C_V\} \tag{34}$$

Correspondingly, there are $4 \times 4 = 16 (= 2^{2m})$ demand combinations at the junction, in the centre pipe and at the root.

For the binary operated appliances n_1, n_2, the centre pipe demand n_1, n_2, is, with matrix elements of v, φ are given in Table 1,

$$v = \{v_{11}, v_{12}, v_{13}, v_{14}, v_{21}, ..., v_{44}\} \tag{35}$$

$$i, j = 1 + n_1 + 2n_2; \; n_{1,2} = [0,1] \tag{36}$$

$$\varphi = \{\varphi_{11}, \varphi_{12}, ..., \varphi_{44}\}; \; \varphi_{ij} = \varphi_i \varphi_j; \; \varphi_{i,j} = \vartheta_1^{n_2}(1-\vartheta_1)^{1-n_2} \vartheta_2^{n_1}(1-\vartheta_2)^{1-n_1} \tag{37}$$

Table 1. Matrix elements for general equations of order $m = 2$

ij	$i : n_2, n_1$; $j : n_2, n_1$	$C_v^{-1}[v_i, v_j]$	$C_v^{-1} v_{ij}$	$\varphi_{ij} = \varphi_i \varphi_j$	$\Delta P_0 : C_v^{-1} C_0^{-1} p_{ij}$	$\Delta P_1 : C_v^{-1} C_0^{-1} p_{ij}$	$C_v^{-1} C_0^{-1} e_{ij}$	c_{ij}
11	0000	0,0	0	$(1-\vartheta_1)^2 (1-\vartheta_2)^2$	0	0	0	$2^{3/7}$
12	0001	0,1	1	$(1-\vartheta_1)^2 \vartheta_2(1-\vartheta_2)$	$L_0 r_0^{-5}$	$L_1 r_1^{-5}$	$L_0 r_0^{-5} + L_1 r_1^{-5}$	$2^{1/7}$
13	0010	0, C_V	C_V	$(1-\vartheta_1) \vartheta_1(1-\vartheta_2)^2$	$C_V^2 L_0 r_0^{-5}$	$C_V^2 L_1 r_1^{-5}$	$C_V^3 \left(L_0 r_0^{-5} + L_1 r_1^{-5} \right)$	$2^{1/7}$
14	0011	0, $1+C_V$	$1+C_V$	$\vartheta_1(1-\vartheta_1) \vartheta_2(1-\vartheta_2)$	$(1+C_V)^2 L_0 r_0^{-5}$	$(1+C_V)^2 L_1 r_1^{-5}$	$(1+C_V)^3 \left(L_0 r_0^{-5} + L_1 r_1^{-5} \right)$	$2^{1/7}$
21	0100	1,0	1	same with $ij = 12$				
22	0101	1,1	2	$(1-\vartheta_1)^2 \vartheta_2^2$	$L_0 r_0^{-5}$	$4 L_1 r_1^{-5}$	$2 \left(L_0 r_0^{-5} + 4 L_1 r_1^{-5} \right)$	$2^{3/7}$

23	0110	$1, C_V$	$1+C_V$	$\vartheta_1(1-\vartheta_1)$ $\vartheta_2(1-\vartheta_2)$	$C_V^2 L_0 r_0^{-5}$	$(1+C_V)^2 L_1 r_1^{-5}$	$(1+C_V)\left(C_V^2 L_0 r_0^{-5} + (1+C_V)^2 L_1 r_1^{-5}\right)$	$\dfrac{2^{1/7}(1+C_V)^{2/7}}{C_V^{2/7}}$
24	0111	$1, C_V$	$2+C_V$	$\vartheta_1(1-\vartheta_1)\vartheta_2^2$	$(1+C_V)^2 L_0 r_0^{-5}$	$(2+C_V)^2 L_1 r_1^{-5}$	$(2+C_V)\left((1+C_V)^2 L_0 r_0^{-5} + (2+C_V)^2 L_1 r_1^{-5}\right)$	$\dfrac{2^{1/7}(2+C_V)^{2/7}}{(1+C_V)^{2/7}}$
31	1000	$C_V, 0$	C_V	same with $ij=13$					
32	1001	$C_V, 1$	$1+C_V$	same with $ij=23$					
33	1010	C_V, C_V	$2C_V$	$\vartheta_1^2(1-\vartheta_2)^2$	$C_V^2 L_0 r_0^{-5}$	$4C_V^2 L_1 r_1^{-5}$	$2C_V^3\left(L_0 r_0^{-5} + 4L_1 r_1^{-5}\right)$	$2^{3/7}$
34	1011	$C_V, 1+C_V$	$1+2C_V$	$\vartheta_1^2 \vartheta_2(1-\vartheta_2)$	$(1+C_V)^2 L_0 r_0^{-5}$	$(1+2C_V)^2 L_1 r_1^{-5}$	$(1+2C_V)\left((1+C_V)^2 L_0 r_0^{-5} + (1+2C_V)^2 L_1 r_1^{-5}\right)$	$\dfrac{2^{1/7}(1+2C_V)^{2/7}}{(1+C_V)^{2/7}}$
41	1100	$1+C_V, 0$	$1+C_V$	same with $ij=14$					
42	1101	$1+C_V, 1$	$2+C_V$	same with $ij=24$					
43	1110	$1+C_V, C_V$	$1+2C_V$	same with $ij=34$					
44	1111	$1+C_V, 1+C_V$	$2(1+C_V)$	$\vartheta_1^2 \vartheta_2^2$	$(1+C_V)^2 L_0 r_0^{-5}$	$4(1+C_V)^2 L_1 r_1^{-5}$	$2(1+C_V)^3\left(L_0 r_0^{-5} + 4L_1 r_1^{-5}\right)$	$2^{3/7}$

The pressure loss required at the network ΔP and the corresponding pressure loss are given by, where the matrix elements p_{ij}, e_{ij} are summarized in Table 1 for easy reference,

$$\Delta P[p,\varphi] = \Delta P_0[p,\varphi] + \Delta P_1[p,\varphi] \; ; \; p = \{p_{11}, p_{12}, p_{13}, p_{14}, p_{21}, \ldots, p_{44}\} \tag{38}$$

$$E : e = \{e_{11}, e_{12}, \ldots, e_{44}\} \; ; \; e_{ij} = (\Delta P : p_{ij})(V_1 : v_{ij}) \tag{39}$$

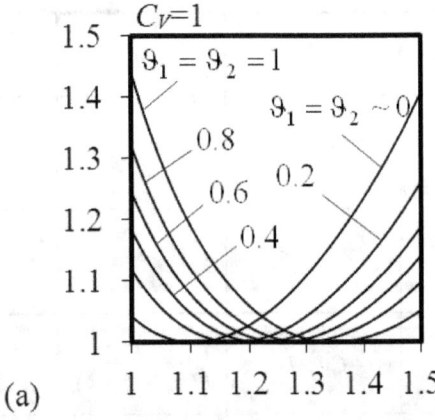

(a)

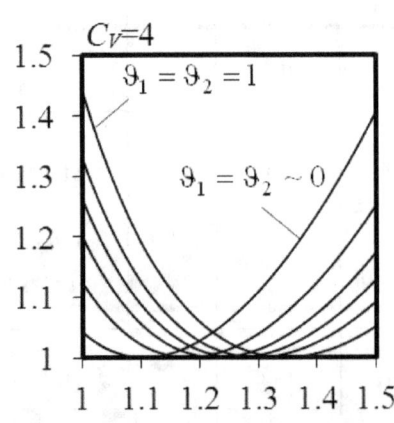

(b)

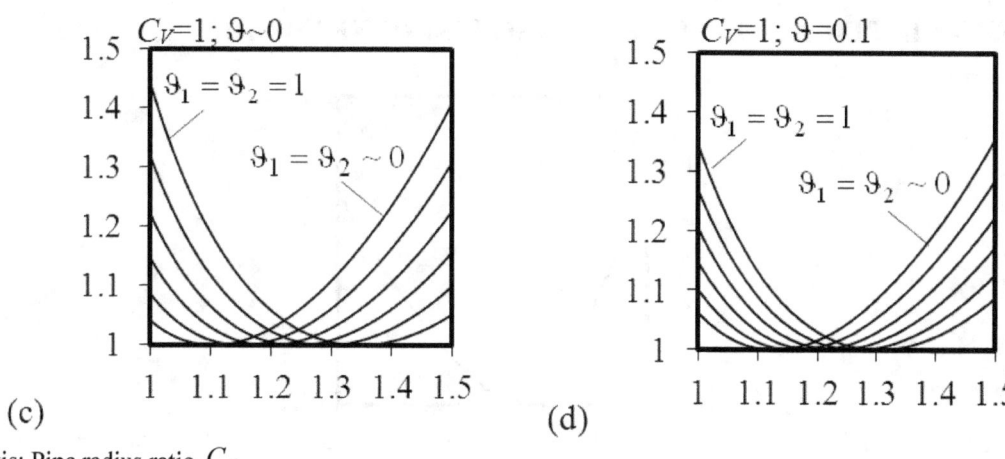

x-axis: Pipe radius ratio C_R

y-axis: Relative pumping energy E/E_{opt}

Figure 5. Relative energy loss for a T-shaped water supply piping network when $m = 2$

Figure 5 presents the relative energy loss for some demand cases while Figure 6 exhibits the optimal pipe radius ratios for various sets of demand flow rates and demand probability combinations. Again, they confirm the validity of the $C_{R,opt}$ range and boundary conditions as in case $m = 1$. It can be seen that demand flow rates $(C_V > 1)$ have some influences on the middle range of demand probabilities (e.g. $\vartheta = 0.1$ to 0.9), but not on $\vartheta > 0.9$ (almost steady flow) or $\vartheta < 0.1$ (almost minimal flow).

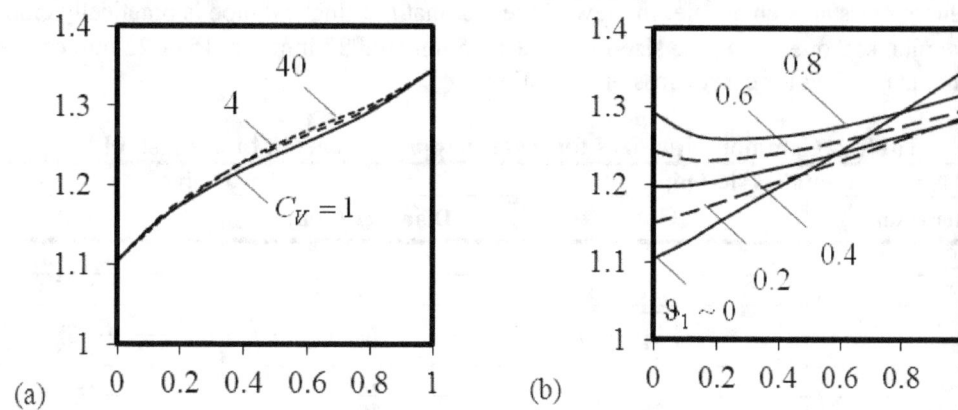

x-axis: Probability ϑ_2

y-axis: Optimum pipe radius ratio $C_{R,opt}$

Figure 6. Optimum pipe radius ratios, $m = 2$

In Figure 6(a), the optimal pipe radius ratios for $\vartheta = 0.5$ are 1.247 and 1.251 when $C_V = 4$ and 40 respectively. It is noted that typical appliance flow rates are in the range 0.08 Ls^{-1} (shower) to 0.3 Ls^{-1} (kitchen sink) corresponding to a $C_V < 4$ [10]. In Figure 6(b), the $C_{R,opt}$ values for fixed demand probabilities $\varphi_1 = 0.2, 0.4, 0.6$ and 0.8 are 1.218, 1.239, 1.268 and 1.304 respectively if $\dfrac{\partial C_{R,opt}(\varphi_1, \varphi_2)}{\partial \varphi_2} = 0$. As expected, more frequent demands (i.e. 'larger' flow rates) lead to higher $C_{R,opt}$ values. The findings suggest that $C_R = 2^{2/7} \approx 1.22$ should be the optimal choice for the design of water supply piping networks that serve probabilistic demands (at uniformly distributed probabilities).

6. ENERGY IMPLICATIONS OF EXISTING PIPING NETWORKS

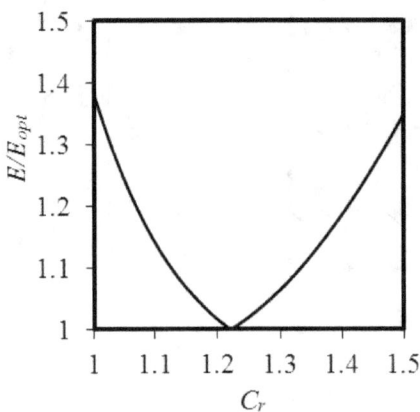

Figure 7. Maximum relative pumping energy at $C_{R,opt} = 2^{2/7}$

Figure 7 shows the maximum relative energy loss for probabilistic demands with $C_R = 1$ to 1.5, illustrating a potential reduction of energy loss up to 38% at $C_{R,opt} = 2^{2/7}$. Table 2 exhibits the common pipe sizes available for water supply systems in buildings [10]. The pipe radius ratios C_R shown are in the range of 1.13-1.47 and many of them are very close to the optimal ratio value $\left(C_{R,opt} = 2^{2/7} \approx 1.22\right)$ proposed in this work. If the optimal value 1.22 substitutes 1.13 or 1.47, 9-30% savings in energy loss are achievable. In view of the fact that the smallest pipe is practically employed in end-use appliances, one more pipe size in between 15 mm and 22 mm, i.e. 15×1.22 mm or 22÷1.22 mm = 18 mm, is required for energy loss optimization.

Table 2. Common pipe sizes for water supply systems in buildings[10]

Copper and stainless steel pipes		Plastic pipe	
Diameter (mm)	C_R	Diameter (mm)	C_R
15		16	
22	1.47	20	1.25
28	1.27	25	1.25
35	1.25	32	1.28
42	1.2	40	1.25
54	1.28	50	1.25
67	1.24	63	1.26
76	1.13	75	1.19
108	1.42	90	1.20
133	1.23	110	1.22
159	1.2	160	1.45

It should be noted that for a wide range of sanitation appliances operated at a demand probability typically lower than 0.1 [11,12], a pipe radius ratio based on steady flow conditions $\left(C_{R,opt} = 2^{3/7}\right)$ leads to an additional energy loss of 14% and 10% respectively, and thus the choice is not optimized for many water supply systems in buildings. Sizing pipes with $C_R = 2^{2/7}$ will be a better choice corresponding to a less energy loss of 2.6% and 1.5% as compared with the cases of known single set of demand probabilities.

Typical water supply systems are designed to cope with a design condition of the probable maximum demand that sufficient pressure is available at all appliance outlets at its design flow rate. The outlet pressure control is achieved by the user through regulating the flow control valve of the appliance.

However, this over-provided pressure relates to energy wastage. The significance of this paper is to understand the required pressure of probabilistic demands. With proper demand control on the appliance outlet pressure at probabilistic demands, potential pumping energy savings for water supply networks can be studied.

7. CONCLUSION

This paper presented the general equations for solving energy loss optimization problems associated with a common basic T-shaped water supply piping network that serves infinite probabilistic demands, and established a mathematical model for energy loss optimization under a fixed pipe volume constraint. Potential reduction of energy loss through proper pipe size was investigated and the optimal pipe radius ratios were found to be in between $2^{1/7}$ and $2^{3/7}$. The findings suggested that $2^{2/7}$ should be the optimal choice for the design of water supply pipe networks that serve probabilistic demands. They also showed that existing piping designs are not optimized for probabilistic demands and reduction of energy loss up to 38% can be made at supply networks, with proper demand control of pumping system.

Acknowledgements

The work described in this paper was partially supported by a grant from the Research Grants Council of the HKSAR, China (PolyU527213E).

References

[1] A.F.E. Wise and J.A. Swaffield. "*Water, sanitary and waste services for buildings*", Fifth ed., Butterworth Heinemann, 2002, London.
[2] K.W. Mui and L.T. Wong. "*Modelling occurrence and duration of building drainage discharge loads from random and intermittent appliance flushes*", Building Service Engineering Research and Technology, 34(4), pp. 381-392, (2013).
[3] L.T. Wong and K.W. Mui. "*Modeling water consumption and flow rates for flushing water systems in high-rise residential buildings in Hong Kong*", Building and Environment, 42(5), pp. 2024-2034, (2007).
[4] L.S. Galowin. "*"Hunter" fixture units development*", Proceedings of the 34th CIBW062 International Symposium on Water Supply and Drainage for Buildings, 8-10 September, The Hong Kong Polytechnic University, Hong Kong, pp. 58-80, (2008).
[5] K.W. Mui and L.T. Wong. "*A comparison between the fixture unit approach and Monte Carlo simulation for designing water distribution systems in high-rise buildings*", Water SA, 37(1), pp. 109-114, (2011).
[6] L.T. Wong and K.W. Mui. "*Stochastic modelling of water demand by domestic washrooms in residential tower blocks*", Water Environment Journal, 22(2), pp. 125-130, (2008).
[7] C.L. Cheng. "*Study of the inter-relationship between water use and energy conservation for a building*", Energy and Buildings, 34(3), pp. 261-266, (2002).
[8] A. Bejan and S. Lorente. "*Design with constructal theory*", John Wiley & Sons, 2008, USA
[9] M. Franchini and S. Alvisi. "*Model for hydraulic networks with evenly distributed demands along pipes*", Civil Engineering and Environmental Systems, 27(2), pp. 133-153, (2010).
[10] *Plumbing Services Design Guide*. Second ed., The Institute of Plumbing, 2002, Essex.
[11] K.W. Mui, L.T. Wong and H.S. Lam. "*Modelling sanitary demands for occupant loads in shopping centres of Hong Kong*", Building Service Engineering Research and Technology, 30(4), pp. 305-318, (2009).
[12] K.W. Mui, L.T. Wong and Yeung MK. "*Epistemic demand analysis for fresh water supply of Chinese restaurants*", Building Services Engineering Research and Technology, 29(2), pp. 183-189, (2008).

Insights and Improvements Based on Updates to Low Power and Shutdown PRAs

J. F. Grobbelaar, J. A. Julius, K. D. Kohlhepp, & M. D. Quilici
Scientech, a Curtiss-Wright Flow Control Company, Tukwila, WA, U.S.A.

Abstract:

In several countries, the requirements for probabilistic risk assessments have increased beyond a Level 1 internal events PRA to add or address spatial and external hazards. In a growing number of countries the requirements have further increased to address all Level 1 hazards in all plant operating modes. Scientech developed its first shutdown probabilistic risk assessment (PRA) in the early 1990s for a European nuclear power plant. Since then several additional low power and shutdown PRA models were developed in the United States following the same approach. The original shutdown PRA model was expanded to evaluate hazards challenging fuel in the reactor vessel and fuel in the spent fuel pool; modeling Level 1 core damage for all hazards in all plant operating modes, with corresponding Level 2 (release) and Level 3 (consequence) models. This complete PRA of all hazards and all modes was incorporated into the European plant's licensing basis, and in 2010 a peer review was conducted.

In the last three years, the shutdown PRA model was updated and a follow-on peer review conducted. Plant operational state definitions were revised to better agree with technical specifications governing the plant operating modes. Additional initiating events were modeled for the fuel pool plant operational states as well as the refueling plant operational states. Initiating event frequencies have been updated to reflect recent operating experience. Success criteria and accident sequence development were revised based on insights from new thermal-hydraulic analyses. New shutdown procedures and "FLEX" strategies were considered in the accident sequence development. New operator actions were credited and human reliability analyses were performed. During the same period, additional model changes and refinements were developed on the USA shutdown PRA models.

This paper presents the insights and improvements made in the PRA modeling of low power and shutdown states, and also presents a summary of insights and benefits that the plant obtained during the development and updates of the underlying shutdown PRA models.

Key Words: Shutdown PRA, Shutdown PSA, Low Power and Shutdown

1. INTRODUCTION

In several countries, the requirements for probabilistic risk assessments (PRA) have increased beyond a Level 1 internal events PRA to add or address spatial and external hazards. In a growing number of countries the requirements have further increased to address all Level 1 hazards in all plant operating modes. Scientech developed its first shutdown probabilistic risk assessment in the early 1990s for a European nuclear power plant. Since then, several additional low power and shutdown PRA models were developed in the United States following the same approach. The original shutdown PRA model was expanded to evaluate hazards challenging fuel in the reactor vessel and fuel in the spent fuel pool; modeling Level 1 core damage for all hazards in all plant operating modes, with corresponding Level 2 (release) and Level 3 (consequence) models. This complete PRA of all hazards and all modes was incorporated into the European plant's licensing basis, and in 2010 a peer review was conducted.

In the last three years, the shutdown PRA model was updated and a follow-on peer review conducted. Plant operational state definitions were revised to better agree with technical specifications governing the plant operating modes. Additional initiating events were modeled for the fuel pool plant operational states as well as the refueling plant operational states. Success criteria and accident sequence development were revised based on insights from new thermal-hydraulic analyses. New shutdown procedures and "FLEX" strategies were considered in the accident sequence development. New operator actions were credited and human reliability analyses were performed. During the same period, additional model changes and refinements were developed on the USA shutdown PRA models.

This paper presents the insights and improvements made in the PRA modeling of low power and shutdown states, and also presents a summary of insights and benefits that the plant obtained during the development and updates of the underlying shutdown PRA models. Section 2 of this paper provides background information about the scope and development of the shutdown PRAs. Section 3 of this paper presents insights into the plant operational state modeling, and Section 4 presents insights into the PRA elements.

2. BACKGROUND

In 2012 a complete shutdown probabilistic risk assessment (PRA) update was developed for the Borssele nuclear power plant (NPP) to update the original 1990's shutdown PRA. The update was conducted to address findings and observations from an International Atomic Energy Agency (IAEA) International Probabilistic Safety Review Team (IPSART) mission, and to improve the as-operated modeling of the plant in accordance with recent modifications and development of shutdown emergency operating procedures (EOPs) [1]. These plant improvements were conducted as part of the plant's continuing process to review and improve plant safety, and also to consider insights from the Fukushima stress test [2] and guidance provided in NEI 12-06 [3]. A summary of the scope of the PSA-2013 paper describing the extent of the Borssele Shutdown PRA changes is provided below.

> "The scope of the shutdown PSA update is a Level 1, 2 and 3 PSA including internal events and all hazards. The scope of this paper is limited to the Level 1 PSA. This paper provides some background in section 2, discusses the plant operational state (POS) definitions in section 3, initiating events in section 4, system modeling in Section 5, success criteria in section 6, accident sequence development in section 7, human reliability analysis in section 8, results and conclusions in section 9, and further work in section 10......
>
> The Borssele NPP is a single unit, 485 MW_e, 2 loop pressurized water reactor of Siemens KWU design. It is located in The Netherlands and has been operational since 1973. Notable features of this plant are a digital control room and "bunkered systems", which are additional systems - beyond the conventional standby engineered safeguard systems - designed to mitigate external events like large scale flooding. These systems include 380 VAC emergency power, 24 VDC power, reserve auxiliary feedwater, reserve decay heat removal, reserve cooling water, reserve high pressure injection, and a reserve spent fuel pool (SFP) cooling train. Unique to the Borssele NPP is shutdown EOPs. Pertinent to the shutdown analyses is that the SFP is located inside containment while the SFP cooling system is outside containment."

Also during the 1990's, shutdown PRA models were developed in the United States in order to address outage risk management considerations. Shutdown PRA models were developed for configuration risk management. These models were developed as an extension of the full power PRA model. Typically, full power probabilistic risk or safety assessments (PRAs/PSAs/IPEs) were already in existence for the nuclear power plant to be modeled. Additionally, the longer range goal was to be able to evaluate the risk in conducting plant maintenance in different plant states, such as to evaluate if the risk was less to remain at power rather than shutdown to cold shutdown to conduct maintenance. The general methodology used was consistent with the International Atomic Energy Agency guidance for the modeling of accident sequences during shutdown and low power conditions [4]. Further, other than the definition of outage types and plant operational states, the underlying methodology (small event tree, large linked fault trees), system models (fault trees) and data were the same as that typically used in the full power PRAs/IPEs with the necessary considerations made for differences in plant response due to shutdown conditions.

3. PLANT OPERATIONAL STATE INSIGHTS

One of the major issues for a shutdown PRA address is the selection and modeling of plant operating states since this task sets the scope of the shutdown PRA. A full power PRA consists of one plant operating state (POS). Because plant system alignments (both operating equipment alignment and also the equipment out of service) change frequently throughout the outage, there are potentially thousands of plant operational states to model. Additionally, the POS requirements vary depending on the type of outage. For example, a quick shutdown to hot standby consists of different POS's than a refueling outage. Similar to the initiating event task in a PRA, the goal of this task is to identify, group and quantify the characteristics, frequency and duration of the modeled POS's.

The recommended modeling practice is to select POSs for fuel in the reactor pressure vessel (RPV) based on plant Technical Specification (TS) operating modes. The primary benefit is to be able to easily link the PRA to the outage schedule, and to improve the understanding when talking to plant operations and maintenance staff. Whereas the POS concept is typically limited to PRA practitioners, the concept of Technical Specification operating mode is widely known to plant staff. This advantage is summarized in the draft USA LPSD PRA Standard [5]:

> "Several plant conditions could be used to define a POS, but model understanding and configuration control are facilitated when the set of plant conditions chosen is consistent with those used by plant personnel to govern LPSD operations (i.e., plant operating modes or operating conditions as defined in plant technical specifications)".

The 2012 update of the European model eliminated "low power" from the shutdown analysis, as the power POS was extended to apply and provide a bounding analysis whenever the core is critical. In the previous version of the PRA, "low power" was included in hot steaming POS, which was the part of the shutdown/startup evolution between criticality and approximately 15% reactor power while the turbine generator is not synchronized to the grid and house loads are powered from the startup transformers and not the generator transformer. The new definitions of the power and hot steaming POSs resulted in simplified event trees for the hot steaming POSs, since reactor trip and RCS pressure control were not needed. In both the European and USA shutdown models, the "low power" portion of the model was typically not used, as the success criteria for safe shutdown components during this state were bounded by those for full power.

Also during the 2012 update of the European model, the POS boundaries of the shutdown states using residual heat removal shutdown cooling (RHR-SDC) were revisited and refined to better match refueling outage practices and recent changes to the plant's shutdown emergency procedures. Thermal hydraulic calculations that were conducted with the RPV head on, RPV head off, and RPV vents open and closed were reviewed to understand the impact on Level 1 plant response success criteria. The original 1990's model considered RPV with vents open to be equivalent to RPV with head off (based on thermal hydraulic analysis done in the 1990s), which is good for Level 1, but may have different Level 2 consequences. At that time, vents were assumed to stay open once opened, but since then new procedures have been developed that instruct closing of the vents on loss of RHR-SDC, so RPV vent status is now questioned in midloop sequences with RPV head on. Additional POSs model the state with RPV head off.

In the USA, similar POS refinements have been conducted based on insights from thermal-hydraulic analyses. For example, for some Westinghouse PWRs when the refueling basin is flooded and the RPV upper internals are installed, there are limited connections between the refueling basin and the RPV such that the RCS behaves as if the extra volume of water in the refueling basin is not available for cooling using natural circulation flow. During this portion of the refueling basin flooded POS, the

PRA success criteria have been reviewed to ensure natural circulation is not available as a long term success state for decay heat removal.

In both Europe and the USA, following the Fukushima Daiichi event, the Spent Fuel Pool (SFP) has received increased attention. A SFP POS has been part of the European and USA models since the early 1990s. In the 2012 European update, the SFP POS was explicitly modeled as two states, one defined as fuel pool early (FE) and for fuel pool late (FL). While POS FE corresponds directly with a specific Technical Specification mode, the Technical Specification does not define a mode for SFP following completion of core reload (POS FL). POS FL was defined to address the spent fuel in the SFP while the "active" core is in the RCS until the next core offload. A new POS to model reshuffling outages (outage without a complete core unload where approximately one third of the core is replaced and the core reconfigured while the SFP and reactor cavity remain connected) was investigated, but it was concluded that the reshuffling outage is adequately represented by POS's for Core Offload and Core Load.

In both Europe and the USA, the strategy and management of plant outages has changed, requiring review in order for the PRA to properly reflect current plant practices. The trend has been for longer intervals between refueling outages and shorter refueling outages. In a PRA that captures an "average" or typical year, the historical data for the preceding 5-10 years should be reviewed in order to check and update (if appropriate) the calculated POS durations and annual fractions. In the application of the shutdown PRA to configuration risk management, such as the usage of the USA models as well as the European model, then the duration and configuration of the plant states is provided by the outage schedule.

4. PRA TASK INSIGHTS

4.1 Internal Initiating Events

The internal initiating events typically identified and considered in the PRA modeling of shutdown POSs are summarized in Table 1 below.

Table 1 - Shutdown Initiating Events in Each POS

Category	Initiating Event POS:	Hot Steaming	Cold Shutdown (RHR)	Mid loop (RHR)	Core Offload Load	Spent Fuel Pool
Transients	Loss of Offsite Power	X	X	X	X	X
	Single SGTR (PWR)	X				
	Steam/Feedwater Line Break	X				
	Loss of Main Feedwater	X				
	Loss of SFP Cooling				X	X
	Loss of RHR-SDC		X	X	X	
Loss of Support Systems	Loss of DC Buses	X				
	Loss of High Voltage AC Bus	X	X	X	X	X
	Loss of Instrument Buses	X	X	X		
	Feedwater Tank Rupture	X				
	Loss of component cooling water system/s	X	X	X	X	X
	Loss Low Voltage AC buses				X	X
LOCAs	ISLOCA via RHR Suction	X				
	ISLOCA via LPSI Inj. Line	X				
	Flow diversion from RHR		X	X		
	RHR pipe LOCAs		X	X	X	
	SFP LOCA				X	X

Table 1 - Shutdown Initiating Events in Each POS

Category	Initiating Event POS:	Hot Steaming	Cold Shutdown (RHR)	Mid loop (RHR)	Core Offload Load	Spent Fuel Pool
	Large LOCA/RPV rupture	X				
	Intermediate LOCA	X	X			
	Small LOCA	X	X	X	X	X
	Very Small LOCA	X	X	X	X	X
Special	Extreme grid disturbance				X	X

In developing the list of internal initiating events, the following insights should be noted.

- Transient losses of decay heat removal apply to all POSs, but are typically not significant during shutdown with the refueling basin flooded (core offload/load).
- Support systems affect decay heat removal, and should be considered, especially where the support system potentially affects fuel in the RPV as well as fuel in the Spent Fuel Pool.
- Losses of inventory (shutdown LOCAs) apply to all POSs, and reduce the advantage of the additional inventory that is initially available at the start of the refueling basin flooded (core offload/load) and Spent Fuel Pool POS's.
- Losses of inventory outside of containment can occur such as with certain RHR-SDC breaks, and thus contribute to interfacing systems LOCA (where the RCS sump is not filled and a pathway exists that breaches containment).
- Initiating event frequencies have been updated to reflect recent operating experience. Since the early 1990's there has been a significant reduction in the frequencies associated with a loss of decay heat removal.

4.2 External Hazard Events

A significant comment from the Peer Review of the European shutdown PRA was that the hazard screening had not been reviewed and updated for many years. After a screening process, the internal and external hazards in Table 2 were included in the modeling of the European shutdown PRA.

Table 2 - Hazards Included in the European Shutdown PRA

Internal Hazards	Internal flooding
	Internal fire
	Failure of pressure parts, supports or other structural components, including HELB
	Disruptive failure of rotating machinery or other equipment
	Dropped or impacting loads
External Hazards	Tornado/high winds
	Meteorite
	High water level
	Explosion after transportation accident (ships transporting LNG and ammonia)
	Chemical release after transportation accident: toxic clouds
	Missiles from off-site activity (e.g. military, (wind / steam turbines)
	Aircraft crash
	Earth quake (analysis on going)

The following insights were developed in Reference 1.

> "Given the Fukushima accident, special attention was given to combinations of hazards. Looking at the (screened in) hazards the following can be concluded on credible combinations. High wind and flooding are always combined, as high wind (storm) is a precondition to cause the high water levels that could threaten the plant. The consequences of storms during a flood are bounded by those for the flooding itself.
>
> High winds could cause shipping incidents. However, shipping incident statistics include those caused by storm. The possible consequences (flash fire, toxic clouds, explosion) envelope the possible damage by high winds. Furthermore, the high wind will dilute the chemicals that can be released from the ships; stable weather is needed for a shipping incident to develop into a threat to the plant.
>
> External flooding caused by an earth quake has been screened while tsunamis can be excluded. So there is no need to consider the impact of a flood wave on a plant that is already possibly damaged by an earthquake.
>
> Also the combination of the plant damaged by an earthquake and a flooding that is made possible by a dike that is damaged by the same earthquake is beyond consideration as there is still a storm surge needed to create a flooding. A damaged or failed dike in itself will not lead to site flooding as the site is 3 m above mean sea level (NAP), a normal tide is between – 2 m NAP and +2 NAP, and the all equipment that is susceptible for flooding is placed at a minimum elevation of +5m NAP.
>
> The combination of seismic damage to the plant and gas clouds originating from seismically damaged industry might need investigating, depending on the outcome of the seismic analysis that is being performed.
>
> The consequences of possible (internal) fires caused by explosions or aircraft impact safety are bounded by the consequences of the explosion or impact on the buildings: the complete loss of the building and the equipment inside."

4.3 Success Criteria

This is an element where all shutdown PRA models benefit in model review and refinement. Understanding the plant behavior following an initiating event is important to the development of the accident sequence (plant response) models as well as for the timing of operator recovery actions.

4.4 Accident Sequence Development

The insights from this PRA task are summarized below.

- New Shutdown Procedures. With increasing emphasis on the losses of RHR shutdown cooling in the USA and internationally, plant response procedures are being reviewed and updated. The shutdown PRA model should check current plant procedures to ensure the modeled plant response reflects the "as-operated" plant. In the European PRA, shutdown procedures included a shutdown diagnosis procedure, or in the Westinghouse terminology a shutdown E-0 (S-E-0) applicable during POS's with RHR-SDC. The S-E-0 procedure is unique to the Borssele NPP (to the best of our knowledge). Additionally, some modifications were also made to the SAMGs.

- Event Tree Updates for Internal Events. In many of the USA plants, the shutdown models were initially created by copying and adapting the full power PRA event trees. Thus, the shutdown event tree models require review and update to re-synchronize with the full power models. This insight applies to both internal events, spatial events and external hazards.

- Defense in Depth Modeling. Many USA shutdown PRA models developed and focused on defense-in-depth models, especially for functions such as Reactivity Control that are not well suited to PRA quantification. These models work well for outage risk management as they match well with shutdown Technical Specifications.

- Post-Fukushima Insights. One of the insights of the Fukushima stress test of the European plant was that the available time to restore core cooling is very limited during mid-loop operation following a SBO. This was not really a new insight of course, but what is new is the requirement for a measure to reduce these consequences, irrespective of their likelihood. The measure had to be twofold: provision of the necessary procedure/s and training of the operators to perform these procedures.

4.5 Systems Modeling

The insights from this PRA task are summarized below.

- Additional systems such as Fire Protection Back-Up to Spent Fuel Pool are often added to the shutdown PRA models.

- Component test and maintenance modeling, including "disallowed maintenance" combinations should be reviewed to ensure they are consistent with the current outage practices.

- Unavailability of automatic actuation. The reactor protection system (RPS) modeling should be reviewed to ensure correct availability/unavailability of RPS signals according to the Technical Specifications applicable to each POS. RPS signal availability can vary, for example into one of three groups: 1) available at power and hot shutdown only, 2) available at power, hot shutdown, cold shutdown, and midloop, 3) available in all POSs. The enabling/disabling of RPS signals for different POSs is typically accomplished using house events.

- Consistency with full power system fault trees. In many of the USA plants, the shutdown models were initially created by copying and adapting the full power PRA fault trees. Thus, the shutdown fault tree models require review and update to re-synchronize with the full power models (or merge the two models). In the European PRA update, for example, the common cause event identification update for full power was propagated into the shutdown PRA model.

4.6 Human Reliability

The shutdown HRA models developed for the USA PRAs as well as the European shutdown PRA included detailed analysis for many HFEs including fire HFEs. These studies showed that that in most cases that current HRA methods are appropriate, but it also confirmed that a weakness of available HRA methods is their ability to model recovery actions with very long time windows (available time). The insights from the shutdown HRA task are summarized below.

- Identification. A shutdown HRA update starts with identification of operator actions modeled as human failure events (HFEs) in the shutdown PRA. The identification task starts with a review of the current identification and grouping of all human failure events already modeled by POS. The HRA task then reviews any new procedures and incorporates new actions into the PRA model on an as needed basis. If applicable, the review should also include identification of all instrumentation required for each shutdown action credited in the fire PRA.

- Quantification. The quantification of each HFE followed the "EPRI HRA Approach" as implemented in the EPRI HRA Calculator [6]. The CBDTM and HCR/ORE [7] quantification methods were used. However, these methods were developed for at-power, EOP driven actions which are typically required within the first couple of hours of an initiating event. In the shutdown model there are many actions in which the time available can be on the order of days and the lower bound quantification limits of the methods are reached. The shutdown HRA applied a lower bound of 1E-6 for individual actions. In some cases these lower bound HEPs are risk significant and there is currently no available HRA method which can be used to systemically justify lower HEP values. Additionally, the currently available HRA methods do not provide guidance on how to apply recoveries from off-site given the long time available. It would seem reasonable to take credit for these additional long term recovery factors, but there is no available guidance on the consistent treatment for such long term recovery factors.

- Spatial Events. For the shutdown fire HRA of the European PRA, the guidance in NUREG-1921 [8] was followed. Although this guidance was intended for at-power applications, in general it remained applicable for shutdown fire HRA. The insight from the fire HRA was that in many cases the fire was extinguished hours before the actions were required, and the fire would have little to no impact on the operators' performance. For each fire HFE, the instrumentation cabling was verified or traced to ensure that the fire impact on cues and indications was understood. Due to the redundancy and diversity in cues and indications, very few shutdown actions were significantly impacted by fire.

- Dependency After the individual HEPs were quantified and implemented into the PRA model, a dependency analysis was performed by review all combinations of HFEs, for each POS. Common cognitive HFEs were developed for actions which are based off the same procedure transfer or same cue. In the Spent Fuel Pool POS, almost all operator actions share a common cognitive for diagnosis of loss of SFP cooling. In this plant state, there is at least 6 hours available for diagnosis and multiple redundant cues and indications including SFP cameras. This risk significant common cognitive applies the lower bound HEP and due to the exceptionally long time window this action was considered to be independent from all other actions in the combinations.

5. SHUTDOWN PRA RESULTS AND CONCLUSIONS

In several countries, the requirements for probabilistic risk assessments have increased beyond a Level 1 internal events PRA to address all Level 1 hazards in all plant operating modes. Scientech developed its first shutdown probabilistic risk assessment in the early 1990s, starting with a European nuclear power plant. This European PRA was used to communicate average, annual risk as well as to provide the plant a tool for outage risk management. Several shutdown PRA models were then developed in the USA, primarily for use in outage risk management.

In the last several years, many of these shutdown PRA models have undergone review and update. Plant operational state definitions were revised to better agree with technical specifications governing the plant operating modes. Additional initiating events were modeled for the fuel pool plant operational states as well as the refueling plant operational states. Initiating event frequencies have been updated to reflect recent operating experience. Success criteria and accident sequence development were revised based on insights from new thermal-hydraulic analyses. New shutdown procedures and "FLEX" strategies were considered in the accident sequence development. New operator actions were credited and human reliability analyses were performed.

The updated shutdown PRAs have been used to evaluate risk-significance and for outage risk assessment. The improvements in the shutdown PRA models have been used to provide the following benefits to the plant.

- Defense-in-depth (qualitative) modeling facilitates outage risk management, especially for functions such as Reactivity Control that are not well suited to PRA quantification. These models work well for outage risk management as they match well with shutdown Technical Specifications.

- Ability to evaluate risk-significance of shutdown events and issues (e.g. significance determination process in the USA).

- Ability to quantify and evaluate risk trade-offs of conducting maintenance in different POSs (e.g. the risk of staying online for maintenance as opposed to shutting down).

- Ability to conduct outage risk assessment and outage risk management.

- Ability to evaluate proposed plant improvements (hardware and procedure changes).

Current PRA methods, especially HRA, do not lend themselves to modeling very long term scenarios. In the case of loss of SFP cooling, the time to core uncover can be several days following a transient loss of decay heat removal. During this time, there will be many opportunities for repair and/or recovery from both onsite and offsite sources in accordance with the FLEX concept, but the offsite sources are not currently credited due to the shortcomings of the PRA methodology. It would be useful in future if the PRA methodologies can be developed or strengthened so that long term scenarios can be modeled with less uncertainty, or eliminated from the model.

6. REFERENCES

[1] *Shutdown Probabilistic Safety Assessment Update*, Erik Roose et al, presented at PSA 2013, American Nuclear Society sponsored Probabilistic Safety Assessment Conference, Columbia, SC, September, 2013.
[2] *Final Report Complementary Safety Margin Assessment NPP Borssele*, EPZ, October 31, 2011.
[3] *Diverse and Flexible Coping Strategies (FLEX) Implementation Guide*, NEI, Suite 400, 1776 Street NW, Washington, D.C., May 2012
[4] *Working Material, Guidelines for Shutdown Risk Assessment*, Report of a Consultants Meeting Organized by the International Atomic Energy Agency, Vienna Austria, 1994.
[5] ANSI/ANS-58.22-2012, American National Standard Low Power and Shutdown PSA Methodology
[6] *The EPRI HRA Calculator® Software Users Manual*, Version 4.2, EPRI, Palo Alto, CA, and Scientech, a Curtiss-Wright Flow Control company, Tukwila, WA, EPRI Software Product ID #: 1021230: 2007.
[7] *An Approach to the Analysis of Operator Actions in Probabilistic Risk Assessment*, EPRI TR-100259, EPRI, Palo Alto, CA: 1992.
[8] EPRI/NRC-RES Fire Human Reliability Guidelines, NUREG-1921 Final Report, US NRC and EPRI, July 2012.

When is it justified to delay the implementation of safety improvements after they have been approved?

Patrick Momal [a]

[a]IRSN, Fontenay-aux-Roses, France

Abstract: After safety improvements have been approved, actual implementation can often be delayed significantly in the nuclear sector. This situation can be unsatisfactory for safety experts conscious of the safety benefits foregone during such a delay. They rightly assimilate this to a cost in terms of safety. The present paper proposes a cost-benefit analysis of this question. Two different types of delay benefits are distinguished: the time value of delaying the implementation on the one hand; and possible reductions in implementation costs. These two benefits are first approached separately after which a general formula is proposed and discussed. Delays generally appear difficult to justify, except when cost reductions are substantial and delays are limited.

Keywords: Economics of Safety, cost-benefit analysis, safety modifications

Implementation delays can be quite substantial after safety improvements have been decided upon. This is challenging for safety experts, especially those who have been researching and advocating the improvements. They may feel that protracted delays carry heavy safety costs. On the other hand, managers responsible for carrying out the works face technical constraints and more immaterial competing objectives.

This paper proposes a cost-benefit analysis of such delays — one aspect for consideration in broader complex and delicate discussions. The analysis involves the safety costs of delaying implementation, on the one hand; and on the other hand, two largely independent types of benefits involved in a possible delay:

1. The time value: when implementation delays do not result in lower implementation costs, the potential benefit only lies in postponing the considered expenses — economists use the expression "time value". A first simplified cost-benefit analysis simply compares the time value and the safety cost.

2. Cost optimization: when implementation delays offer the possibility of reducing costs, it can be justified to delay the implementation of a safety modification. Intuitively, the reduction in costs must be attractive and the delay reasonable. In order to examine a "pure" cost optimization, this situation is analyzed when there is no time value (this obtains when the interest rate is zero).

The paper first addresses these two specific cases which provide a good understanding of the trade-offs involved; a general cost-benefit formula is then derived and conclusions are drawn.

1. DELAYS ARE NOT JUSTIFIED WITHOUT COST REDUCTIONS

This result is obtained without difficulty, using minimal formalism with the following notations:

G is the yearly safety benefits: the considered safety modification typically yields a gain in terms of accident probabilities, for instance a reduction in yearly failure probability of 10^{-8}/year. This safety gain is translated into monetary terms resulting in a value G.

N is the number of years of production left for the plant.

i is the interest rate or discount rate; from a public point of view, i would be the social discount rate while from a private point of view, it would be the financial yield the operator can obtain by disposing of the funds made available by the delay.

n is the number of years of delay between the decision to go forward with the considered safety modification and its actual implementation in the plant.

C is the cost of the modification under scrutiny.

The total safety benefit in case of immediate implementation is the discounted sum of benefits from now until year N that is:

$$\sum_{1}^{N} \frac{G}{(1+i)^t}$$

And the value of the safety modification is:

$$V_0 = \sum_{1}^{N} \frac{G}{(1+i)^t} - C$$

Similarly, for a delay of n years, safety benefits would be:

$$\sum_{n+1}^{N} \frac{G}{(1+i)^t}$$

And the value of the safety modification becomes:

$$V_n = \sum_{n+1}^{N} \frac{G}{(1+i)^t} - \frac{C}{(1+i)^n}$$

Comparing the two options boils down to considering the difference $D = V_0 - V_n$. If the difference D is positive, then implementation should not be delayed because the net benefit of immediate implementation V_0 is larger than the benefit with delayed implementation V_n. In other words, if D is positive the delay would imply safety costs higher than the time value, i.e. higher than the financial benefit (e.g. the return which would be obtained by wisely investing C on financial markets). D is the potential loss involved in the delay. If D is negative, then the delay yields benefits larger than the safety costs.

We have:

$$D = \sum_{1}^{n} \frac{G}{(1+i)^t} - C\left[1 - \frac{1}{(1+i)^n}\right] \tag{1}$$

and simple calculations lead to (appendix 1):

$$D = \frac{C}{i}\left[1 - \frac{1}{(1+i)^n}\right]\left(\frac{G}{C} - i\right) \tag{2}$$

for which a first order approximation, generally acceptable since i is small, is:

$$D = nC\left(\frac{G}{C} - i\right)$$

In both expressions, the sign of D is that of $(G/C - i)$ or $G - iC$, the difference between the annual safety gain and the annual yield of the funds the safety modification requires to be implemented.

This result is remarkable: it does not depend upon n. In other words, if it is justified to delay the safety modification by *one* year, then it is also justified to delay it by n years *whatever the value of n*! This is easily conceived through the following argument: if G – iC is negative, works should be delayed by one year because financial yield iC is larger than safety benefits G. When one asks the question again, one year later, the answer has not changed! One should again postpone implementation by one year; and so forth *ad infinitum*.

In still other words, if it is justified to implement the safety modification, then it should be implemented immediately (remember: no cost reductions are expected).

In practice, usual orders of magnitude suggest that implementation delays are not justified by the time value alone and safety modification should be implemented immediately when the decision is agreed upon if no cost savings can be expected. Safety modifications are often acceptable as long as the cost is lower than 10 times the safety benefit which is roughly equivalent to considering that most safety modifications provide benefits for more than 10 years:

$G/C > 10\%$.

On the other hand,

$i < 10\%$.

Indeed, from the authority point of view, i is the social discount rate which, in most countries is in the range 3 – 4%. From the operator point of view, this is also the case as long as real rates of return are lower than 10%; for example, in France, a figure of 8% has been used in the past (it could be lower nowadays considering general economic conditions).

The time value (iC) is generally low in comparison to the value of safety (G). In other words, the efficiency (G/C) of safety modifications which are agreed upon is generally above the discount rate i.

2. A DELAY CAN BE JUSTIFIED WITH SIGNIFICANT COST REDUCTIONS

For the sake of clarity, we now disregard the modest time values treated in the previous section. The potential loss D then simply boils down to: $D = nG - \Delta C$. indeed there is no discount rate, the non-discounted sum of benefits foregone during the n year delay is simply nG and the benefit of postponing works is the cost reduction ΔC. Looking at things slightly differently, in this simplified case the value of immediate implementation is $NG - C$ while the value of the delay would be $(N-n)G - (C-\Delta C)$ and the difference is $D = nG - \Delta C$.

If D is positive, it is beneficial to implement the modification immediately because the safety gain nG is larger than the foregone cost-reduction ΔC. If D is negative it is preferable to delay the works in order to benefit from the cost-reduction because it is substantial and higher than the safety gains.

It is convenient to write D as:

$$D = nC \left(\frac{G}{C} - \frac{\Delta C}{nC} \right) \qquad (3)$$

As in the previous case, the sign of D features the efficiency ratio G/C; the cost reduction ΔC only intervenes through its ratio $\Delta C/C$; and contrary to the previous case, the number of years of delay n plays a crucial role. Even when a one year delay is beneficial, nothing guarantees that a two-year delay would still be justified. Furthermore, when n is large, delays can no longer be justified because the second term tends towards zero.

In practical terms,

- as we have seen above, G/C > 10% for most economically viable safety modifications; therefore, cost reductions should be larger than 10% for a one-year delay to be acceptable; larger than 20% for a two-year delay; and so forth; a 10 year delay cannot be justified
- if G/C > 1 then even a one year delay cannot be justified; if G/C > 50% then a two-year delay cannot be justified; and so forth.

It therefore appears that the window of opportunity for economically justifiable delays is not immense: it requires that safety benefits be moderate and simultaneously that cost reductions be significant.

Notice that the decision does not depend upon the safety modification cost C. The sign of D is not affected by C; however, the *size* of C directly affects the *size* of D. Therefore indeed, when C is small, the question is irrelevant in practice and delays need not be considered. In contrast, when C is large, the stakes justify an in-depth analysis, but this should involve neither C nor ΔC, but the percentage savings ΔC/C.

Neither does the decision depend upon N, the number of years of potential future operation. Indeed total safety gains are close to NG, but the decision only depends upon the efficiency G/C.

3. A GENERAL FORMULA

Considering the general case allows the two types of above results to be taken into account. Formula (1) becomes:

$$D = \sum_{1}^{n} \frac{G}{(1+i)^t} - C + \frac{C - \Delta C}{(1+i)^n} \tag{4}$$

which yields:

$$D = \frac{1}{r}\left[1 - \frac{1}{(1+i)^n}\right][G - i(C - \Delta C)] - \Delta C \tag{5}$$

The expression $\frac{1}{i}\left[1 - \frac{1}{(1+i)^n}\right] = n^*$ represents n years discounted at rate i[1].

D then becomes:

$$D = n^* \left(G - i(C - \Delta C) - \frac{\Delta C}{n^*} \right) \tag{6}$$

Safety | Time value | Cost savings

This expression directly involves costs measured in currency. The value of cost savings appears as $\Delta C/n^*$ which again suggests that implementation delays cannot be protracted without affecting the economic benefit of the delay.

Expression (6) can usefully be presented using dimensionless ratios:

$$D = n^*C \left[\frac{G}{C} - i\left(1 - \frac{\Delta C}{C}\right) - \frac{\Delta C}{n^*C}\right] \tag{7}$$

This leads to six major conclusions:

1. For many safety modifications, a delay is never justified. If G/C is larger than 100% (the yearly safety gain is larger than the safety modification cost), postponing implementation of the safety modification is never justified in practice because the second term of the expression is low (see hereafter) and the third term is lower than 1.

2. For the majority of other safety modifications, only limited implementation delays can be justified. In practice for a delay of n* years to be justified, it is necessary that G/C < 1/n*; for example, a delay of 10 discounted years can only be justified with G/C < 0.1.

---

[1] It can be developed as: $n^* = n\left(1 - \frac{n+1}{2}r + \varepsilon(r^2)\right)$

3. The benefit of cost savings rapidly declines with the number of years of delay. The third term relative to cost savings is always lower than one, for example with $\Delta C/C = 50\%$, it is worth 50% for $n^* = 1$, but only 25% for $n^* = 2$.

4. The time value is generally small. The second term relative to time values is lower than i; for example for $i = 4\%$ (point of view of the safety authority) and $\Delta C/C = 50\%$, it is only worth 2%; in general, it should play a secondary role in the decision.

5. The cost of the safety modification plays no direct role in the decision. As far as the decision is concerned, it only intervenes through the efficiency G/C and the rate of cost savings $\Delta C/C$. C determines the stakes involved in the decision; if C is large and the decision is incorrect the loss is large.

6. The discount rate plays a minor role.

4. CONCLUSION: THE ROLE OF THE EFFICIENCY G/C

As is clear by now, the efficiency G/C should play a major role in the decision to delay or not delay the implementation of a safety modification. While the authority and the operator can often agree on the cost C, they may well differ in their estimation of the safety gain G.

(i) In legal terms, the utility would only be responsible for a limited portion of total accident costs. While it could be held accountable for on-site costs and for some so-called direct consequences, it is highly improbable that image costs and impacts on electricity production, domestic and worldwide, would be borne by the utility. In practice, it may only have a limited financial capacity to pay for its own damages not to mention compensation of direct victims.

We refer to IRSN estimates of accident costs in France [1] which show that for a severe but not major nuclear accident, image costs and costs related to electricity production (domestic) would represent more than three quarters of the total cost in France. Even for a major accident, these components would account for more than half the cost. If therefore utilities were to consider their limited involvement in compensation for accident losses, they might largely underestimate accident costs and therefore safety benefits.

(ii) There could also be a tendency among utilities to consider failure probabilities limited to initiators which directly involve their own responsibility. Along this line of thinking, they could advocate that "acts of God" are beyond their responsibility and are not to be included. On the contrary, authorities considering the interest of the nation globally should include all causes of failure. Thus there might be another difference in appreciation of safety benefits due to the authority considering failure probabilities higher than those assumed by the utility.

(iii) And there is a third consideration related to risk aversion with respect to nuclear accidents. Risk aversion might be higher among representatives of the nation who would naturally be concerned with the fate of victims. Utility managers could exhibit lower risk aversion due to their familiarity with nuclear and because they might disregard the entire scope of consequences.

For these three reasons, safety gains considered by the nuclear authority could be more than 10 times higher than those considered by the utilities. This would naturally be reflected in different values of the safety efficiency G/C. For instance, while utilities would naturally exhibit a willingness to pay or safety modifications with efficiencies down to 5 to 10%, authorities might consider these same modifications to have efficiencies of 50% to 100% and higher.

Therefore, utilities may argue for delays in a perfectly candid way, on the basis of parameter estimates which are perfectly justified from their own point of view, white authorities should enforce immediate or rapid implementation when keeping in mind the social benefits of safety.

Disclaimer
The ideas presented in this paper are those of the author and may not represent positions held by IRSN.

Appendix

$$\sum_{1}^{n} \frac{1}{(1+i)^t} = \frac{1}{(1+i)} + \frac{1}{(1+i)^2} + \cdots + \frac{1}{(1+i)^n}$$

$$= \frac{1}{(1+i)} \left[1 + \frac{1}{(1+i)} + \frac{1}{(1+i)^2} + \cdots + \frac{1}{(1+i)^{n-1}} \right]$$

$$= \frac{1}{(1+i)} \frac{1 - \frac{1}{(1+i)^n}}{1 - \frac{1}{1+i}} = \frac{1}{i} \left(1 - \frac{1}{(1+i)^n} \right) \quad \text{from which (2) follows immediately.}$$

This represents the number of discounted years. Based on $(1+i)^{-n} = 1 - ni$, the first order development reduces to n; in other words, the first order corresponds to no discounting.

The second order approximation is $n - \frac{n+1}{2} i$, the discounted number of years is lower than n. Here are a few values for this reduction:

Reduction in number of years due to discounting

Number of years	Discount rate i			
	2%	4%	6%	8%
2	-2%	-3%	-5%	-6%
4	-1%	-3%	-4%	-5%
6	-1%	-2%	-4%	-5%
8	-1%	-2%	-3%	-5%
10	-1%	-2%	-3%	-4%

In particular, when n is « large », the percentage reduction is close to r/2.

Reference

[1] Pascucci-Cahen, L., and Momal, P. Massive radiological releases profoundly differ from controlled releases. Eurosafe 2012.

The Underlying Principles and Quantitative Values of Risk Limits

Dennis R. Damon
U. S. Nuclear Regulatory Commission, Washington DC 20555, U. S. A.
dennis.damon@nrc.gov

Abstract: The purpose of this paper is to clarify the principle that risks imposed on individuals should be limited in their magnitude. The paper also discusses implementation of this risk limitation principle by regulation, including quantitative risk limits. This principle of limits arises when an activity, although beneficial to society, nevertheless imposes some risk of harm on any individual without their consent and not for their benefit. In this case, fairness requires that the magnitude of the risk be limited. If the harm is fatality, then compensation is no remedy, and regulation should limit the risk a priori. This principle of limitation of imposed risk is not the same as the idea that it is desirable to reduce risk in general. The principle to be applied for general reduction of risk is optimization. Optimization considers all impacts, both beneficial and adverse, on all persons to improve net benefits to society collectively. The appropriate magnitude of a quantitative risk limit may vary if the individual has some degree of consent, or benefits from the activity having risk. Thus values recommended for limits on risk have typically differed between workers who benefit from the activity, and members of the general public who do not.

Keywords: Risk Limits, Individual Risk, PRA, Decision Analysis, Regulation

1. INTRODUCTION

Risk is the possibility or probability of harm. Risk to individual persons arises from their own activities, from actions of other persons, and from natural phenomena. Some risks to ourselves arise from actions we choose to take because they are beneficial to us. Some harmful or risky actions by others are maliciously intended as such. Some risks come from nature, and cannot easily be affected by human action. There are many types of risk. This paper is only about unintentional preventable risks arising from activities of benefit to society. Risk refers to the future. Quantitative risk assessment and reliability engineering are the recognition that, despite the good intentions of design and operating procedures, harmful events may happen because we cannot entirely control the future. This paper is about one particular type of risk, and what should be done about it. Some risks to us arise from activities under our own control; other risks are imposed on us by activities that are not under our control, nor for our benefit. The principle clarified in this paper is that this latter type of risk should be limited in magnitude, and, when the harm is irrecoverable, may need to be regulated. The views in this paper on regulation are the author's, and not those of the United States Nuclear Regulatory Commission or its staff.

2. THE PRINCIPLE OF LIMITATION OF RISK IMPOSED ON INDIVIDUALS

The main points to be made about the principle of limitation of risk imposed on individuals are: 1) that the principle applies to certain situations and does not apply in others, 2) that it is not just a practical tool for risk management, but embodies a fundamental principle of equity, that each individual is entitled to, 3) that it is not the same as optimizing societal benefits, such as minimizing collective risk, and 4) that it is not the same as the principle of making risk to individuals as low as reasonably practicable, and 5) risk limits are the upper bound of tolerable, not acceptable, risk. The principle is that there is essentially an ethical obligation to limit imposed risks as a matter of equity. This does not mean that the principle should always be implemented by a requirement to quantify risk to individuals, and meet quantitative risk limits. This subject of implementation of the principle will be discussed in section 3 below.

2.1 Decision Situations where Risk Limitation Applies

A decision situation is a situation where a decision-maker must choose from among mutually exclusive options. A number of different groups of persons may be affected by these options, including the decision-maker; and there may be various benefits and adverse impacts on these different groups. Many such situations involve options which will cause risks to some persons. The principle of risk limitation applies only to those situations where the options involve an initiating an activity that is beneficial to society as a whole, but imposes some risk on some individuals without compensating benefits to them. There are two keys that define this situation. First the persons who will receive the risk are not the decision-maker. This is what is meant by the term "imposed". Secondly, the persons who will undergo the risk will not receive commensurate benefits from the option. Other situations exist where this principle does not apply. Reference [1] discusses five archetype decision situations that vary depending on who receives the risk, who receives the benefit, and who has control of the decision. For example, there is the situation arising in medical practice where decisions are made that trade off benefits from medical treatment against their risks. There is no limit on the magnitude of risks that might be appropriately taken in such cases. Similarly there is the rescue situation where a person voluntarily enters a risky situation to provide a benefit to others. The risk taken is thus voluntary, not imposed. Warfare also involves situations where the principle of risk limits does not apply.

The typical situations envisioned here are where risk is imposed on members of the public or workers by an activity of considerable economic value to society. The principle is that, regardless of the magnitude of the economic benefit, there is a magnitude of risk that should not be imposed.

2.2 The Issue of Equity in Imposed Risk

This principle that imposed risks should be limited is not new. For radiation doses, and the risk of radiation doses, that arise from activities beneficial to society, the International Commission on Radiological Protection recommends in Reference [2] (ICRP60) that "constraints" be applied. The following is a quote from ICRP 60 paragraph S18 referring to the "procedure" of optimizing protection:

> "This procedure should be constrained by restrictions on the doses to individuals (dose constraints), or the risks to individuals in the case of potential exposures (risk constraints), so as to limit the inequity likely to result from the inherent economic and social judgments."

Thus ICRP explains the principle of limits as an issue of equity. In reference [3], ICRP discusses extension of this principle to include "potential exposures", that is risks of accidental doses rather than routine exposures.

The idea that limiting imposed risks is a matter of equity or fairness is also found in paragraph 119 of the United Kingdom Health and Safety Executive (document "Reducing Risks and Protecting People" (reference [4]).

The point made in these authoritative documents is that, despite careful evaluation that an activity is beneficial to society, and therefore incurring some collective risks or other adverse impacts is justified by those benefits, there is some magnitude of risk that should not be imposed on any individual person. The reason given is that it is a matter of equity. The UKHSE (Ref. 4) notes that there may be exceptional circumstances where this principle might be violated.

Note that section 2.1 above makes it clear that the principle of risk limitation only applies in the specific situation of imposed risk not for the individual's benefit. Even in this situation, in principle, if the benefits to society were sufficiently great, and the imposed risk not too much above a normal limit, some temporary imposed risk might be considered such an "exceptional circumstance". However, in practice, it is almost never necessary to impose high risks long term. If a facility is of such value to society, and individual residents might be at excessive risk, then they could be moved, perhaps involuntarily.

There is also the unusual circumstance where a choice is made such that risk of harm will occur for a few individuals in order to reduce risk to another group that is either larger or would be subject to greater harm. When the risk is unavoidable, this special situation is called an "intervention", and does not constitute "imposed" risk. Excessive imposed risk occurs only when there is the option of avoiding it.

2.3 Limitation Applies to Individual Risk not Collective

The principle of limitation of risk applies to each individual person. Thus in calculating a risk metric, in principle, the risk to each potentially affected person should be calculated separately to see if it is excessive. In practice, this is not done. Instead one typically identifies a restricted group that clearly is at highest risk, called the "critical group"; or alternatively a "reasonably maximally exposed individual". Either of these may be hypothetical bounding cases that are convenient for the purposes of calculating and managing risk, but the principle itself applies to actual individual persons. A typical metric of individual risk is the estimated annual frequency of fatality for that individual due to accidents in a particular facility. Collective risk is a different quantity. For collective risk of fatality due to accidents, the total number of fatalities is calculated for each scenario, and the mean value is calculated by the probability weighted sum over all scenarios. Thus a typical metric of collective fatality risk is mean fatalities, usually per year.

The point here is that the risk to each individual is the quantity of concern, not a sum or average over all affected persons. This is because the issue is one of fairness to each individual. This is a key to understanding the common situation where the risk arises from unintended releases of radioactive or toxic material. In such release scenarios, a particular individual may not be affected at all unless the wind is blowing in his direction. Even when it is, if dispersion by turbulence is high, the dose may not be enough to cause fatality. Thus the probability of fatality depends on the probability of weather conditions. The equation for the frequency of fatality thus includes several factors: 1) the frequency of the release scenario, 2) the probability distributions on the size of the release given each scenario, and 3) the probability distribution on weather conditions. The probability of fatality for one scenario is then the joint integral of the two probability distributions over the domain where fatality occurs. Thus the total average frequency of the consequence to the individual is the sum over all accident scenarios of the frequency of the scenario times the probability of the consequence given that scenario. Note that the probability distribution on weather conditions may be highly conditional on the scenario. For instance, a high wind event may cause failures that lead to release, but there is a high probability that the weather condition is highly dispersive of any release that may occur. Likewise, the presence of individual residents may be conditional on the scenario. For example, many residents around a facility would be evacuated in most flood scenarios, so that releases caused by the flood would not affect them. Thus the risk assessor needs to model locational distributions of persons, variations in weather, and other factors for each scenario.

Typically, for a release large enough to cause fatality, the probability that the wind is blowing toward a particular individual is about 0.1. If the frequency of the release is 0.001 per year, the individual's frequency of fatality is $(0.1)(0.001/year) = 0.0001$ /year. This point is often misunderstood in that it appears to some inappropriate not to count a release that goes in a direction away from the individual. However, the mathematics is correct, in that all releases are, in fact, considered, but the risk is calculated for each individual person (or group) at a time; and for each individual calculation, risk to other persons is not relevant.

2.4 The Principle of Individual Risk Limitation is not Individual Risk Minimization

The United Kingdom applies a principle that risks to individuals should be made As Low As Reasonably Practicable" (ALARP). ICRP applies a similar concept of optimization or minimization to both routine doses and the risk of accidental doses. The ideas here amount to a recommendation or requirement that individual risk be reasonably reduced. This principle can be implemented by performing an optimization that considers all impacts of each of a set of options to select that yielding the highest net benefit. This principle could be implemented as a regulatory requirement (as it is in the United Kingdom) and applied to minimizing risk to individuals. This is not the same as the principle of risk limitation discussed in this paper. A risk limit would be the highest tolerable value of risk to an individual. Any proposed activity that causes the limit to be exceeded is unacceptable. A requirement to further reduce individual risk below this limit is a separate issue.

2.5 Meaning of Risk Limits

ICRP in Reference [3] and UKHSE in Reference [4] recommend quantitative limits on risk to individuals. The meaning of these values is clearly spelled out as values which are not to be exceeded for the situations to which they apply under almost any circumstances. UKHSE in Reference [4] makes it clear that it regards these risk limits as the boundary between tolerable and unacceptable risk. The term tolerable implies a situation that has some undesirable character, but nevertheless can be allowed. The term "acceptable" is not used by UKHSE until individual risk levels are much lower. In addition, both ICRP and UKHSE endorse the principle of As Low As Reasonably Practicable. Thus one might say that individual risk, as well as collective, is not "acceptable", unless this principle has been satisfied as well. Further, the term "acceptable" raises the question of acceptable to whom. Persons on whom the risk is imposed may not find any imposed risk "acceptable". The principle of risk limits is thus best characterized as a tolerability limit determined by society to prevent large inequities on individuals from activities that are otherwise a net benefit to society.

3. RISK LIMITATION IN DECISION ANALYSIS AND REGULATION

Although the principle of limitation of imposed risk may apply to a regulatory situation, this does not mean that the only way to implement the principle is by quantitative regulatory limits. Both the United Kingdom Health and Safety Executive (UKHSE) and the ICRP in reference [4] provide quantitative risk limits, one for workers and one for the general public. However, the UKHSE limits are not regulatory or statutory requirements. They are guidelines for the regulators. If quantitative risk limits were required, it would imply that the risks be quantified. Regulation involving issues of compliance is better if compliance can be determined accurately and objectively. Accurate objective quantification of risk is a difficult standard to achieve. There are several ways around this difficulty. Risk can be limited by prescriptive measures that have been determined by risk analysis to be reasonably effective in meeting quantitative risk limitation guidelines that are not themselves regulatory requirements. Risk assessment and comparison of results to quantitative risk guidelines can be used in various ways other than as direct regulatory limits. Demonstration that risk limits have been met can also be done not by true risk assessment, but by bounding quantitative risk analysis using conservative assumptions in lieu of data. Another approach is to use standardized pre-approved risk models and data. This would only be practical for a limited range of types of facility and hazards.

Both the ICRP and UKHSE recommend implementation of the principle of risk limitation by two risk limits, one for the public and one for workers. Two limits is not the only possibility, others could be defined. The magnitude of risk that it is fair to impose depends on the degree of control and the amount of benefit that the individual gets from the activity causing the risk (see reference 1). These two factors could vary among different groups of persons. The differences in control and benefit are the reasons that the worker risk limit can be higher than that for the public. The worker derives substantial benefits from the risk taken. The worker also has exercised the choice to undertake the risk to get the benefit. However, worker risk should still be limited in that the degree of control by the worker is limited. In principle, there could be other groups of persons who, analogous to workers, have some amount of control and/or benefit from the risky activity. Thus, in principle, a variety of risk limits could be established to accommodate these variations in control and benefit.

Some risks that are imposed on individuals, such as property damage, are recoverable, typically through insurance. Irreversible serious health effects and fatality are not recoverable. Thus any mechanism used to assure that imposed risk of these irrecoverable consequences is limited is better done by a priori evaluation, regulation, or standards.

The conclusion here is that although imposed risk should be limited, the mechanisms used to provide such limitation could vary. Risk, however, is inherently a quantitative concept; hence, quantitative guidelines and some form of risk analysis is usually part of the process of determining whether limitation is tolerable.

4. QUANTITATIVE VALUES OF RISK LIMITS

UKHSE uses limits of 10^{-4} per year for the public and 10^{-3} per year for workers as the boundary between tolerable and unacceptable risk of fatality imposed on individuals. ICRP in Reference [3] recommends risk limits on potential exposures to radiation doses that are of the same order of magnitude as its annual limits on routine doses of 20 millisieverts to workers and 1 millisievert to the public. The UKHSE fatality risk limits are consistent with the latent cancer fatality risk from these ICRP dose levels. Thus these two recommendations on limits are consistent with each other. No basis for an exact numerical value for risk limits has been identified. To illuminate whether the recommended limits are reasonable, consider a value 10 times the public limit, namely a risk of fatality of 10^{-3} per year. Over an 80 year life span this would be a risk of fatality of 0.08 imposed on an individual. This seems clearly excessive. On the other hand, if the risk limit were 10^{-5} per year, the risk imposed on the average person in the U. S. by other drivers who are at fault would not meet it. This risk of fatality imposed by other drivers is about 4×10^{-5} per year. Thus, to within an order of magnitude, the limits used by UKHSE appear reasonable.

In considering the reasonableness of risk limits, it should be remembered that these limits are maximum tolerable values, not desirable objectives, or negligible risk levels. Negligible individual risk levels are a different concept, and much lower than the limits discussed here.

5. CONCLUSIONS

In conclusion, the principle that risk imposed on individuals not for their benefit should be limited has been recognized by authorities applying it to a wide range of practices. Even though not used as regulatory requirements, quantitative risk limits are recognized as useful. Limits of 10^{-4} per year for risk of fatality to individual members of the public, and 10^{-3} per year for workers, as the boundary between tolerable and unacceptable appear reasonable. Individual risk is not the same metric as the collective risk metric used in cost-benefit optimization to manage risk.

References

[1] D. R. Damon, "*A Risk-Benefit-Control Paradigm for Decision-Making*", Eighth International Conference on Probabilistic Safety Assessment and Management, 2006, New Orleans.
[2] International Commission on Radiological Protection, Publication 60, "*1990 Recommendations of the Commission*", Pergamon Press (Elsevier), 1991, Oxford, paragraphs S16-S18.
[3] International Commission on Radiological Protection, Publication 64, , "*Protection from Potential Exposure: A Conceptual Framework*", Pergamon Press (Elsevier), 1992, Oxford.
[4] United Kingdom Health and Safety Executive, "*Reducing Risks, Protecting People*", www.hse.gov.uk/risk/theory/r2p2.pdf, paragraph 119, pp. 40-41.

Development of A Framework for Establishment of Risk-informed Safety Goals for Nuclear Power Plants Operation in the UAE

Jun Su Ha [a*], Sung-yeop Kim [b], Jamila Khamis Al Suwaidi [c], and Philip Beeley [a]

[a] Khalifa Univ. of Science, Technology and Research, Abu Dhabi, UAE
[b] Korea Advanced Institute of Science and Technology (KAIST), Republic of Korea
[c] Federal Authority for Nuclear Regulation (FANR), Abu Dhabi, UAE

Abstract: A framework for establishment of risk-informed safety goals for nuclear power plants (NPPs) operations in the UAE was developed in this study. The current regulatory circumstance to the safety goals in the UAE was addressed as well. Representative parameters related to the core integrity (Level 1 PSA) and containment integrity (Level 2 PSA) are used as surrogate measures, Core Damage Frequency (CDF) for cancer (latent) fatality and Large Early Release Frequency (LERF) for early (prompt) fatality, for risk-informed safety goals. Under this framework a conservative evaluation of risk-informed safety goals was performed on the basis of conservative assumptions and data which were obtained and/or derived from the PSA results of APR-1400, the same type of the Barakah NPPs which are under construction in the UAE, and public health risk assessments. The current safety targets specified in the regulatory guideline (FANR-RG-004) in the UAE were examined to be appropriately determined with sufficient conservatism from the evaluation results. Limitations of the study and recommendations for appropriate applications of the risk-informed safety goals were provided as well.

Keywords: safety goal, quantitative health objective (QHO), probabilistic safety assessment (PSA)

1. INTRODUCTION

Basically requirements for safe design and operation of Nuclear Power Plants (NPPs) have been based on deterministic principles. Under legal requirements, they are incorporated in some forms of dose limits and performance criteria on the radiation barriers including fuel rods, reactor coolant system, and reactor containment building. In addition to these deterministic requirements based on defense-in-depth (DID) principle, quantitative health objectives (QHOs) for NPPs are or can be set up by either stipulating them in regal requirements or guidelines or policy statements. In the UAE there are regulations for NPP design, radiation dose limits, and the application of Probabilistic Risk Assessment (PRA, the same term as Probabilistic Safety Assessment (PSA)) requirements which should be reviewed for the establishment of risk-informed safety goal. Also regulatory guides regarding the evaluation criteria for probabilistic safety targets and design requirements are provided in a regulatory guidance (FANR-RG-004).

In this study, considering the regulatory circumstance, a framework for establishment of risk-informed safety goals for NPPs operation in the UAE was developed as shown in Figure 1. Risk-informed safety goals are defined in this study as quantitative safety targets evaluated on the basis of available risk insights. Generally QHO values limiting public health risks in the vicinity of the plant site cannot be directly used to check whether an NPP or NPPs satisfies them or not because the health risks cannot be obtained unless Level 3 PSA is performed. Therefore, it is convenient to define surrogate measures for risk-informed safety goals which correspond to QHOs and can be directly compared to the safety level of an NPP or NPPs such as CDF (Core Damage Frequency) and LERF (Large Early Release Frequency). Hence the CDF and the LERF are selected and used as surrogate measures for early (prompt) and cancer (latent) fatalities, respectively. The next step after the selection of surrogate measures for risk-informed safety goals is to determine quantitative values so that they can be used as the safety targets. General risks other than those due to NPPs operation are evaluated with statistical data such as accident and cancer fatalities to estimate the GHOs for early and cancer fatalities, respectively. Public health risks are assessed for the evaluation of conditional probabilities of early and cancer fatalities. Finally, CDF and LERF criteria as risk-informed safety goals are determined by comparing the QHO values and the conditional probabilities of early and cancer fatalities, respectively. Under this framework a conservative evaluation of risk-informed safety goals was performed on the basis of conservative assumptions and data which were obtained and/or derived from the PSA results of APR-1400, the same type of the Barakah NPPs which are under construction in the UAE, and public health risk assessments. Appropriateness of existing safety targets provided in regulatory guides in the UAE was examined by comparing those values with the risk-informed safety goals evaluated in this study.

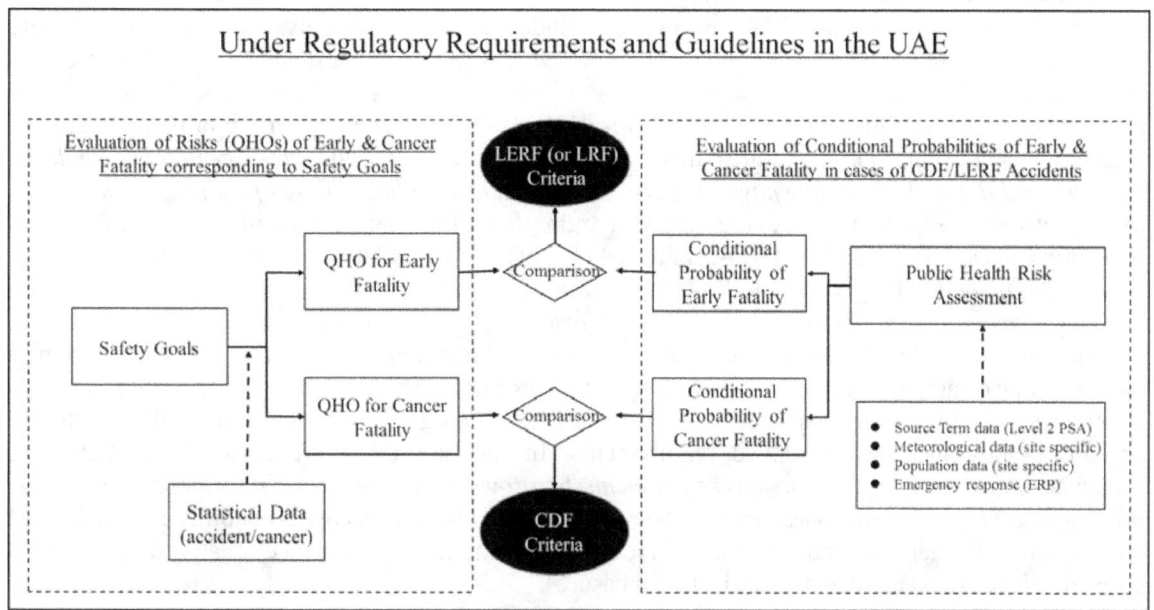

Figure 1: Framework for establishment of risk-informed safety goals

2. LEGAL REQUIREMENTS AND POLICIES TO SAFETY GOALS

2.1. UAE Current Situation

An analysis that carried out by official UAE entities and published in the *Policy of the United Arab Emirates on the Evaluation and Potential Development of Peaceful Nuclear Energy* in 2008 [1] has concluded that, the projections of the energy demand is expected to double in 2020. Based on this feasibility study, an evaluation has been done to find viable options, that capable to fit to the predicted energy demand with environmental and economic considerations. Nuclear power-generation was one of the competitive options that has been adopted. Nuclear policy has set an outlines of highest commitment to nuclear safety, security, and non-proliferation. The Federal Authority for Nuclear Regulation (FANR) as an independent regulator has been formally established with *Federal Law by Decree No 6 of 2009, Concerning the Peaceful Uses of Nuclear Energy* [2]. The federal law has set the responsibilities and requirements for regulating and licensing of nuclear sector in the UAE toward the peaceful purposes. The Emirates Nuclear Energy Corporation (ENEC) has been established with Abu Dhabi Law No. (21) of 2009. ENEC submitted a construction license application to FANR in December 2010, to construct the first two units (Barakah Units 1 and 2) of a nuclear facility at the western region of Abu Dhabi. The applicant submitted a Preliminary Safety Analysis Report (PSAR) which is based on the reference plant of the Shin-Kori Units 3 and 4 facility in Korea, and of the type of Korean APR1400 reactors. The construction license of units 1 and 2 of the Barakah nuclear facility and related regulated activities, is granted in July 2012 which specifies activities authorized and license conditions [3].

2.2. Regulation for Design Requirements

Part of the FANR activities and responsibilities is issuing regulations and regulations guidance, one of these essential regulations was the *Regulation for the Design of Nuclear Power Plants* (FANR-REG-03) [4], which is consistent with International Atomic Energy Agency (IAEA) safety requirements NS-R-1, "*Safety of Nuclear Power Plants: Design*" which superseded by new equivalent publication of IAEA SSR-2/1 [5]. The (FANR-REG-03) aims to establish the design requirements for Structures, Systems and Components (SSCs) important to safety and requirements for safety assessment for different plant states (operational states and incident/accident conditions). The requirements for safety assessment process state that, the complementary techniques of deterministic safety analysis and Probabilistic Risk Assessment (PRA) have to be included. In the safety analysis requirements (Article 44), stipulate "*It shall also be demonstrated that the nuclear facility as designed is capable of meeting any approved limits or criteria for radioactive releases and potential radiation doses for each category of plant operation and that defense-in-depth will be maintained*". In the principal technical requirements, (Article 7), it requires that "*in the design process, the defense-in-depth shall be incorporated to provide multiple physical barriers to the uncontrolled release of radioactive materials to the environment*"; and to provide safety margin to ensure maintaining of safe operation and preventing accidents. The defense-in-depth (Article 7.f), also aims to "*Provide multiple means for ensuring that each of the fundamental safety functions, i.e. control of the reactivity, heat removal, and the confinement of radioactive materials is performed, thereby ensuring the effectiveness of the barriers and mitigating the consequences of any Postulated Initiated Events (PIEs)*". The regulation of the design also include the requirements for safety classification, general design basis, PIEs, internal events and external events, site related characteristics, Design Basis Accidents (DBAs), severe accidents, and other design related requirements. In the part of severe accidents (Article 24), "*considerations shall be given to severe accidents by providing in the design reasonably practicable preventive and/or mitigative measures*". These events which have very low probability but could lead to the core degradation and release of radioactive materials have to be analyzed to provide correspondence preventive and/or mitigative measures.

2.3. Regulation of Radiation Dose Limits

The *Regulation for Radiation Dose Limits and Optimization of Radiation Protection for Nuclear Facilities* (FANR-REG-04) [6] aims to establish the radiation dose limits and the requirements for optimization of radiation protection that are relevant to a nuclear facility during its design, construction, normal operation and decommissioning. The limit for the effective dose during the operation of the nuclear facility does not exceed the dose limits as shown in Table 1:

Table 1: Dose Limits

Category	Worker who is Occupationally Exposed during the normal operation of a nuclear facility	Member of the public
Effective Dose	An average of 20 milli sieverts (mSv) per year averaged over a period of five years (100 mSv in 5 years), and 50 mSv in any one year.	1 mSv (this includes persons working in the nuclear facility other than those categorized under the Worker definition)
The annual Equivalent Dose in the lens of the eye	150 mSv, nor shall the annual Equivalent Dose exceed 500 mSv at any point on the hands, feet or skin.	15 mSv, nor shall the annual Equivalent Dose at any point on the skin exceed 50 mSv.

2.4. Regulation for the Application of Probabilistic Risk Assessment

The objective of the *Regulation for the Application of Probabilistic Risk Assessment (PRA) at Nuclear Facilities* (FANR-REG-05) [7] is to require the applicant or licensee constructing or operating a nuclear facility to conduct a high quality PRA to support the construction and operating licensing. The scope of the PRA has to be defined including internal and external events and all modes of plant operation. The regulatory guide (FANR-RG-004) [8] regarding the *Evaluation Criteria for*

Probabilistic Safety Targets and Design Requirements is defining the evaluation criteria the staff will use in assessing plant Safety assessments associated with probabilistic safety targets and the design requirements in FANR-REG-03.

The probabilistic safety targets – as indicated in the evaluation criteria of regulatory guide (Article 6) are as follows:

1. Core Damage Frequency (CDF) to $< 10^{-5}$/yr (mean value from the PRA considering internal and external events and all modes of Operation).
2. Overall Large Release Frequency (LRF) to $< 10^{-6}$/yr (mean value from the PRA considering internal and external events and all modes of Operation).

3. EVALUATION OF SAFETY GOALS UNDER THE FRAMEWORK

3.1. Risk-Informed Safety Goals

Performance measures for the safety goal must be chosen based on prevention and mitigation of core damage and radioactive material release. Representative parameters of facilities related to the core integrity (Level 1 PSA) and containment integrity (Level 2 PSA) can be used as performance measures. Two selected measures for the risk-informed safety goals are the CDF and the LERF. The CDF is defined as the frequency of an accident which can cause the fuel in the reactor to be damaged. The LERF is defined as the frequency of those accidents leading to significant, unmitigated releases from containment in a time frame prior to effective evacuation of the close-in population such that there is a potential for early health effects (NUREG/CR-6595) [9]. The CDF and LERF have been adopted as performance measures in similar studies performed for other countries [10].

3.2. Quantitative Health Objectives

Qualitative safety goals for securing safety due to NPPs operation were announced in the USA as follows [11]:

- *"Individual members of the public should be provided a level of protection from the consequences of NPP operation such that individuals bear no significant additional risk to life and health."*
- *"Societal risks to life and health from NPP operation should be comparable to or less than the risks of generating electricity by viable competing technologies and should not be a significant addition to other societal risks."*

The qualitative safety goals were supported by quantitative objectives (QHO: Quantitative Health Objectives) used in determining achievement of the qualitative safety goals as follows:

- *"The risk to an average individual in the vicinity of a NPP of prompt fatalities should not exceed one-tenth of one percent (0.1 percent) of the sum of prompt fatality risks resulting from other accidents to which members of the U.S. population are generally exposed."*

- *"The risk to the population in the area near a NPP of cancer fatalities that might result from NPP operation should not exceed one-tenth of one percent (0.1 percent) of the sum of cancer fatality risks resulting from all other causes."*

South-Korea also published *The Policy on Severe Accident of Nuclear Power Plants* in August, 2001 [12], which declared the similar safety goals to the USA's and recommended implementations of Probabilistic Safety Assessments (PSAs) for NPPs.

To develop the quantitative safety goals, risks from other causes than NPPs should be evaluated, which is called "general risk" hereafter. Annual death numbers of accident and cancer per 100,000 people were evaluated 28.7 and 59.2 on average, respectively from literature survey on death statistics

[13-15]. The corresponding general risks due to accident and cancer were calculated 2.87×10^{-4}/year and 5.92×10^{-4}/year, respectively. If the 0.1% rule from the QHOs is applied to the general risks, the values of 2.87×10^{-7}/year and 5.92×10^{-7}/year could be calculated as crude quantitative safety goals (QHOs) for early (prompt) and latent (cancer) fatalities due to NPPs operations, respectively.

3.3. Public Health Risks Assessments

Code: The WinMACCS code has been used to calculate plume dispersion and dose risk assessment. The code includes three modules of ATMOS, EARLY and CHRONC which evaluate atmospheric dispersion, emergency phase impact, and intermediate/long-term impact, respectively [16].

Source Term: The most important information in consequence analysis is source term data which can be released to the environment. The source term obtained from the Level 2 PSA of APR-1400 consists of seventeen source term categories (STCs). The accident set of the early containment failure includes 9 STCs and that of the late containment failure contains the other 6 STCs (see Table 4). In addition to the release quantity of radioactive material, the source term information includes the timing of the radioactive material release, the amount of energy associated, the release height and the predicted frequency of the release. 72 hours release duration after accidents was applied and the frequencies of internal accidents and external accidents including fire and earthquake were considered. For more conservative calculation, radioactive sources were assumed to be released from 0 m height and thermal plume rise was not considered. The radionuclides included in the analysis have been categorized according to their chemical properties. Table 2 shows nine categories of chemical group. The code library of radionuclides and decay chain data is "Indexr.dat" which can handle 825 kinds of radionuclides and six generation decay series. Database has been provided by the Radiation Shielding Information Center (1994) as a part of the FGR-DOSE/DLC-167 data package.

Meteorological Data: Hourly meteorological data of 2012's from National Center of Meteorology and Seismology (NCMS) have been sampled in the dispersion calculation. Due to the atmospheric stability class was not provided in the meteorological data from NCMS, it was derived from the categorization method of modified Pasquill stability classes [17], which is described in Table 3. The meteorological data consist of wind direction, wind speed, atmospheric stability, and accumulated precipitation. The ground surface roughness length representing the desert environment was applied as 0.03 cm to calculate some parameters used in plume dispersion modeling.

Population Distribution: The recent and projected population distributions near NPP site have been obtained from local government statistics available during the period of years 2001 through 2100. The 80km radius area around the plant was divided into sixteen directions that are equivalent to a standard navigational compass rosette. This rosette was further divided into 10 "inner" rings, each with sixteen azimuthal sections. The projected population in 2050 has been applied in this assessment considering the operating reactor and assuming high population density conservatively. Usually, the range of 1.6 km (1 mile) has been applied to calculate early fatality risk in Korea and USA. However, there is no resident within 2 km in the Barakah site and the area of 4 km distance from the center point of the NPP was considered by carrying out distance-population depended sensitivity analysis. The range of 8 km was applied to the evaluation of latent cancer fatality risk.

Countermeasures: As planned in the radiological emergency response plan, countermeasures include such activities as sheltering, evacuation, and dose-dependent relocation. Sheltering and evacuation have been excluded to consider more conservative situations. Dose-dependent relocation and KI ingestion were considered.

Dose Calculation: The WinMACCS code includes five pathways in early exposure scenario: (1) direct external exposure to radioactive material in the plume (cloudshine), (2) exposure from inhalation of radionuclides in the cloud (cloud inhalation), (3) exposure to radioactive material deposited on the ground (groundshine), (4) inhalation of resuspended material (resuspension inhalation), and (5) skin dose from material deposited on the skin [18]. Dose Conversion Factors (DCFs) in Federal Guidance Report 13 (FGR-13) issued by the Environmental Protection Agency (EPA) have been applied to the dose calculation [19].

Results: Early and cancer fatalities have been calculated and then corresponding risks have been obtained with population data. Table 4 shows the results of each STC and the averaged values.

Table 2: Chemical Group of Radionuclides

Chemical Group	Radionuclides
1. Inert Gases	KR-85, Kr-85m, Kr-87, Kr-88, Xe-133, Xe-135
2. Iodine	I-131, I-132, I-133, I-134, I-135
3. Cesium	Rb-86, Cs-134, Cs-136, Cs-137
4. Tellurium	Sb-127, Sb-129, Te-127, Te-127m, Te-129, Te-129m, Te-132, Te-131m
5. Strontium	Sr-89, Sr-90, Sr-91, Sr-92
6. Ruthenium	Co-58, Co-60, Mo-99, Tc-99m, Ru-103, Ru-105, Ru-106, Rh-105
7. Lanthanum	Y-90, Y-91, Y-92, Y-93, Zr-95, Zr-97, Nb-95, La-140, La-141, La-142, Pr-143, Nd-147, Am-241, Cm-242, Cm-244
8. Cerium	Ce-141, Ce-143, Ce-144, Np-239, Pu-238, Pu-239, Pu-240, Pu-241
9. Barium	Ba-139, Ba-140

Table 3: Modified Pasquill Stability Classes

Wind Speed [m/s]	Daytime Incoming Solar Radiation [W/m^2]				Within 1 hr before Sunset or after Sunrise	Nighttime Cloud Amount [oktas]		
	Strong (> 600)	Moderate (300-600)	Slight (< 300)	Overcast		0-3	4-7	8
≤ 2.0	A	A-B	B	C	D	F or G	F	D
2.0-3.0	A-B	B	C	C	D	F	E	D
3.0-5.0	B	B-C	C	C	D	E	D	D
5.0-6.0	C	C-D	D	D	D	D	D	D
> 6.0	C	D	D	D	D	D	D	D

Table 4: Number and Risk of Early and Cancer Fatalities

STC	Early/Late Containment Failure	Early Fatality (4 km)		Cancer Fatality (8 km)	
		Number (Mean)	Risk (Mean)	Number (Mean)	Risk (Mean)
1	Early	0.00E+00	0.00E+00	4.58E+00	3.34E-03
2	Early	1.23E+00	8.95E-04	2.05E+01	1.50E-02
3	Early	0.00E+00	0.00E+00	3.89E+00	2.84E-03
4	Early	0.00E+00	0.00E+00	3.53E+00	2.57E-03
5	Early	5.71E-03	4.16E-06	1.17E+01	8.56E-03
6	No Failure	0.00E+00	0.00E+00	4.99E-06	3.65E-09
7	No Failure	0.00E+00	0.00E+00	9.44E-05	6.89E-08
8	Late	0.00E+00	0.00E+00	1.84E-01	1.34E-04
9	Early	0.00E+00	0.00E+00	9.18E-01	6.70E-04
10	Early	0.00E+00	0.00E+00	1.96E+00	1.43E-03
11	Early	0.00E+00	0.00E+00	5.40E+00	3.94E-03
12	Early	0.00E+00	0.00E+00	2.06E+00	1.51E-03
13	Late	0.00E+00	0.00E+00	6.81E-02	4.97E-05
14	Late	0.00E+00	0.00E+00	1.94E-01	1.42E-04
15	Late	0.00E+00	0.00E+00	1.75E-03	1.28E-06
16	Late	0.00E+00	0.00E+00	1.77E-01	1.29E-04
17	Late	0.00E+00	0.00E+00	1.73E-03	1.27E-06
Average		*1.37E-01	*9.99E-05	**3.68E+00	**2.69E-03

*the value was averaged only for early containment failure cases
**the value was averaged only for both early and late containment failure cases

3.4. Development of Risk-Informed Safety Goal Criteria

Early risk: Individual Early Risk (IER) is calculated based on data and phenomena associated with the large early release of radioactive materials, as follows [20]:

$$IER = \sum_{n=1}^{N} LERF_n \times CPEF_n = \sum_{n=1}^{N} LERF_n \times \frac{EF_n}{TP} \quad (1)$$

where,
$LERF_n$: occurence frequency of source - term release category (STRC) - n
$CPEF_n$: conditional probability of early fatality near the site in the release case of STRC - n
EF_n : the number of fatality near the site in the release case of STRC - n
TP : the number of population near the site

$CPEF_{AVG}$ can be evaluated with weight-averaged $LERF_n$.

$$CPEF_{AVG} = \frac{\sum_{n=1}^{N} CPEF_n \times LERF_n}{\sum_{n=1}^{N} LERF_n} \quad (2)$$

Then, Equation (1) can be reduced into:

$$IER = CPEF_{AVG} \times \sum_{n=1}^{N} LERF_n = CPEF_{AVG} \times LERF \quad (3)$$

where,

$$LERF = \sum_{n=1}^{N} LERF_n$$

Cancer risk: Individual Latent Risk (ILR) is calculated based on data and phenomena associated with STRC, as follows [20]:

$$ILR = \sum_{n=1}^{N} F_n \times CPLF_n = \sum_{n=1}^{N} F_n \times \frac{LF_n}{TP} \quad (4)$$

where,
F_n : occurence frequency of source - term release category (STRC) - n
$CPLF_n$: conditional probability of latent fatality near the site in the release case of STRC - n
LF_n : the number of cancer fatality near the site in the release case of STRC - n
TP : the number of population near the site

$CPLF_{AVG}$ can be evaluated with weight-averaged F_n:

$$CPLF_{AVG} = \frac{\sum_{n=1}^{N} CPLF_n \times F_n}{\sum_{n=1}^{N} F_n} \quad (5)$$

Then, Equation (4) can be reduced into:

$$ILR = CPLF_{AVG} \times \sum_{n=1}^{N} F_n = CPLF_{AVG} \times CFF$$

where, (6)

$$CFF \text{ (Conditional Failure Frequency)} = \sum_{n=1}^{N} F_n$$

In Equation (6), if it is conservatively assumed that all the radioactive materials released into containment are released into the environment, that is, CCFP (Conditional Containment Failure Probability)=1, the ILR can be evaluated as follows:

$$ILR = CPLF_{AVG} \times CDF$$

where, (7)

CDF : Core Damage Frequency

The early and cancer risks to public in the UAE NPP site were estimated based on the Level 2 PSA results and site-specific data described in the previous section. WinMACCS code was used for the evaluation of conditional probability of early and cancer fatality (CPEF and CPLF). The CPEF and the CPLF are then compared with the QHOs for early and cancer risks to determine the corresponding performance criteria for CDF and LERF, respectively, as follows:

$$IER = CPEF_{AVG} \times LERF < QHO_{early} \qquad (8)$$

$$ILR = CPLF_{AVG} \times CDF < QHO_{cancer} \qquad (9)$$

If the crude QHO values of 2.87×10^{-7}/year for early fatality and 5.92×10^{-7}/year for cancer fatality calculated in Section 3.2 and the conservative values of 9.99×10^{-5} for $CPEF_{AVG}$ and 2.69×10^{-3} for $CPLF_{AVG}$ calculated in Section 3.3 were applied, the risk-informed safety goals for CDF and LERF should less than 2.20×10^{-4}/yr and 2.87×10^{-3}/yr, respectively.

3.5. Discussions on Risk-Informed Safety Goal Criteria

In this study, conservative assumption and data were used for the evaluation of the risk-informed safety goals as follows:

- Conservative source term assumptions: the release height, thermal plume rise (see Section 3.3)
- Assumed high population density near the plant
- Assumptions of no evacuation and sheltering for the countermeasures
- Assumption of CCFP=1: usually it is required that CCFP < 0.1, especially newly constructed plants as a requirement or recommendation.

The values of the CDF ($< 10^{-5}$/yr) and the LRF ($< 10^{-6}$/yr) as probabilistic safety targets in the UAE [8] had been set lower than those considered in other countries such as USA (CDF< 10^{-4}/yr and LERF< 10^{-5}/yr; for new plants LERF< 10^{-6}/yr) and South-Korea (CDF< 10^{-4}/yr and LERF< 10^{-5}/yr) [10]. The LRF includes LERF and LLRF (Large Late Release Frequency). The large late release doesn't contribute to the early fatality risk (the risk contribution to early fatality = 0) but contributes to the cancer fatality risk, as shown in Table 4. In the $CPEF_{AVG}$ calculation, averaging only for LERF cases makes more conservative results, which is the reason why the LERF is adopted as a surrogate measure for the early fatality risk instead of the LRF. In the APR-1400 PSA, the value of LERF was estimated almost same as that of LLRF. Hence the doubled value ($2 \times 2.87 \times 10^{-3}$/yr = 5.74×10^{-3}/yr) of the risk-informed safety goal for LERF can be used for the comparison with the safety target for LRF. Finally, the probabilistic safety targets for CDF ($< 10^{-5}$/yr) and LRF ($< 10^{-6}$/yr) in the UAE [8] are concluded to be set sufficiently lower in comparison with the conservatively evaluated criteria of the CDF (2.20×10^{-4}/yr) and the doubled LERF (5.74×10^{-3}/yr) in terms of engineering judgment.

4. LIMITATIONS AND RECOMMENDATIONS

4.1. Limitations of the study

Quality of PSA: Data and information from current PSAs for APR-1400 were used in this study. The PSAs were prepared for the licensing of Barakah NPPs as construction PSAs. A lot of generic data and assumptions were used for the PSAs. As living PSAs are available with the operation of the plant, this study should be updated with quality data and assumptions.

Uncertainty and Sensitivity Analysis: Probabilistic analyses are always associated with uncertainty. To cope with uncertainty, sufficient conservatism or high quality uncertainty and sensitivity analyses are widely adopted for safety analyses in nuclear industries. Even though this study is based on sufficient conservatism for determining the safety goals, a wide range of uncertainty and sensitivity analyses was not performed in this study. If a wide range of uncertainty and sensitivity analyses is performed in this framework, more reasonable evaluation can be accomplished, which is remained as a further study.

Statistical Data for General Risk Calculation: Data used for the calculation of general risk other than NPPs operation were obtained from literature and website surveys to competent authorities of relevance. However only several years' data were collected. More well-organized data base should be developed as a further study as well.

Data and Assumptions used in Public Health Risk Assessments: Usually public health risk assessments require a huge amount of data and assumptions some of which are directly collectable and others inferable from some rational calculations or logical approaches. US Nuclear Regulatory Commission (NRC) derives the atmospheric stability class by correlating with vertical temperature gradient but some required data were not available in this study. Instead, modified Pasquill stability classes were applied. NRC use 1.6 km (1 mile) for the consideration of early fatality. However, there is no resident in 1.6 km in the Barakah site so 4.0 km distance has been chosen to calculate early fatality by the sensitivity analysis of the population weighted distance. Evacuation and sheltering was not considered conservatively and ingestion scenario was excluded in this study because of absence of farmland, few residence areas, and the lifestyle near the Barakah site.

4.2. Recommendations for Application

Limits or Objectives: due to the uncertainty and insufficient analysis scope of PSA, the risk-informed safety goals should be recommended as objectives rather than strict limits, even though some countries which have sufficient PSA experience and technologies adopt the quantitative safety goals as regulatory limits.

Future NPPs and Multi-unit Site: to get public acceptance for future reactors, the increase in risk due to the addition of new NPPs should be as low as possible. Also there have been issues regarding multi-unit sites. The total risk of a multi-unit site is not always expressed as the numerical sum of all reactors. Hence, it is desirable for future NPPs and/or multiple modular NPPs to have lower safety goals compared to those of operating NPPs. The question on "how low it should be" should be studies in a comprehensive way considering type and number of NPPs to be constructed, constructions and operating experience, quality of PSA technologies applied, and so on.

Scope of Analysis: current PSA technologies have been evolved to quantify most of the risk factors reasonably. However, in reality, risks from external events or low power and shutdown states still have large uncertainties in their analyses. Therefore, consideration should be taken into account for level of analysis technology and uncertainties.

5. CONCLUSION

In this study, a framework for establishment of risk-informed safety goals for NPPs operations in the UAE was developed and under this framework a conservative evaluation of risk-informed safety goals was performed on the basis of conservative assumptions and data. The current safety targets specified in the regulatory guideline (FANR-RG-004) in the UAE were examined to be appropriately determined with sufficient conservatism from the evaluation results. However, considering the limitations coupled with insufficient quality and scope of PSAs considered, lacks of uncertainty and sensitivity analyses, insufficient statistical data for the general risk calculation, and data and assumptions in public health risk assessments, the conservative evaluation done in this study should be considered as a first attempt and further studies should be performed to generate more reasonable evaluation on risk-informed safety goals (or targets) under the developed framework.

References

[1] UAE Government, Policy of the United Arab Emirates on the Evaluation and Potential Development of Peaceful Nuclear Energy, 2008: http://www.fanr.gov.ae/En/Documents/whitepaper.pdf

[2] UAE Government, Federal Law by Decree No 6 of 2009, Concerning the Peaceful Uses of Nuclear Energy, 2009: http://www.fanr.gov.ae/En/Documents/20101024_nuclear-law-scan-eng.pdf

[3] Federal Authority for Nuclear Regulation (FANR), Safety Evaluation Report of the Construction Licence Application for Barakah Units 1 and 2, 2012: http://www.fanr.gov.ae/en/AboutFANR/OurWork/Documents/Safety%20Evaluation%20Report%20of%20an%20Application%20for%20a%20Licence%20to%20Construct%20Barakah%20Unites%201%20and%202.pdf

[4] Federal Authority for Nuclear Regulation (FANR), Regulation for the Design of Nuclear Power Plants (FANR-REG-03), 2010: http://www.fanr.gov.ae/En/RulesRegulations/RegulationsGuides/Documents/20110629_fanr-reg-03-npp-design-with-cover.pdf

[5] International Atomic Energy Agency (IAEA), Safety of Nuclear Power Plants: Design, No. NS-R-1, 2012: http://www-pub.iaea.org/MTCD/publications/PDF/Pub1534_web.pdf

[6] Federal Authority for Nuclear Regulation (FANR), Regulation for Radiation Dose Limits and Optimisation of Radiation Protection for Nuclear Facilities (FANR-REG-04), 2010: http://www.fanr.gov.ae/SiteAssets/PDF/REg/20101128_fanr-reg-04-dose-limits-version-0-english-13.07.10.pdf

[7] Federal Authority for Nuclear Regulation (FANR), Application of Probabilistic Risk Assessment (PRA) at Nuclear Facilities (FANR-REG-05), 2010: http://www.fanr.gov.ae/SiteAssets/PDF/REg/20110227_fanr-reg-05-eng.pdf

[8] Federal Authority for Nuclear Regulation (FANR), Evaluation Criteria for Probabilistic Safety Targets and Design Requirements (FANR-RG-004): http://www.fanr.gov.ae/SiteAssets/PDF/20110801_fanr-rg-004-final-approved-version.pdf

[9] U. S. Nuclear Regulatory Commission, An Approach for Estimating the Frequencies of Various Containment Failure Modes and Bypass Events, NUREG/CR-6959 (Rev.1), 2004.

[10] Do Sam Kim, Taesuk Hwang, Key Yong Sung, Jun Su Ha and Ingoo Kim, "*Status of Safety Goal Establishment for Nuclear Power Plants in Korea*," Technical Meeting on Safety Goals in Application to Nuclear Installations, IAEA, Vienna, Austria, April, 2011.

[11] U. S. Nuclear Regulatory Commission, Policy Statement: Safety Goals for the Operations of Nuclear Power Plants, August 1986.

[12] Korean Nuclear Safety Commission, Policy on Severe Accident, August 2001.

[13] Ministry of Health, UAE, Cancer Incidence Report of UAE from 1998 to 2001.

[14] Ministry of Health, UAE, http://www.moh.gov.ae/en/OpenData/Pages/2011.aspx.

[15] World Health Rankings website, http://www.worldlifeexpectancy.com/cause-of-death/all-cancers/by-country/.

[16] McFadden, K., Bixler, N.E., Cleary, V.D., Eubanks, L., Haaker, R., WinMACCS, a MACCS2 interface for calculating health and economic consequences from accidental release of radioactive materials into the atmosphere, User's Guide and Reference Manual WinMACCS Version 3, Sigma Software LLC, Sandia National Laboratories, Gram Inc., AQ Safety, Inc., 2007.

[17] Manju Mohan and T. A. Siddiqui, "*Analysis of Various Schemes for the Estimation of Atmospheric Stability Classification,*" Atmospheric Environment Vol. 32, No 21, pp. 3775-3781, 1998.

[18] Chanin, D., Young, M.L., Randall, J., Jamali, K. Code Manual for MACCS2: Volume 1, User's Guide. Technadyne Engineering, Sandia National Laboratories, US Nuclear Regulatory Commission, US Department of Energy, 1998.

[19] Keith F. Eckerman, Richard W. Leggett, Christopher B. Nelson, Jerome S. Puskin and Allan C. B. Richardson, Cancer Risk Coefficients for Environmental Exposure to Radionuclides - Federal Guidance Report No. 13, Oak Ridge National Laboratory and United States Environmental Protection Agency, 1999.

[20] U. S. Nuclear Regulatory Commission, Feasibility Study for a Risk-Informed and Performance-Based Regulatory Structure for Future Plant Licensing, NUREG-1860, 2007.

Insights from PSA Comparison in Evaluation of EPR Designs

Ari Julin[*a], Matti Lehto[a], Patricia Dupuy[b], Gabriel Georgescu[b], Jeanne-Marie Lanore[b], Shane Turner[c], Paula Calle-Vives[c], Anne-Marie Grady[d], Hanh Phan[d]

[a] Radiation and Nuclear Safety Authority (STUK), Finland
[b] Institute of Radiological Protection and Nuclear Safety (IRSN), France
[c] Office for Nuclear Regulation (ONR), United Kingdom
[d] Nuclear Regulatory Commission (USNRC), United States of America

Abstract: The paper describes the outcome of a limited probabilistic safety assessment (PSA) comparison on the following EPR designs: Olkiluoto 3 Nuclear Power Plant (NPP) in Finland, Flamanville 3 NPP in France, UK EPR design, and U.S. EPR design. The objective of this PSA comparison was to identify differences in the modeling aspects and results of EPR PSAs, as well as to assess the rationale for these differences. The comparison covered various types of initiators challenging a broad scope of safety functions. Insights from the EPR PSA comparison and rationale for the differences originated from modeling assumptions, applied reliability data, designs, and operational aspects. The EPR designs chosen for comparison represents various design and licensing stages, as well as level of detail, which gives the main rationale for the identified differences. The outcomes and lessons learned from the EPR PSA comparison have been used to facilitate the regulatory reviews and assessment work of various EPR designs and to enhance the scope, level of detail, and quality of EPR PSA models and documentation.

Keywords: PSA, EPR, Licensing, Regulation, Design Evaluation

1. INTRODUCTION

The EPR is an Evolutionary Pressurized Water Reactor (a.k.a. European Pressurized Water Reactor), whose design takes benefit from operating experience especially in France and Germany. Design improvements have been introduced to enable more reliable prevention and mitigation of severe accidents. EPR PSA development was initiated from the beginning of the conceptual design stage. At the end of the basic design phase, Level 1 PSA for internal initiating events as well as the so called Level 1+ PSA, to estimate the frequency of potential failures of the containment, taking into account measures for severe accident mitigation were completed. Later, Level 2 PSA and hazards PSA were developed. PSA has been used during the design process in order to optimize the design with respect to safety and availability [1].

EPR PSA comparison was performed by the Radiation and Nuclear Safety Authority of Finland (STUK), Institute of Radiological Protection and Nuclear Safety (IRSN) of France, Office for Nuclear Regulation (ONR) of the United Kingdom, and United States Nuclear Regulatory Commission (USNRC) within the Multinational Design Evaluation Program (MDEP) design specific working group on the Evolutionary Power Reactor (EPR). The comparison was conducted on the following EPR designs: Olkiluoto 3 Nuclear Power Plant (NPP) in Finland, Flamanville 3 NPP in France, UK EPR design, and U.S. EPR design, respectively.

MDEP was established in 2006 as a multinational initiative to develop innovative approaches to leverage the resources and knowledge of the national regulatory authorities who are currently or will be tasked with the review of new reactor power plant designs. The Organization for Economic Co-Operation and Development (OECD) Nuclear Energy Agency (NEA) facilitates MDEP's activities by acting as technical secretariat for the program.

---

[*] ari.julin@stuk.fi

The objective of this PSA comparison was to identify differences in the modeling aspects and results of EPR PSAs, as well as to assess the rationale for these differences. The PSA comparison exercise was aimed to provide support for safety evaluations and PSA reviews in MDEP member countries.

The scope was limited to the following four initiating events (IEs): medium loss-of-coolant accident (LOCA), loss of offsite power (LOOP), steam generator tube ruptures (SGTR), and loss of cooling chain (LOCC). The selection covered various types of initiators challenging a broad scope of safety functions. The comparison focused on the IE definition, modeling of accident sequences (i.e., timing, safety functions, automatic and manual actions, etc.), minimal cut sets, importance measures, and quantitative results.

2. LICENSING AND PSA REQUIREMENTS

The licensing process is country specific, but it contains many similarities. PSA is a licensing document and a full scope PSA is required at the latest in the operating license phase. Licensing steps, status of licensing process and the role of PSA in France, UK, USA and Finland are described in more detail in the following subsections.

France

In accordance with the "Technical Guidelines" [2], the safety demonstration for the nuclear power plants of the next generation has to be achieved in a deterministic way, supplemented by probabilistic methods. In the frame of the construction license application of Flamanville 3 (FA3) reactor (2006) EDF provided a Level 1 PSA for internal events for the reactor and fuel pool, and a Level 1+ PSA and simplified analysis for internal and external hazards. For the FA3 operating license application, EDF will provide, according with the French safety requirements, a full scope Level 1 and Level 2 PSA (internal events and hazards). Some of these PSAs are already being reviewed by IRSN in the frame of so called "anticipated instruction" of the operating license.

UK

ONR has developed a process of generic design assessment (GDA) [3] for new reactor designs. Under the GDA process ONR assesses the safety case for the generic design of a specific type and make of reactor. ONR expects that the submission for design acceptance should include a full scope Level 1 and Level 2 PSA. The PSA should be used to help show that the design satisfies the requirement to reduce risk as low as reasonably practicable (ALARP). A Level 3 PSA relevant to the generic site will also be expected. The PSA for the UK EPR was assessed by ONR as part of GDA [4].

Prior to start of nuclear safety-related construction of a new reactor the responsible body (the licensee) would have to hold a nuclear site license [5]. ONR will then ordinarily use the primary power provided by License Condition (LC) 19 (4) [6] to specify that the licensee should not commence nuclear safety-related construction without a regulatory Consent. Throughout construction and installation, ONR may employ LC19 (4) to identify further 'hold points' where ONR Consent is required before the licensee may proceed from one stage to the next. For each stage, a safety case would be submitted to support the licensee's request to move from one stage to the next. Safety cases commonly produced include: pre-construction safety case, pre-inactive commissioning safety case, pre-active commissioning safety case, pre-operational safety case and operational safety case. For each safety case ONR expects that a full scope, site specific Level 1, 2 and 3 PSA would be included. This PSA would need to be aligned to the relevant reference design for the specific stage. The licensee has submitted to ONR an initial pre-construction safety report (PCSR) for the construction of two EPRs at Hinkley Point C [7]. However, this PCSR will be updated by the licensee prior to requesting consent to start construction of the nuclear island at Hinkley Point C.

Ultimately it is ONR's expectation that a full scope site specific Level 1, 2 and 3 symmetric PSA is produced to support operation that is consistent with international good practice and is capable of supporting a risk monitor application.

ONR expectations relevant to PSA can be found in its Safety Assessment Principles [8] (SAPs) and in the ONR technical Assessment Guide (TAG) on PSA, TAG 030 [9]. ONR is also guided in its safety case assessments by certain numerical targets in the SAPs. In assessing against these, ONR will seek sufficient information for it to be able to judge that the targets are likely to be achieved and that the overall risk is ALARP.

USA

The PSA is performed to support the design certification of the U.S. EPR. The principal objectives of this analysis are:
- to demonstrate that the design poses an acceptably low risk of core damage accidents;
- to identify opportunities for effective and timely improvements during the design phase, through a systematic assessment of the design;
- to provide the foundation for a plant-specific PSA for the combined operating license (COL) and operating phases.

The COL applicant that references the U.S. EPR design certification will either confirm that the PSA in the design certification bounds the site-specific design information and any design changes or departures, or update the PSA to reflect the site-specific design information and any design changes or departures.

This PSA is a Level 1 and Level 2 PSA and addresses the risks associated with nominal full-power operation, low-power operation, and shutdown conditions. The PSA assesses both internal and external events (except acts of sabotage).

The PSA assesses risk for comparison against the Commission's safety goals: core damage frequency (CDF) less than 1E-4/year and large release frequency (LRF) less than 1E-6/year; and containment performance goals: containment integrity be maintained for approximately 24 hours following onset of core damage for the more likely severe accident challenges and the conditional containment failure probability (CCFP) less than approximately 0.1 for the composite of all core damage sequences assessed in the PSA.

The design certification application for U.S. EPR is still under review. USNRC has completed phase three out of six in the safety review process. It is not known when the final safety evaluation report will be available.

Finland

The foundation for the risk informed safety management is laid in the nuclear safety legislation. Detailed regulations called YVL Guides are issued by STUK. As a necessary complement to the deterministic safety design, a PSA is required to verify the reliability of all vital safety functions and the balance of the design features.

A plant specific, design phase Level 1 and 2 PSA is required as a prerequisite for issuing the construction license, and a complete Level 1 and 2 PSA for issuing the operating license. The plant specific PSA includes internal initiators, internal hazards (fires, floods, missiles, etc.) and external hazards (harsh weather conditions and seismic events, etc.) analyzed in all operating modes. In each licensing phase PSA has to be used to demonstrate that the following probabilistic design objectives, specified in the Regulatory Guide YVL A.7 [10], will be met:
- mean value of CDF is less than 1E-5/year; assessed and verified in full scope Level 1 PSA;
- mean value of LRF is less than 5E-7/year; assessed and verified in full scope Level 2 PSA.

PSA will be complemented during construction as the detailed design of the plant unit will be finalized. Design has to be modified unless these objectives are met. If dominant risk factors are identified after issuing a construction license, all reasonable efforts have to be taken to reduce the risk.

During construction, PSA shall be updated to comply with the detailed design information of systems, structures and components (SSC) and more detailed modeling of plant response to various initiating events. The fulfillment of the aforementioned numerical criteria for CDF and LRF has to be demonstrated as well.

In addition, several PSA applications have been required in Regulatory Guides as a condition for construction and operating licenses. Examples of required risk informed PSA applications include Pre- and In-Service Inspection (RI-PSI/ISI), In-Service Testing (RI-IST), Technical Specifications (RI-TS), Safety Classification of SSCs (RI-SC), staff training, and identification of potential design changes and/or plant modifications.

In Olkiluoto 3 (OL3) project, risk informed approach has been applied in a large scale for the first time in the design, construction and commissioning of a new NPP unit in Finland.

3. EPR PSA COMPARISON

3.1. Development of EPR PSAs

The first Level 1 PSA for internal initiating events was completed at the end of the basic EPR design in 1999. This PSA model and documentation has been utilized in the further development of the first versions of EPR PSAs for Olkiluoto 3 and Flamanville 3 NPPs. Since then, the OL3 construction license PSA (2004) has been updated several times in the course of the detailed design process more or less independently from other EPR PSAs. OL3 PSA (2004) was used in the development of U.S. EPR PSA for Design Certification process in 2007. PSA for UK EPR GDA process was at least partially based on the three aforementioned PSAs: OL3 (2004), FA3 (2006) and U.S. EPR (2007). Although EPR PSA developers are exchanging PSA information and findings, each EPR PSA has been extended and updated in accordance with his own project specific requirements while the licensing and/or the detailed design processes have progressed.

3.2. Information on Selected EPR PSAs and Documentation

The analysis of internal initiating events constitutes the backbone of any plant specific PSA. EPR designs under the review represent various stages of the design process, licensing process, as well as level of modeling detail. Some PSAs are more or less so called full scope PSAs in terms of the coverage of initiating events i.e. internal IEs, and internal and external hazards are included in the analyses. The others include somewhat limited analysis of hazards. Therefore internal events PSAs were selected for EPR PSA comparison effort. The following subsections provide more information on the status and details of PSAs and related documentation chosen for the comparison. The source of the background information on the EPR PSAs is summarized in Table 1.

Table 1: EPR PSA Models and Documentation

	PSA information source (design stage)
FA3	Final Safety Analysis Report (FSAR) (2010)
UK EPR	GDA step 4 (2011) [4], GDA PCSR (2011) [11]
OL3	pre-Operating License Application (pre-OLA, v104, 2010)
U.S. EPR	Design Certification (DC) rev. 5 + PSA (2013)

Flamanville 3 NPP

The FSAR 2010 version of the Level 1 PSA internal events is an update of the PSA version provided by EDF and analyzed by IRSN in the frame of the construction license application in 2006. It considers the conclusions of the 2006 instruction and the design evolution until 2009. The updated version of this PSA provided by EDF in the frame of the operating licensee, will include the results of the "anticipated instruction" by IRSN (done in 2010 and 2013) of the FSAR 2010 version and of the subsequent updates as well as the final design and operation. However, due to inherent difficulties in developing a design PSA, some aspects will be finalized later, before starting commercial operation (like for example, the detailed human reliability analysis (HRA) based on the finalized procedures or the detailed modeling of maintenance).

UK EPR

The UK EPR GDA PCSR 2011 [11] version of the PSA model was considered as part the comparison exercise. This was assessed by ONR during the GDA process. This is a Level 1, 2 and 3 PSA that considers both internal events, and internal and external hazards. The Level 1 PSA also includes consideration of all non-power operating states. The scope of this PSA excluded any requirement on the PSA modeling that needed detailed design information or site specific data beyond the scope of GDA.

Updates have since been made to the PSA model to account for site specific features at Hinkley Point C, including, for example, site specific heat sink modeling, site specific loss of ultimate heat sink frequency and site specific loss of off-site power frequencies. The revised PSA has been provided to ONR to support the licensee's initial site specific pre-construction safety report [7]. Further updates to the PSA are anticipated as the detailed design progresses and as procedures are developed.

U.S. EPR

The U.S. EPR Level 1 PSA 2013 is a revision of the original 2009 Level 1 PSA, and is based on the design input through September 2012 and its supporting documentation. The 2013 PSA also reflects some modeling and data changes. The 2009 and 2013 PSA significant initiating event contributions to CDF for internal events are quite similar, with LOOP and small LOCA (SLOCA) still among the most important contributors.

Design changes resulting from earlier PSA insights include: increased safety related chiller capacity; each component cooling water header can provide seal cooling to all four reactor coolant pumps (RCPs); and, the four emergency feed water system (EFWS) storage tanks cross tie valves have been closed.

Pending design changes will be assessed against the PSA model periodically and the cumulative impact on the CDF will be determined and documented. If the impact on the cumulative CDF is less than 10 percent (positive or negative), then no further action will be taken. If the impact on the cumulative CDF is greater than 10 percent (positive or negative), then further impact on the PSA will be evaluated.

Olkiluoto 3 NPP

OL3 PSA has been updated several times during the construction and detailed design process. Hundreds of design changes ranging from minor to major have been implemented since the start of the construction in 2005. The PSA documentation and model chosen for EPR PSA comparison is based on the situation around the end of 2010 and the so-called pre-operating license application FSAR documentation.

Changes in the OL3 risk profile are foreseen due to the finalization of the detailed instrumentation and control (I&C) design and more detailed and realistic modeling of internal hazards, especially fires.

3.3. Main Results of EPR PSAs

Table 2 presents the results of four different EPR designs' internal events PSAs for power operating modes. The total CDFs are fairly similar but the risk profiles are not identical. Based on the experience from previous PSA comparisons performed e.g. in France and Finland, it was evident that the comparison should not focus on only those IEs, which CDF differs the most. Even with similar CDFs, significant difference may be identified related to IE frequencies, most important cut sets, modeling details, most important basic events, assumptions etc. Therefore the selection of candidate IEs was focused on those initiators challenging a broad scope of safety functions. Finally, the following four initiating events were chosen for comparison: medium loss-of-coolant accident, loss of offsite power, steam generator tube rupture(s), and loss of cooling chain.

Table 2: EPR PSA internal events CDF (1/a)

IE	DESCRIPTION	FA3	UK EPR[A]*	U.S. EPR*	OL3*
LOOP	**Loss of Offsite Power**	**1,40E-07**	**2,97E-07**	**1,23E-07**	**1,33E-07**
LOCA	Loss of primary coolant accident	5,70E-08	1,06E-07	4,48E-08	7,08E-08
MLOCA	*Medium LOCA*	*(3,6E-08)*	*(9,2E-09)*	*(9,1E-10)*	*(3,1E-08)*
V-LOCA	LOCA leading to containment bypasses	6,50E-10	3,70E-09	2,99E-09	1,50E-08
Prim-Tr	Primary circuit transients	2,00E-08	5,25E-08[D]	-	1,07E-08
Sec-Tr	Secondary circuit transients	4,60E-09	1,63E-08	1,37E-08	8,37E-08
Sec. Br.	Secondary circuit breaks	1,80E-08	1,3E-08	1,66E-09	8,88E-09
SGTR	**Steam Generator Tube rupture(s)**	**1,10E-08**	**1,02E-08**	**2,63E-08**	**2,21E-08**
LOCC	**Loss of cooling chain or heat sink**	**8,80E-08**	**9,46E-08**	**3,61E-08**	**1,94E-08**
ATWS	Anticipated Transient w/o Scram	1,00E-07	2,14E-08	8,95E-09	(1,84E-08[B])
LV-bus	Loss of low voltage busbars	2,50E-09	-	-	-
I&C	Spurious I&C actions	3,50E-08	-	-	-
IND SGTR	Induced SGTR	-	4,35E-09	8,50E-09	-
BDA	Loss of 6.9kV Power from Bus BDA	-	-	1,14E-08	-
GT	General Transient (Includes Turbine Trip and Reactor Trip)	-	-	2,02E-08	-
CCI-SAC	Loss of SAC divisions 3 & 4 due to CCF	-	-	-	2,50E-09
PSD	Planned Shutdown (pseudo IE) [C] - I&C passive CCFs dominate the result	-	-	-	1,18E-07
	TOTAL	**4,8E-07**	**6,2E-07**	**3,0E-07**	**4,8E-07**

[A] *PSA for a UK EPR[TM] at Hinkley Point C [7]*
[B] *Modeled together with related transients, not as a separate IE*
[C] *Event sequences which may occur (only) during a planned shutdown maneuver*
[D] *This includes some contribution for non at power operating states*
* *At power operating states*

3.4. Medium LOCA Comparison

IE definitions and plant response

Following a LOCA there are several scenarios possible depending on the break size. The definition of the LOCA categories is based on the accident mitigation means required depending on the impact of the break size on the reactor and given by the results of thermal-hydraulic analysis performed to support the PSA.

During at-power states, the LOCA scenarios typically modeled in the EPR PSAs studied result in a depressurization of the reactor coolant system, a decrease of pressurizer level and an increase in the pressure in the containment. This results in a reactor and turbine trip, and the initiation of the safety injection systems. In particular, the partial cooldown is initiated in order to decrease the reactor coolant system pressure to allow the required safety injection. Cooldown is performed by releasing the steam from the steam generators (SG) via the steam dump to the condenser or to the atmosphere.

Typical accident sequences and progression

The accident scenarios are similar in all the compared PSAs. Following a medium LOCA, partial cooldown is initiated in order to allow medium head safety injection into the cold legs. If partial cooldown fails, actuation of the primary circuit feed and bleed function is necessary.

The safety injection system accumulators discharge cold water to ensure complete quenching.

If the medium head safety injection system trains are unavailable, fast secondary cooldown can be manually actuated to reduce reactor coolant system pressure sufficiently to allow low head safety injection into the cold legs.

IRWST cooling is performed with one residual heat removal system train or containment heat removal system (CHRS) train.

Main differences

The main differences identified were the assumed medium LOCA frequencies. There is a significant difference between OL3/FA3 and UK/US medium LOCA frequencies. The UK and US medium LOCA frequencies' data source is NUREG 1829. The medium LOCA frequencies used in the FA3 and OL3 medium LOCA models have their origin in studies developed in France or in Germany respectively.

The conditional core damage probability is similar in all the PSAs. However, the following important differences have been identified:
- There are differences in the range of break size considered in the medium LOCA category. The main reason for this difference seems to be the thermal-hydraulic support studies available to the PSA analysts. The support studies are specific to each project except for the UK EPR PSA. The OL3 and the UK EPR PSAs studied in this comparison share the same support studies. However, some small differences exist in LOCA category definition between both PSAs. According to the information provided by the "EPR family", these differences seem to be due to modeling assumptions regarding the break spectrum.
- There are differences in the reliability data and human error probabilities. The impact in the overall medium LOCA results is negligible. However, the impact of these differences on the overall CDF may be more significant, but it was not in scope of the comparison exercise.
- There seems to be important differences in the treatment of digital I&C amongst the different models. Some of these differences may be due to modeling assumptions, the detailed design information available during the development of the PSA and potential design differences.

- The main difference identified in the success criteria is that the four steam generators (no additional feed) are required to ensure the cooldown functions in the U.S. EPR PSA as opposed to one steam generator (with additional feed) required in the other PSAs studied. Although this is different, it is not necessary contradictory as the same objective could be achieved by different means. However, it is important to note that differences in the supporting analysis may result in a different pressure in the containment after a medium LOCA. Depending on the protection system design, differences in the containment pressure may have an impact on the state of the main steam isolation valves (open or close). This would have an impact on the number of steam generators required to ensure the cooldown function. On the basis of the information available for the study it was not clear if there are different settings in the logic to control the main steam isolation valves between the different EPRs. The requirements for secondary feed would depend on the number of steam generators claimed in the PSA. As indicated previously, a review of the supporting analyses was considered as out of scope of the comparison exercise. *NOTE!: The success criteria for the U.S. EPR has been changed. Similarly to the other EPRs PSA studies, one steam generator is required, a difference in the U.S. EPR PSA is that no additional feed is required.*
- There are also differences regarding the need for primary bleed (during primary feed and bleed) in case of the partial cooldown failure, between the UK PSA and the rest of the PSA models.

Potential reason(s) for differences

The main reasons for the differences identified seem to be differences in modeling assumptions and thermal-hydraulic analyses. There was insufficient information available to realize any significant design differences that could impact the results.

Areas and topics that require additional information

Assumptions regarding the thermal-hydraulic analyses were not considered in this comparison exercise. These studies are not fully representative of all the EPR designs at this stage so a comparison at this stage may not be meaningful.

There was insufficient information to understand the differences in reliability data and human error probabilities. It is recommended to study these differences for the dominant contributors to the overall PSA results. Furthermore, there was insufficient information regarding I&C modeling and design differences. In view of the I&C significance for most of the CDF scenarios, it is recommended to study specific differences in the I&C modeling and design.

3.5. Loss of Off-site Power Comparison

IE definitions and assumptions

The external electrical power supply of the EPR plant is provided by two electrical grids: main grid designed for the normal operating conditions; auxiliary grid in case of "main grid" failure. In general, three types of LOOP initiating events can be analyzed in the EPR PSAs:
- loss of main grid: this initiator is defined as the loss of the external main power supply only;
- short term loss of offsite power: total failure of the main and auxiliary grid for a short term;
- long term loss of offsite power: total failure of the main and auxiliary grid for a longer term.

All compared PSAs consider these initiating events, but the frequency and grouping of initiating events are slightly different. It was decided to continue the comparison only for the initiating events Short LOOP and Long LOOP. In all PSAs the recovery times considered for these two initiating events are 24 hours for the long LOOP and 2 hours for short LOOP. The origin of 2 hours border between short and long term LOOP is that, for EPR reactor, the diesel generators are not needed before 2 h (if no RCP seals LOCA). The IE frequencies used in the different PSAs are very similar, although coming from different specific data sources.

Typical accident sequences and progression

The accident scenarios are similar in all compared PSAs. Following the loss of the main and auxiliary grids the plant will be transferred to House Load Operation. In case of unavailability of the house load, the reactor trip is triggered. The turbine trip and the closure of Main Feed Water (MFW) large flow lines are also triggered. The Emergency Diesel Generators (EDGs) are started and connected to the safety busbars automatically. Since the MFW and the Startup and Shutdown System (SSS) pumps are not supplied by the EDGs, the SG level decreases leading to EFWS automatic actuation. The SG regulation is automatic. In case of EFWS unavailability primary Feed and Bleed (F&B) is necessary to avoid core damage. As the Component Cooling Water System (CCWS) and Chemical and Volume Control System (CVCS) are supplied by the EDGs, the reactor coolant pumps seals injection and the thermal barriers cooling is maintained (Note! In the U.S. EPR, CVCS is supplied from the station blackout DGs). In case of failures of these systems or theirs support systems, the Stand Still Sealing System (SSSS) will be automatically actuated in order to maintain the primary circuit integrity.

Similarities and differences

There is a large consistency among the four PSAs as regards to:
- the initiating events considered: short (2h) and long (24h) LOOP;
- the main CD sequences: in the four studies the main functional sequences are either a loss of heat removal (EFWS and F&B failure) mainly due to the failure of all the diesel generators, or a seal LOCA followed by a total loss of water injection;
- the overall results are very similar: about 1.3E-07 /year for the CDF relating to LOOP (as presented in Table 3 below) and a dominant contribution of LOOP to the total CDF.

Table 3: CDF (1/a) related to LOOP IE

	FA3	UK EPR	U.S. EPR	OL3
Short LOOP (<2h)	3,4E-08	1,4E-07	1,2E-07	1,1E-07
Long LOOP (< 24h)	9,5E-08			2,1E-08
Total	**1,3E-07**	**1,4E-07***	**1,2E-07**	**1,3E-07**

* *Result from PSA for GDA 2011 PCSR [11] inclusive of consequential LOOP. Note, PSA for Hinkley Point C, LOOP CDF ≈ 3E-07 /a*

However, with a more detailed analysis, differences were identified which could have an impact on results, especially in case of risk informed decision making (e.g. design optimization, TechSpecs, maintenance programs):
- Differences in success criteria and strategy in degraded situations: for example F&B is considered as possible with LPSI only in U.S. EPR PSA, the actuation of SBO diesel generators is possible only after the loss of all EDGs for FLA3. These differences seem to be due to procedures and to support calculations.
- Differences in the level and detail of modeling: for example modeling of ventilations, of electrical interconnections, of manual alignment of headers or other manual actions could lead to significant contributions.
- Differences in modeling and role of batteries, especially the need for 2h batteries for SBO diesels actuation and the CCFs between batteries.
- There are some differences in the reliability data and human errors probabilities, but it seems that they have not an important impact on the comparison.
- Significant differences appear in modeling and quantification of I&C: the CCFs identified and quantified, the account for diversified means (non-computerized) are not similar. The differences seem to arise from different modeling, assumptions and design.

It can also be underlined that some dominant results rely on similar assumptions in the four studies, especially the treatment of the seal LOCA risk and the CCFs between EDGs. For this last point all the PSAs consider a CCF between the four main diesels and a CCF between the SBO diesels, but no CCF between the two categories (assumption of adequate diversity), and since the loss of the six diesel generators is a dominant cut-set for all the studies, this assumption of diversity is very important.

Areas and topics that require additional information

Some of the identified differences between the compared PSAs, like the I&C, ventilations and CCF modeling may have an important impact on the PSA results and applications. However the available information was not sufficient for a detailed comparison. Especially I&C, which is an important and cross-cutting issue, needs a particular attention.

3.6. Steam Generator Tube Rupture Comparison

IE definitions and assumptions

The initiating event is a steam generator tube rupture (SGTR), for: the double ended rupture of a single tube; the double ended rupture of two tubes; and an induced rupture of multiple tubes, following a secondary side break. The multiple tube rupture is modeled as ten ruptured tubes in a single steam generator. The SGTR causes a loss of inventory from the primary to the secondary side of the SG, leading to a primary pressure decrease and a level increase in the affected SG.

Typical accident sequences and progression

The accident scenarios are similar for all four PSAs. The case of a single tube rupture, which was evaluated by all four PSAs, is described below.

Upon diagnosing the SGTR, operators trip the reactor, then isolate the faulted SG and initiate cooldown with the intact SGs. Failure to isolate the faulted SG means there is a LOCA outside containment and the operators must cool down, depressurizing the reactor cooling system (RCS). If the faulted SG is not isolated, an automatic signal initiates MHSI, which ensures primary circuit makeup and extends the time available for operators to cool down. Secondary cooldown can be accomplished by utilizing either the startup and shutdown system (SSS) or the emergency feed water system. If secondary cooldown fails, the operator initiates feed and bleed using MHSI, opening all the pressurizer safety valves (PSV) or one of the severe accident depressurization valves (PDV). Long term cooling (LTC) is either provided by one train of LHSI or a single train of the severe accident heat removal system (SAHRS i.e. CHRS in FA3/OL3/UK EPR) and the IRWST. Operator aligns and initiates residual heat removal via secondary circuit before IRWST is lost.

Similarities and differences

The four PSA models are largely similar, with respect to:
- the initiating events considered a single tube leak (all), a double tube leak (all but US), and multiple tube leaks (US and FA3);
- the main core damage sequences are failure to initiate fast secondary cooldown and failure of primary feed and bleed;
- the overall results for the single tube rupture, with the exception of the UK EPR, are similar ~ 2E-08 /year and are presented in Table 4 below. For the UK EPR, the CDF resulting from single tube ruptures has increased to 6,24E-09 /year in the PSA for a UK EPR^{TM} at Hinkley Point C as a result of a correction to the PSA model [7].

Table 4: CDF (1/a) related to SGTR at power

	FA3	UK EPR[*]	U.S. EPR	OL3
Single tube rupture	5,5E-09 (1,0E-08 incl. small SGTR)	2,2E-10	2,6E-08	2,2E-08
Double tube rupture	7,7E-10	4,0E-09	N/A	9,0E-12
Multiple tube ruptures (following sec. side break)	4,4E-09 (SGTR SSB)		8,5E-09 (IND SGTR)	

[*] *Results from PSA for GDA 2011 PCSR [11]*

Areas and topics that require additional information

Potential design changes from I&C, especially in the U.S. EPR, need to be reflected in the PSA.

3.7. Loss of Cooling Chain Comparison

The Loss of Cooling Chain (LOCC) initiating events cover several failure modes, including pipe breaks and leaks, of the Component Cooling Water System (CCWS) and of the Essential Service Water System (ESWS). There are two common user headers (CH) in CCWS, each connected to two CCWS trains. There is also an interconnection between CHs for RCP thermal barrier cooling in all other EPR designs, except for OL3.

IE definitions and assumptions

In EPR PSAs, partial or total loss of cooling chain is treated as a so called common cause initiator (CCI). A common cause initiator is defined as an event which either causes a reactor trip or requires a shutdown of the reactor, and which at the same time degrades one or several of the safety functions which are required for the shutdown.

LOCC events are categorized in following main groups, presented in Table 5 below. Note, loss of one CCWS/ESWS alone do not lead to a CCI if the switchover to the standby train is successful. The number of LOCC IE groups in EPR PSAs varies from one to seven. For example, in OL3 PSA, seven IE groups were modeled in the construction license PSA (2004). Later, six of these were screened out based on frequency screening, more realistic modeling and taking into account that many of the event combinations are modeled in the fault trees i.e. there is no need to model each combination as a separate IE. According to EPR vendors, the modeling of LOCC IEs in EPR PSAs may evolve towards similar direction than in OL3 PSA.

Table 5. IE group frequencies (1/a) for LOCC events

	FA3	UK EPR	U.S. EPR[**]	OL3[*]
Loss of one train	4,7E-1	***	2,7E-3	-
Loss of one common header	5,8E-3[2]	***	2,0E-1[1]	3,5E-3[3]
Loss of all trains	1,8E-7	***	2,4E-6	-

[*] *Seven IE groups were modeled in construction license PSA (2004)*
[**] *The U.S. EPR values for these IE group frequencies are based the older model; in the 2013 revision all the LOCCW IEs are integrated into the model through a single initiating event fault tree*
[***] *Not published*
1. *spurious openings of safety valves contribute 93% and leaks contribute 5%*
2. *Spurious opening of safety valves and leaks contribute each other to about 50% (EDF operating experience)*
3. *includes mechanical failure of one train and failure of switchover to the standby train*

Typical accident sequences and progression

Loss of one CCWS/ESWS train leads to unavailability of corresponding train in following safety systems (through lack of cooling):
- Medium Head Safety Injection
- Residual Heat Removal
- Low Head Safety Injection *(valid for pumps in safety trains 2 and 3, pumps in trains 1 and 4 have diversified cooling)*

The consequences of losing one common user header are:
- for OL3: loss of cooling for two RCP (thermal barrier and motor, motor bearing, pump thrust bearing) which leads to an automatic trip of these two RCP and consequential automatic reactor & turbine trip; for other EPRs either common header can cool all four RCPs (interconnection)
- shutdown of the operating charging pump (make-up) and automatic startup of the standby charging pump.

The consequences of losing two common user headers are:
- loss of cooling for all RCP (thermal barrier and motor, motor bearing, pump thrust bearing) which leads to an automatic trip of the RCP and consequential automatic reactor & turbine trip;
- loss of all charging pumps.

A failure of RCP trip may lead to a RCP seal LOCA. Residual heat removal would in both aforementioned cases be performed automatically via the secondary side feed & bleed with all secondary systems available with the exception that emergency feed water system train(s) may be affected via loss of room cooling in the corresponding trains in which the CCWS/ESWS failure(s) occur.

Following the LOCC the pressure in the main steam lines increases until the main steam by-pass (MSB) is automatically opened. If the MSB is not available the main steam release trains (MSRT) are opened. If the main feed water is not available, the start-up and shutdown (SSS) feed water pump is actuated. If both the MFW and the SSS fail the emergency feed water system (EFWS) pumps are automatically actuated.

Similarities and differences

In general, the plant response to LOCC IEs is fairly similar in all EPR PSAs. Significant differences exist in the grouping of IEs. The number of LOCC IEs varies from one IE group in OL3 up to seven IE groups e.g. in UK EPR. There are also differences in the exact definition of LOCC IEs, their frequencies, as well as in data sources and in the se of operating experience (pipe breaks and leaks).

Some of the differences in IE groups and frequencies may be explained by conservative modeling, choice of modeling approach (use of fault trees and/or calculation of IE specific frequencies) or data source.

In some cases the consequences of losing one or two common user headers in CCW system may vary due to design differences e.g. in air conditioning and ventilation systems. In OL3 NPP, the room cooling in the safeguard buildings was diversified by adding new heat exchangers cooled by CCWS. Examples of other design differences are given below.
- CCWS common user header (CH) valves
  - U.S. EPR: CCW CH valves need two divisions (trains) to open/close these valves (Div1 and 2 "OR" Division 3 and 4); specific combinations of double failures could fail all valves;

- o FA3, OL3, UK EPR: The solenoid valves are power supplied from the division the main valve belongs to. Thus the main valve closes (common user header will be isolated) if the power supply of the respective division is lost.
- There is an interconnection between common user headers for RCP thermal barrier cooling in all EPR designs, except for OL3 NPP. Adding this design feature in OL3 would have no significant impact on the risk.

Insufficient information was available for more detailed comparison of LOCC events in EPR PSAs.

Areas and topics that require additional information

The treatment of software failures and spurious signals of I&C systems as well as their impact on the results and the most important cut sets is still under review by some regulators. Another important modeling issue is the RCP seal LOCA. Based on the comparison, there are clear differences in the modeling. Complexity of potential failure combinations and assumption related to the leakage potential of the RCP seals need to be studied in more detail before drawing any definitive conclusions on identified differences.

4. EPR DESIGN DIFFERENCES

The MDEP EPR specific PSA working group has held joint meetings with EPR vendors exchanging information related to regulatory review findings, modeling details, design differences and potential new design changes. The work is still on-going, especially related to the identification of design differences affecting the risk. The aim is to find rationale for differences in EPR PSAs, whether their origin is in design, PSA modeling or data.

Examples of known differences, which are implemented due to regulations, site, operator, industry or project timing (not all of these are directly related to the PSA comparison exercise):
- Different SGTR management strategy in OL3: aim is to minimize steam release into environment, i.e. faulty steam generator automatically isolated at the end of partial cooldown (time < 10min). In other EPR designs isolation is done around 60 minutes post fault.
- Differences in system design, e.g. air conditioning and ventilation systems, extra boration system, fuel pool cooling system, EDG size and cooling, fire design, and some of the I&C systems.
- Full rupture (2A LOCA) of reactor cooling systems (RCS) is not the design basis for ECCS in all EPR designs,
- There are differences in RCS insulation material (mineral vs. glass wool).
- There are design differences related to severe accident management, for example:
  - o Fulfillment of single failure criterion in severe accident systems is required in OL3.
  - o Diversity between severe accident and design basis accident equipment is not required in all EPR designs.
  - o Redundancy in severe accident depressurization is not required in all EPR designs.
  - o Severe accident containment filtered venting is not required in all EPR designs.

5. CONCLUSIONS

One of the most important reasons for the identified differences is due to the fact that compared EPR PSAs represent various stages of the design process, licensing process, as well as level of modeling detail. Some PSAs are so called full scope PSAs in terms of the coverage of initiating events, i.e. internal IEs, and internal and external hazards are included in the analyses. The others include somewhat limited analyses of hazards.

Comparison of the numerical results of different EPR design PSAs is not straightforward. Firstly, each PSA represents various phases of licensing and detailed design processes. Secondly, there are differences in EPR designs, which affect the risk. Thirdly, studying the numerical results alone does not reveal the definitions and assumptions related to the modeling of IE groups and the accident progression.

The differences in the details and assumptions related to the modeling of I&C systems explain some of the identified differences. In addition, the detailed design of the OL3 I&C system is still under development and some changes are foreseen. The treatment and assumptions concerning software failures and spurious actions of I&C systems as well as their impact on results and most important cut sets is to be reviewed. Comprehensive fault analyses are needed for more detailed and realistic modeling of I&C systems.

The outcomes and lessons learned from the EPR PSA comparison have been used to facilitate the regulatory reviews and assessment work of various EPR designs and to enhance the scope, level of detail, and quality of EPR PSA models and documentation.

References

[1] J-L. Caron et al. *"The Use of PSA in Designing the European Pressurized Water Reactor (EPR)"*, Proceedings of the 5th International Conference on Probabilistic Safety Assessment and Management (PSAM-5), November 27–December 1, 2000, Osaka, Japan.
[2] Letter ASN, *"Options de sûreté du projet de réacteur EPR" (2004) endorsing the "Technical guidelines for the design and construction of the next generation of nuclear power plants with pressurized water reactors"*
[3] ONR, New nuclear reactors: Generic Design Assessment Guidance to Requesting Parties, ONR-GDA-GD-001 Revision 0, August 2013, (www.hse.gov.uk/newreactors/ngn03.pdf).
[4] ONR, Generic Design Assessment – New Civil Reactor Build, Step 4 Probabilistic Safety Analysis Assessment of the EDF and AREVA UK EPR™ Reactor, ONR-GDA-AR-11-019, Revision 0, 10 November 2011, (www.hse.gov.uk/newreactors/reports/step-four/technical-assessment/ukepr-psa-onr-gda-ar-11-019-r-rev-0.pdf).
[5] ONR, Licensing Nuclear Installations, Second edition: August 2013, (www.hse.gov.uk/nuclear/licensing-nuclear-installations.pdf).
[6] ONR, Licence condition handbook, Issue Date: October 2011, (www.hse.gov.uk/nuclear/silicon.pdf).
[7] Hinkley Point C Pre-Construction Safety Report 2012, (http://hinkleypoint.edfenergyconsultation.info/public-documents/hinkley-point-c-pre-construction-safety-report-2012/).
[8] ONR, Safety Assessment Principles for Nuclear Facilities, 2006 Edition, Revision 1 SAPs, (www.hse.gov.uk/nuclear/saps/saps2006.pdf).
[9] ONR, Nuclear Safety Technical Assessment Guide, Probabilistic Safety Analysis, NS-TAST-GD-030 Revision 4, June 2013, (www.hse.gov.uk/nuclear/operational/tech_asst_guides/ns-tast-gd-030.pdf).
[10] Regulatory Guide YVL A.7, *"Probabilistic Risk Assessment and Risk Management of a Nuclear Power Plant"*, Radiation and Nuclear Safety Authority (STUK), 2013.
[11] UK EPR pre-construction safety report, Chapter 15, probabilistic safety analysis, 2011, (www.epr-reactor.co.uk/scripts/ssmod/publigen/content/templates/show.asp?P=290&L=EN).

OECD WGRISK – Challenges and Recent Tasks

Marina Roewekamp[a*], Jeanne-Marie Lanore[b], Kevin Coyne[c],
Milan Patrik[d], Abdallah Amri[e], Neil Blundell[e]

[a] Gesellschaft für Anlagen- und Reaktorsicherheit (GRS) mbH, Köln, Germany
[b] Institut de Radioprotection et de Sûreté Nucléaire (IRSN), Fontenay-aux-Roses, France
[c] U.S. Nuclear Regulatory Commission, Washington, DC USA, United States of America
[d] UJV Rez, Rez, Czech Republic
[e] OECD Nuclear Energy Agency (NEA), Issy-les-Moulineaux, France

Abstract: The overall objective of the Working Group on Risk Assessment (WGRISK) of the OECD Nuclear Energy Agency (NEA) Committee on the Safety of Nuclear Installations (CSNI) is to advance the understanding of Probabilistic Safety Assessment (PSA) and to facilitate its utilization for enhancing the safety of nuclear installations. To accomplish this mission, WGRISK continuously performs a variety of activities to exchange information on PSA between member countries. This paper presents a brief overview on the actually on-going WGRISK activities and perspectives.
In addition to on-going tasks covering more traditional PSA challenges (e.g. tasks relating to human reliability analysis (HRA) and digital instrumentation and control (I&C)), new challenges for PSA have arisen from the recent nuclear power plant operating experiences and the insights from the post-Fukushima stress tests[†]. In response to these new challenges, WGRISK conducted an international workshop on "PSA of Natural External Hazards Including Earthquakes" in June 2013. This workshop revealed valuable insights on challenges associated with external events such as scope consideration for PSA, the need to consider combinations of external hazards, and multi-unit impacts. Another ongoing WGRISK activity is the second follow-up workshop on "Fire PRA" to be held in April 2014. The Fire PRA workshop will address many of the technical challenges associated with including fire hazards, which typically provide a non-negligible contribution to the overall core or fuel damage frequency, in PSA.
WGRISK recently initiated a task focused on obtaining insights from PSA related to the loss of electrical power sources. This task will collect examples of PSA insights related to a loss of electrical power sources, including those insights identified as a result follow-up activities to the Fukushima Dai-ichi reactor accidents. It is expected that this task will also highlight the capabilities of PSA as a tool for providing insights related to the potential consequences of the loss of a safety function, such as core damage frequencies or frequencies of radioactive releases. The use of PSA in this manner may provide a measure of defense-in-depth in case of loss of a safety function, which will augment more traditional analysis approaches that emphasize identification of failures that can lead to loss of system function.

Keywords: Risk, External Hazards PSA, WGRISK, fire, electrical power.

1. INTRODUCTION

The main objective of the Working Group on Risk Assessment (WGRISK) of the OECD Nuclear Energy Agency (NEA) Committee on the Safety of Nuclear Installations (CSNI) is to advance the understanding of probabilistic safety assessment (PSA) and to enhance its utilization for improving the safety of nuclear installations. Due to its disciplined, integrated and systematic approach, PSA is considered as a necessary complement to traditional deterministic safety analysis.

To accomplish this mission, WGRISK performs a number of activities to exchange PSA-related information among NEA member countries. WGRISK provides a forum for exchange of information

---

[*] Corresponding author: Marina.Roewekamp@grs.de
[†] For a summary of post-Fukushima Dai-ichi accident activities by OECD/NEA member countries see http://www.oecd-nea.org/nsd/fukushima/

and experience related to risk assessment in member countries. This exchange is not only limited to technical discussions on questions regarding risk analysis approaches, results, insights, applications and interactions with other disciplines and analysis techniques, but also includes identifying and prioritising important issues requiring additional research. WGRISK also prepares technical reviews (such as state-of-the-art reports, technical opinion papers, compilations of ongoing efforts, comparison studies etc. as appropriate) of work in all phases of risk assessment to assist further developments and the application of PSA in risk-informed decision making. This information sharing assists member countries in ensuring adequate safety of existing and future nuclear installations in their respective territories.

The scope of the activities carried out by WGRISK may involve, for current and future nuclear installations under the purview of CSNI, any or all of the two broad sets of activities pursued in managing risk:

- Risk assessment (including risk characterization as well as technical assessment) and
- Risk management (including the development and evaluation of options).

WGRISK provides timely, high-quality work products addressing, to the extent practical, a broad range of risk management needs identified and be forward looking in the identification of risk management issues that may need to be addressed by CSNI and the working group thus being sufficiently flexible to respond to emerging risk management issues, appropriately coordinated with the risk management programmes of the member countries as well as of other international organizations. It also serves as an internationally recognized, authoritative source on risk-related matters and as an important resource for risk-related knowledge management activities.

CSNI, in collaboration with the Committee on Nuclear Regulatory Activities (CNRA), maintains a joint strategic plan and mandates [1] identifying main challenges and focus areas. One main challenge identified in the CSNI/CNRA Strategic Plan is the safe operation of current, new, and advanced nuclear facilities. This paper presents a brief overview of ongoing or recently finished WGRISK activities (hereafter called "Tasks") and perspectives, e.g. the recommendations resulting from these for future WGRISK activities that address this CSNI/CNRA main challenge.

2. RESULTS OF RECENTLY FINSIHED TASKS

2.1. Use and Development of Probabilistic Safety Assessment - An Overview of the Situation at the End of 2010

In the recent past, the results of the continuous information exchanges among member countries related to PSA have been compiled in a standalone CSNI report entitled "The Use and Development of Probabilistic Safety Assessment", first issued in 2002 [2], then updated in 2007 [3] and most recently updated in 2011 [4]. This report provides a description of the PSA activities in member countries at the time when the report was written. The latest report presents an analysis of the position on the use and development of PSA in WGRISK member countries as of the end of 2010. As previously, the corresponding Task was carried out in cooperation with the International Atomic Energy Agency (IAEA), which led to more information and thus provided a better overview on PSA worldwide. The expected readers and "end users" of this report are PSA professionals and generalists dealing with risk and safety management. The current version of the report includes information from twenty-one member and non-member countries and covers a range of topics including national PSA frameworks, numerical safety criteria, PSA standards and guidance status, scope of PSA programs, PSA methodology and data, applications results and insights from, and future development and research activities.

2.2. Use of OECD Data Project Products in Probabilistic Safety Assessment

The OECD/NEA joint Database Projects and information exchange programmes enable interested countries, to pursue research or the sharing of data with respect to particular areas or problems. The following Database Projects have direct relevance to PSA activities:

- International Common Cause Failure Data Exchange (ICDE),
- OECD/NEA Fire Incidents Record Exchange (FIRE) Project,
- Component Operational Experience, Degradation and Ageing Programme (CODAP) having subsumed the former OECD Piping Failure Data Exchange (OPDE), and
- OECD/NEA Computer-based System Important to Safety (COMPSIS) Project.[‡]

These data projects can, in principle, support the collection and analysis of data that is highly relevant to PSA, particularly in the areas of material degradation and aging, common cause failures, fire risk, and digital instrumentation and control systems. All of these projects collect qualitative information that can be useful in the development and review of PSA models. Moreover, several of these projects include specific objectives to support quantification activities. However, to date, WGRISK members, particularly those who are not members of the database projects, have made little use of the data project products (principally reports). To address this challenge, and based on needs expressed by a number of member countries, the CSNI WGRISK initiated a Task on "Use of OECD Data Project Products in Probabilistic Safety Assessment" in NEA member countries in 2011. This task was coordinated with representatives from ICDE, FIRE, OPDE/CODAP, and COMPSIS and benefitted greatly from the perspectives offered by the data project members.

The major objectives of this data project task were the following:

- Identification and characterization of the current uses of OECD/NEA data project products and data in support of PSA. In this context, the term 'products' refers to data analysis results, technical reports, and other project outputs.
- Identification and characterization of technical and programmatic characteristics that either support or impede use of data project products in PSA. This includes an assessment of which PSA parameters could be potentially estimated from the various data project products and gaps between available product information and PSA data needs.
- Identification of recommendations for enhancing the usefulness of data project products and the coordination between WGRISK and the data projects.

An additional objective of this task was to strengthen the relationships between the data project and PSA communities.

The data project task included three main activities:

- **Questionnaire/Survey** – A survey instrument was developed in collaboration with representatives from WGRISK and each of the data projects. Two surveys were developed, one that was distributed to members of the PSA community, and a second that was distributed to each of the data projects. The surveys focused on the task objectives and requested information pertaining to project participation, data access, uses of data project products for PSA, challenges in data collection and use, and best practices in use of data project products. The surveys were distributed in the spring of 2012. Good participation completing the survey was noted, with 22 organizations representing 14 member countries providing survey responses. Survey responses were also obtained from the ICDE, FIRE, OPDE/CODAP, and COMPSIS data project representatives.
- **Task Meeting** – After the survey responses were analysed, a two day task meeting was held in October 2012 at OECD headquarters in Paris, France. Fourteen participants attended the task

---

[‡] The COMPSIS project ended in December 2011 but was an active project when this task was initiated.

meeting, representing eight NEA member countries, the NEA secretariat, and the FIRE, ICDE, and CODAP projects[§]. The task meeting agenda included a review of survey results from each data project, open discussions on enhancing participation in data project activities and identification of new data needs, and identification of conclusions and recommendations.
- **Final task report** – The final task report [5] provides the survey responses and associated analysis, along with a detailed description of the key attributes of each of the data projects. The report also includes recommendations for strengthening collaboration between the PSA community and the joint data projects. Best practices for the use of data project products for PSA are identified, along with a summary of success factors for data project activities. The final report was coordinated with representatives from ICDE, FIRE, OPDE/CODAP, COMPSIS, and WGRISK and is intended to represent a consensus view among each of these organizations.

In general, the OECD/NEA joint data projects represent mature data collection efforts and have enjoyed substantial support from the NEA membership. These projects have endeavoured to ensure that data collection activities have a high level of completeness and quality. This commitment to quality has resulted in the development of project-specific programmatic requirements intended to ensure quality. However, there remain some challenges when attempting to apply data project products to PSA activities (e.g., data completeness and exposure information needed to calculate PSA parameters). As such, data applicability and completeness should be fully assessed prior to applying data project products to a specific application. Despite these challenges, experience has been developed by a number of NEA members in applying ICDE, FIRE, and ODPE/CODAP data to PSA initiatives. Examples include CCF parameter estimation, fire frequency calculation, and estimation of piping rupture frequencies. Overall, the data projects are an important OECD/NEA activity, particularly for member states with a small number of nuclear installations and limited national databases.

This task identified a number of challenges and opportunities for further improvement:

- Enhancing participation in data project activities
- Striving for continual improvement in operating experience data collection efforts
- Increased sharing of data with national organizations including industry and standards organizations (as appropriate)
- Consideration of new data collection needs (e.g., new and advanced reactors, human reliability analysis, external hazards)
- Consideration of success factors for application of data project products to PSA when developing new activities

In order to support wider dissemination of the lessons learned from this activity, a summary of the task and results were presented at the recent American Nuclear Society PSA 2013 topical meeting [6].

3. RESULTS OF ONGOING OR RECENTLY STARTED TASKS

WGRISK has a number of ongoing tasks that are nearing completion, including comparisons of human reliability analysis (HRA) methods to desirable attributes[**] and identification of failure taxonomies to support the digital instrumentation and control (I&C) PSA. In addition to these more traditional PSA-related activities, recent nuclear power plant operating experience and the insights from the post-Fukushima stress tests have highlighted new challenges for PSA. WGRISK has identified several new activities to address these challenges, including conducting international workshops on PSA for naturally occurring external hazards and fire, and using PSA as a tool to identify insights related to electrical power sources.

[§] At the time the workshop was held (October 15-16, 2012), both the COMPSIS and ODPE Projects had ended.
[**] This HRA-related task is being performed in collaboration with the CSNI Working Group on Human and Organizational Factors (WGHOF).

3.1. International Workshop on PSA of Natural External Hazards including Earthquakes

Motivation for the task of an international workshop on PSA for natural hazards was that the 2011 reactor accidents of Fukushima Dai-ichi triggered discussions about the significance of external hazards and their treatment in probabilistic safety analyses and assessment. In addition, the results of the stress tests performed as a result of these accidents have shown vulnerabilities and potential cliff-edge effects in plant responses to external hazards resulting in identifying possibilities of and priorities for improvements and safety measures' implementation at specific sites and for particular designs.

In order to address these issues and provide relevant conclusions and recommendations to CSNI and CNRA, the WGRISK directed, together with the CSNI Working Group WGIAGE, an "International Workshop on PSA of Natural External Hazards Including Earthquakes". The workshop was hosted by UJV Rez and took place in Prague, Czech Republic in June 2013.

Key objectives of this workshop were to collect information from OECD member states on methods and approaches being used and experience gained in PSA for natural external hazards. In addition, the workshop was used to identify new potential topics, such as improving the PSA treatment of the different levels of defense-in-depth, for further WGRISK and WGIAGE activities.

The focus of the workshop was on external hazards PSA for nuclear power plants (NPP), including all modes of plant operation. The workshop scope was limited to natural external hazards including those ones for which the distinction between natural and man-made hazards is not sharp (e.g., external floods caused by dam failures). The participation was open to experts from regulatory authorities and their technical support organizations, research organizations, operators, NPP designers and vendors, industry associations and observers.

The following conclusions have been drawn based on workshop presentations, discussions during particular sessions and two facilitated discussions.

Regulatory framework

Lessons learned from the Fukushima Dai-ichi reactor accidents and related actions at national, European and global level have emphasized the importance to assess risks associated with external hazards (including combinations of these hazards) and their impacts on a plant site (possibly with several units).

Regulators in most countries have taken actions to include seismic and flooding risk, and, to some extent, some other specific external hazards in national PSA practices and safety regulations. The development of systematic approaches for addressing external hazards completely in PSA practices is still ongoing.

The current role of external hazards PSA in the regulatory framework varies from country to country depending on the local conditions, operating experiences and the type of relevant hazards. In some countries adequate deterministic requirements for protection against earthquakes or other external hazards did not exist when the operating reactors were built and the external hazards have been later analyzed in the PSA framework. In other countries the emphasis has been on deterministic design requirements.

Models, methods, tools and data

Useful hazard estimates can be determined with current methods and used in applications in the processes of risk oriented decision making.

Development of methods and preparation of studies aiming to obtain realistic risk assessments, neither too optimistic nor too much conservative, is a key issue. These more realistic evaluations would provide a better view on the real problems and also a better view on the interest of safety improvements.

Standards and guidance

Recently developed methods and guides are available for seismic hazard determination, identification of external hazards and screening of external hazards for detailed consequent analysis. Several lists of screening criteria are available. The methods of Probabilistic Seismic Hazards Assessment (PSHA) have been developed and used in practice for several decades and they have been well documented and described in relevant Standards.

Good practices and applications

The following good practices in external hazards PSA were demonstrated by the presentations made during the workshop (and are applicable to PSA in general, not just for external hazards PSA):

- challenging assumptions,
- calibrating models,
- accounting for underlying physical processes,
- treating dependencies ,
- involving multidisciplinary teams, and
- disseminating information promptly and broadly.

Methods for external hazards analysis have been recently used to evaluate operating NPP units and to identify needs for modification of plant systems and procedures as well as to support new plant designs. The risk contribution (at least for some external hazards) has included events occurring during shutdown and low power operation and examples of plant reactor (and non-reactor) improvements following the results of the analyses were given.

Challenges and opportunities for further enhancements

The task on PSA for Natural External Hazards including earthquakes identified a number of challenges and opportunities for further enhancements.

In general, there are a number of significant technical challenges for external hazards PSA covering various areas of PSA, which include, for example:

- multi-unit impacts,
- combinations of external hazards,
- fragility analysis of non-seismic external hazards,
- correlation effects and consequent damage scenarios,
- HRA for external hazards PSA, including organizational and managerial aspects,
- mission times for long-term scenarios,
- effects of climate change on the derivation of hazard frequencies and magnitudes.

Data analysis, particularly estimation of the initiating event frequency and identification of correlations between external hazards represents another significant challenge.

The broad scope and organizational challenges appear to be:

- increasing the scope of external hazards PSA to match internal events (recognizing resource limitations);
- ensuring appropriate interactions with the appropriate scientific/technical communities;

- ensuring appropriate use in safety-related decision making, including challenges related to quality and acceptance of external hazards PSA.

The workshop provided valuable input for strengthening the role of WGRISK in supporting the development and application of probabilistic safety assessment and risk-oriented decision making methods in the area of external hazards.

Recognizing the impetus for action provided by actual operational events, it has appeared that WGRISK can provide stronger (and better-focused) cases for action by increasing its use of operating experience feedback. Among others, this could imply strengthening ties with associated international working groups, particularly the NEA Committee of Nuclear Regulatory Authorities (CNRA) Working Group on Operating Experience (WGOE).

An additional action for WGRISK suggested by this task concerns the tracking of recommendations from completed tasks. Increased efforts by the WGRISK leadership to systematically track, prioritize, and appropriately resolve past recommendations would improve the group's strategic planning process. Specifically, such tracking would help ensure that each task performed by the group more strongly supports the group's overall objectives and ensure that timely action is taken on risk-significant recommendations.

In general, due to high importance of external hazards risk analysis, WGRISK should consider initiating further activities in this area. For example, a future task to cover (partly or completely) the area of man-induced external hazards, which has been shown in some plant specific studies to be an important contributor to risk, could be considered.

Moreover, in the CSNI framework, WGRISK could provide a contribution to the newly created CSNI Task Group on Natural External Events for including a risk aspect.

3.2. International Workshop on Fire PRA

Another ongoing WGRISK activity is the second follow-up workshop on "Fire PRA" addressing the most recent challenges for Fire PRA, which typically provides a non-negligible contribution to the overall core or fuel damage frequency. The aim of the workshop, to be held in April 2014, is development of recommendations regarding a potential future update of the State-of-the-art Report on fire risk analysis (NEA/CSNI/R(99)27 [7]). This includes providing insights in probabilistic fire risk analysis and the corresponding methods, the collection of operating experience and processing of data to be used in Fire PRA applications.

More specifically, the workshop will cover the following objectives:

- to support assessment of current state of probabilistic analyses of fire hazards for nuclear installations at the design stage and during all plant operational states from start of operation up to the longer lasting post-commercial operating phases,
- to support re-evaluation of Fire PSA, in particular as a tool to address the lessons learned from the post-Fukushima investigations and stress tests with respect to fire events,
- to share methods and good practices and experiences among member states on probabilistic risk assessment of fire hazards and event combinations with fires, and
- to identify new potential topics for further WGRISK activities in this area, including potential update of the State-of-the-Art Report (SOAR) on fire risk analysis.

The workshop builds upon previous, relevant WGRISK work in the field of fire probabilistic risk analysis. In addition, ongoing activities at the OECD/NEA in the field of fire risk analysis in the frame of the OECD FIRE Database project, the PRISME and PRISME2 experimental fire research projects, and the fire related OECD High Energy Arcing Fault (HEAF) experimental program are intended to form a sound basis for discussions in the frame of this workshop.

The workshop will include plenary and technical sessions in addition to facilitated discussion sessions. It has also been organized to encourage active participation of the attendees in the discussions, as well as in the formulation of conclusions and recommendations. Information obtained as a result of the workshop should provide better understanding and interpretations of subjects, topics and issues connected with fire risk analysis. A report covering workshop proceedings with summarized results and some conclusions and recommendations for follow-on activities on good practices and experiences in member states, including lessons learned from operating experience, experimental research and actual applications of probabilistic analyses in the frame of regulatory activities will be prepared based on material presented at the at the workshop.

3.3. Probabilistic Safety Assessment Insights Relating to the Loss of Electrical Sources

Most recently, the activities and related WGRISK tasks focus on insights from probabilistic safety assessment related to the loss of electrical sources. This task aims on collecting good examples of PSA insights related to a loss of electrical sources identified as a result of several activities after the Fukushima Dai-ichi reactor accidents. While in the analysis of the robustness of safety functions the emphasis is generally on the causes of potential functional failures, particularly when using operating experience, PSA is a tool for providing insights related to the potential consequences of the loss of a safety function, such as damage frequencies or frequencies of radioactive releases, and relating to the provisions aiming to avoid such inadmissible consequences of the loss of the safety function. The use of PSA results may provide a measure of defense-in-depth in case of loss of a safety function.

The task intends to illustrate the PSA capabilities with an outstanding practical example. Two types of insights will be gained:

- Insights for plant safety related to results and applications of risk calculations: overall risk as well as relative results relating to dominant contributions, potentially weak points in the defenses, balance between core damage prevention and mitigation, comparison between the contributions of internal initiating events and internal or external hazards, key sources of uncertainty (where available), safety benefits brought by modifications already implemented or planned (including possible post-Fukushima modifications).
- Insights on PSA methodology: identification of good practices, potential gaps, differences in the methodologies used or developed. Potential interesting points are the treatment of common cause failures (CCF) and the treatment of long-lasting scenarios which are not currently introduced in the PSAs.

4. INTENDED FUTURE ACTIVITIES

WGRISK has developed and maintained a comprehensive integrated plan that describes the working group's vision, technical goals, and working methods. The operating plan is reviewed updated annually in order to help ensure that WGRISK remains well aligned with CSNI priorities and is focused on topics of current interest to its members. The operating plan provides a systematic approach of reviewing WGRISK's progress in addressing main CSNI challenges and evolving needs of the membership. Emerging issues of interest to members include human reliability analysis for seismic and other external events and multi-unit integrated risk assessment. WGRISK is currently evaluating proposals for new tasks in both of these areas. A future activity to enhance the tracking and prioritization of past WGRISK recommendations is expected to better inform future task planning.

5. CONCLUSION

WGRISK strives to provide timely, high-quality work products addressing, to the extent practical, the broad range of risk assessment and management needs identified by CSNI and the working group members. In addition, WGRISK serve as an internationally recognized, authoritative source on risk-related matters and as an important resource for risk-related knowledge management activities. It is expected that recently completed and ongoing task activities associated with HRA, digital I&C,

External Hazards PSA, Fire PSA, and electrical power source risk insights will serve to enhance the state of knowledge in the broad PSA community. Finally, it should be noted that WGRISK reports are generally available to the public (including non-members of NEA) and can be found on the OECD/NEA website (e.g., http://www.oecd-nea.org/nsd/docs/indexcsni.html).

Acknowledgements

The author's wish to acknowledge the outstanding support provided to WGRISK by the OECD/NEA secretariat, particularly Abdallah Amri and Neil Blundell. Additionally, special thanks are due to Dr. Nathan Siu, who strategically led and coordinated the activities of WGRISK as its chairman during the period of 2007-2013.

References

[1] Organisation for Economic Co-operation and Development (OECD), Nuclear Energy Agency (NEA), *Joint CSNI/CNRA Strategic Plan and Mandates*, NEA/CSNI/R(2011)1, Paris, France, (February 2011).

[2] Organisation for Economic Co-operation and Development (OECD), Nuclear Energy Agency (NEA), *The Use and Development of Probabilistic Safety Assessment in NEA Member Countries*, NEA/CSNI/R(2002)18, Paris, France, (July 2002).

[3] Organisation for Economic Co-operation and Development (OECD), Nuclear Energy Agency (NEA), *Use and Development of Probabilistic Safety Assessment*, NEA/CSNI/R(2007)12, Paris, France, (November 2007).

[4] Organisation for Economic Co-operation and Development (OECD), Nuclear Energy Agency (NEA), *Use and Development of Probabilistic Safety Assessment - An Overview of the situation at the end of 2010*, NEA/CSNI/R(2012)11, Paris, France, (December 2012).

[5] Organisation for Economic Co-operation and Development (OECD), Nuclear Energy Agency (NEA), *Use of OECD/NEA Data Project Products in Probabilistic Safety Assessment*, NEA/CSNI/R(2014)2, Paris, France (in publication, 2014)

[6] K. Coyne, M. Tobin, N. Siu, and M. Roewekamp, "*Use Of OECD Data Project Products In Probabilistic Safety Assessment*", Paper 70, in: Proceedings of ANS PSA 2013 International Topical Meeting on Probabilistic Safety Assessment and Analysis, Columbia, SC, September 22-26, 2013, on CD-ROM, American Nuclear Society, LaGrange Park, IL, USA, (2013).

[7] OECD Nuclear Energy Agency (NEA) Committee on the Safety of Nuclear Installations (CSNI), *Fire Risk Analysis, Fire Simulation, Fire Spreading and Impact of Smoke and Heat on Instrumentation Electronics, State of the-Art Report (SOAR)*, NEA/CSNI/R(99)27, Paris, France, (March 2000).

RAPP: Method for risk prognosis on complex failure behaviour in automobile fleets within the use phase

Stefan Bracke[a] and Sebastian Sochacki[a]

[a] University of Wuppertal, Chair of Safety Engineering and Risk Management, Wuppertal, Germany

Abstract: The increasing complexity of product functionality and manufacturing process parameters often leads to complex failure modes during the product life cycle. These field information are the basis for risk analyses and damage case prognosis with the goal of an early risk detection and leads to the possibility of nearby interactions e.g. product and manufacturing optimisation or recall action. This paper outlines the essential procedure of the new developed method "Risk Analysis and Prognosis of complex Products (RAPP)". The main focus of the RAPP method is the detection, visualisation and prognosis of risks and damage cases depending on their life span variables regarding to a product fleet - based on a risky production batch - in field. The RAPP method contains multiple steps: First steps include the mapping/prognosis of the failure behaviour and the mapping of product field load profiles. Next step is focusing on the estimation of the critical area regarding the life span variable (e.g. critical kilometer range). Based on these steps, it is possible to perform the risk analysis and risk prognosis regarding the product fleet. Finally, the last step of the RAPP method is the verification of risk analysis and –prognosis. The theory and application of the RAPP method is explained within an automotive case study oil tube leakage.

Keywords: Product reliability, risk analysis, risk prognosis, product fleet risks, statistical methods

1. INTRODUCTION

The development and manufacturing of technically complex products as well as capital goods are confronted to central challenges: The increasing complexity of product functionality and manufacturing process parameters often leads to complex product damage symptoms and failure modes during the product life cycle. In case of this paper technically complex products are meant to be electronic goods, home appliances, consumer electronics or automobiles. These products are manufactured in series with an amount of middle up to a high production batches (statistical population), for example $N \geq 100$ products per day. After the sales of such production batches the products are passed into the use phase. The term product fleet denotes each production batch with the same construction stage (same grade of upgrade/ innovation stage) of a component.

During the usage phase a huge amount of operational data (operating time, operating temperature, load characteristics etc.) regarding each product is generated. A special type of operational data collection can be found in case of failure occurrence during the usage phase of such components. For instance the time of damage appearance, the failure mode as well as the life span at the failure occurrence is recorded. These field informations are the basis for risk analysis and damage case prognosis with the goal of an early risk detection, visualisation and prognosis. Finally, it leads to the possibility of nearby interactions e.g. product and manufacturing optimisations or recall actions.

As part of the research work the RAPP approach (Risk Analysis and Prognosis of technical complex Products) has been already developed [1]: In this paper, the RAPP method is developed regarding an extended procedure. The overarching goal of the RAPP method is the analysis of product fleet risks in the use phase depending on their life span variables regarding to certain failure modes.

2. GOALS OF THE RAPP APPROACH

The challenges and requirements of a risk analysis and risk prognosis based on the RAPP method with regard to product fleets in the use phase are as follows:

*) bracke@uni-wuppertal.de

a) Detection of specific risks regarding to a product fleet in field, which is based on a risky production batch.
b) Prognosis of the future amount of units at risk depending on their life span variables (e.g.: operating/switching cycles, operation time or operation parameters).
c) Visualisation of the determined risk as a basis for further decisions (e.g.: recall actions, garage actions).

This paper demonstrates on the one hand the developed theoretical background of the RAPP method and on the other hand an application of the shown RAPP steps due to the automotive engineering case study oil tube leakage. The case study is based on a synthetic field data set. The creation of these data and information was based on characteristics with respect to real damage case data sets. The result of the RAPP method application is the estimation of the risk regarding the product fleet during the usage phase based on a detected failure mode. The data set includes e.g. field operating times, kilometrage and observation time based on the car-level.

3. RAPP APPROACH AND CASE STUDY OIL TUBE LEAKAGE: BASE OF OPERATIONS

The basis of every risk analysis constitutes two parts [4]: On the one hand the knowledge of the damage case at a certain observation period and on the other hand the empirical product use phase behaviour. The observation of both data base is independent from each other.

The knowledge of the damage case includes data and information regarding the amount of damaged components as well as data of the life span variables regarding the point of observation, when the damage of each component appeared in the use phase (e.g.: operating hours, operation cycles, start and shut down activities). In addition, external conditions (e.g.: temperature, air humidity) have to be analysed and documented [10].

The knowledge of the empirical product usage phase behaviour includes data and information regarding the determination and mapping of the products' life span variables. This data is based on independent observations regarding the damage cases: The data base is collected from the whole - not damaged - product fleet in the field during the usage phase. This means that the data are not censored.

The precondition with regard to the RAPP approach is the correlation between the progression of the analysed failure mode variable and the chosen life span variable. This can be verified with the help of the Spearman's rank correlation coefficient [12].

An excerpt of the data and information of the case study oil tube leakage regarding the damage case and the empirical product use phase behaviour is shown in table 1.

Table 1: Excerpt of data and information of the automotive case study oil tube leak regarding damage cases and product fleet (synthetic field data set)

Nr.	Data array	Format / content
1. Information regarding the product fleet		
1.1	Production batch (automobile fleet)	$N_{PF} = 6,259$
1.2	Production period	01.2009 - 03.2009
1.3	Kilometrage profile of the product fleet regarding the life span of the product	Log-Normal distribution (parameters: $\mu; \sigma$)
2. Field damage case information		
2.1	Observation time of concerned automobile fleet in field	11 months
2.2	Failed oil tube leaks up to observation time	n = 254

4. RAPP METHOD: OVERVIEW, PRINCIPALS AND STATISTICAL TOOLS

The essential procedure of the RAPP method (Risk Analysis and Prognosis of complex Products) includes five steps A – E, which are shown in figure 1 (cf. [1]). The single steps of the method and the used statistical tools are summarised in an overview in this chapter and explained in table 2. A detailed application of the RAPP method based on the theoretical steps is explained within the automotive case study oil leakage in chapter 5.

Figure 1: Overview of the RAPP method [1]

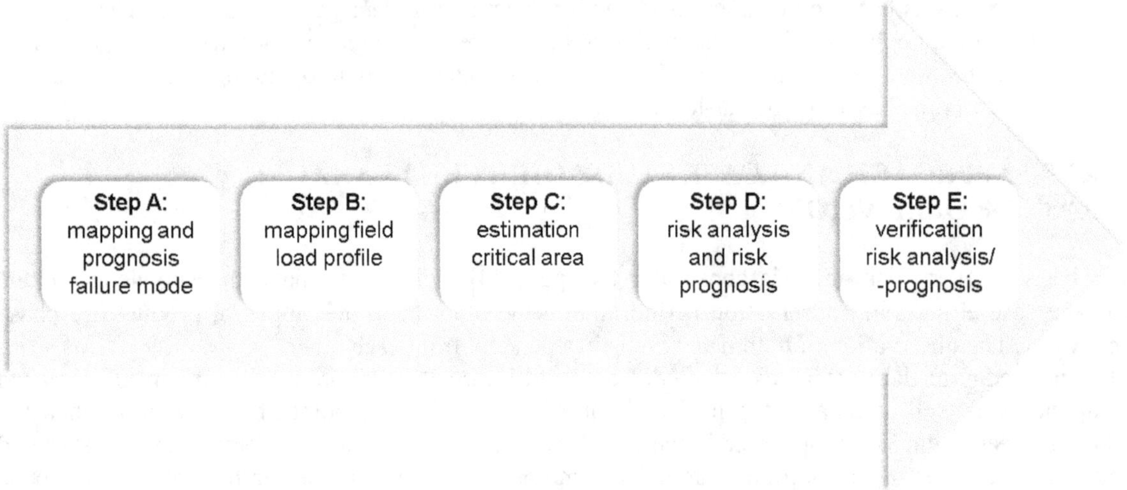

Step A: mapping and prognosis failure mode	Step B: mapping field load profile	Step C: estimation critical area	Step D: risk analysis and risk prognosis	Step E: verification risk analysis/-prognosis

The main goal of step A is the detection of the failure mode related to the product life span variables and the statistical mapping and prognosis of the failure mode (FM). Base of operations is data out of the technical analysis of damaged components and field data (e.g. guarantee/goodwill database). The focus on step B is the empirical mapping of the product fleet (PF) behaviour based on the failure mode related life span variables (cf. step A). The goal of step C is the estimation of the critical area regarding the failure mode related life span variables by definition of a range using the borderlines of important quantiles of the failure mode and the product fleet distribution models. The step D of the RAPP method focuses on the determination of the risk in product fleets in terms of a detected failure mode. Furthermore, step D considers the estimation of the retrospective and the expected risk development within the product fleet. The last step E of the RAPP method is the verification of the risk analysis and prognosis (cf. step D) based on field observation and Monte Carlo simulation methods. The application of step E regarding the case study oil tube leak is part of future research works.

Table 2: Essential procedure of the RAPP method

	Step	Data Bases, statistical tools, procedure
Step A: Mapping and prognosis of the failure behaviour		
No.	Action	Statistical tools and data bases
A1	Analysis of failure mode related life span variables	Technical analysis of damaged components
A2	Determination of life span variable's correlation	Spearmans ς; Kendalls τ
A3	Analysis of the behaviour development of the failure mode	Distribution models; e.g. Weibull distribution (parameters b, T)
A4	Prognosis of the parameters of the statistical model of the failure mode (FM)	Regression analysis; coefficient of determination r^2
A5	Analysis and mapping of clearing effects: e.g. replacement part operations	Consideration of market charging via distribution models
Step B: Mapping field load profile		
No.	Action	Statistical tools
B1	Analysis of the life span variables of the product fleet (PF)	Correlation analysis between technical analysis and product fleet
B2	Empirical mapping of the life span variable development	Distribution models; e.g. log-normal (parameters μ, σ)
B3	Analysis and mapping of clearing effects: e.g. product fleet reduction	Consideration of market scrap influences via distribution models
Step C: Estimation of the critical area		
No.	Action	Statistical tools
C1	Determination of critical areas (quantiles)	E.g.: 50%, 95% 99%-quantile of FM and PF models
C2	Mapping of life span variable quantiles regarding failure mode and product fleet	E.g.: $l_{CLL-CUL} = \vert Q_{PF-LL;0.05} - Q_{FM-UL;0.95} \vert$
Step D: Risk analysis and risk prognosis		
No.	Action	Statistical tools
Risk analysis		
D1	Risk probability in product fleets regarding actual point of time, based on critical areas	Overlap of FM and PF distribution models results in risk probability P_{FM}
D2	Determination of function "damage appearance"	Overlap of FM and PF distribution models results in density function f_{DA}
D3	Analysis of risk probability and damage appearance peaks development [retrospective]	Regression analysis at different points of time t_x leads to damage appearance function
D4	Analysis of significant changes: risk in-/decrease	Significance tests (e.g.: Mann-Whitney U-Test; Siegel-Tukey-Test, Fisher-Test) and confidence intervals
Risk Prognosis		
D5	Risk probability regarding future points of time [future]	$P_{Prog-PF}$
D6	Prognosis of risk probability and damage appearance peaks development [future]	f_{DA}
D7	Determination of clearing effects: e.g. product fleet reduction / replacement part operations	Consideration of market charging distribution models
Step E: verification of risk analysis and –prognosis		
No.	Action	Statistical tools
E1	Field observation regarding to the risk point of time	Determination of failure rate λ and products at risk n
E2	Simulation of the failure and the product fleet behaviour	Monte Carlo simulation

5. RAPP: RISK-ANALYSIS AND -PROGNOSIS WITHIN THE CASE STUDY

5.1 Step A: Mapping and prognosis of the failure mode

Out of the field data base a density function is generated to represent the statistical area of the empirical failure mode (FM). A common distribution model to approximate such failure modes is the three parameter Weibull distribution (cf. density function (1)). An advantage to approximate such failure modes using a three parameter Weibull distribution is the interpretation of the utilised parameters (t_0 = failure-free time; T = characteristic lifetime; b = shape parameter; [2]). The estimation of the parameters is done by state of the art methods such as Trust-Region method [6] or Maximum-Likelihood-Estimator [9].

$$f(x)_{FM} = \frac{b}{T-t_0}\left(\frac{x-t_0}{T-t_0}\right)^{b-1} \cdot \exp\left(-\left(\frac{x-t_0}{T-t_0}\right)^b\right) \qquad (1)$$

Regarding the case study, the point of time of field data analysis is early (11 month in service (MIS); cf. table 1), when the failure behaviour is not comprehensive known. Based on this special case, the use of an approach, which considers an additional prognosis of changing failure behaviour, is useful: The given damage cases within a limited observation period can be splitted into a certain number of classes (e.g. quarter division: 3, 6, 9, ..., 24 months in service). With the use of each location parameter T, the mean value as well as the shape parameter b, the prospective parameters of the failure behaviour regarding the second year (24 month in service) can be calculated by using non-linear regression methods (cf. figure 2). A paradigmatic example is shown in figure 2. The abscissa represents the month in service (MIS) regarding to different field observation times, whereas the ordinate reflects the estimated values of the parameters regarding the Weibull distribution models. If the gradient of the regression functions converges to zero, the parameters are not changing and therefore it is assumed, that the failure behaviour is constant. According to the risk prognosis point of time (MIS point, cf. RAPP method step D), it is possible to choose the predicted parameters based on figure 2.

Figure 2: Estimation of Weibull parameters T, b, $t_0 = 0$ and arithmetic mean value x, using non-linear regression methods for prognosis of the comprehensive failure mode, product fleet field observation time 24 months in service; exemplified by the case study oil tube failure

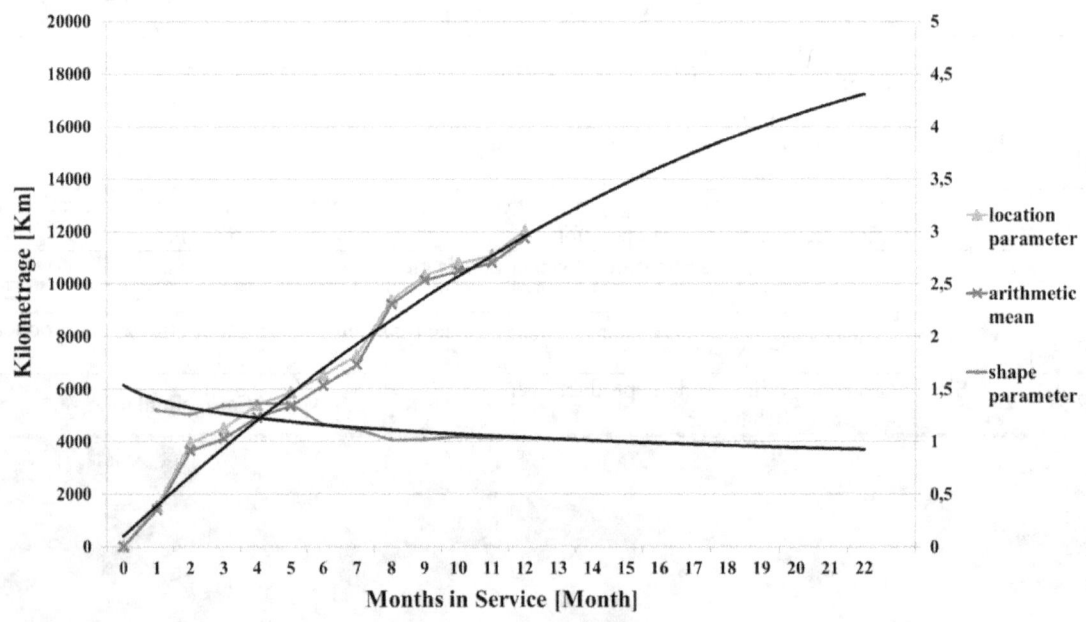

Figure 3: Example for a failure mode using the knowledge after two years of field observation of the product fleet (Weibull distribution fit), case study oil tube leakage

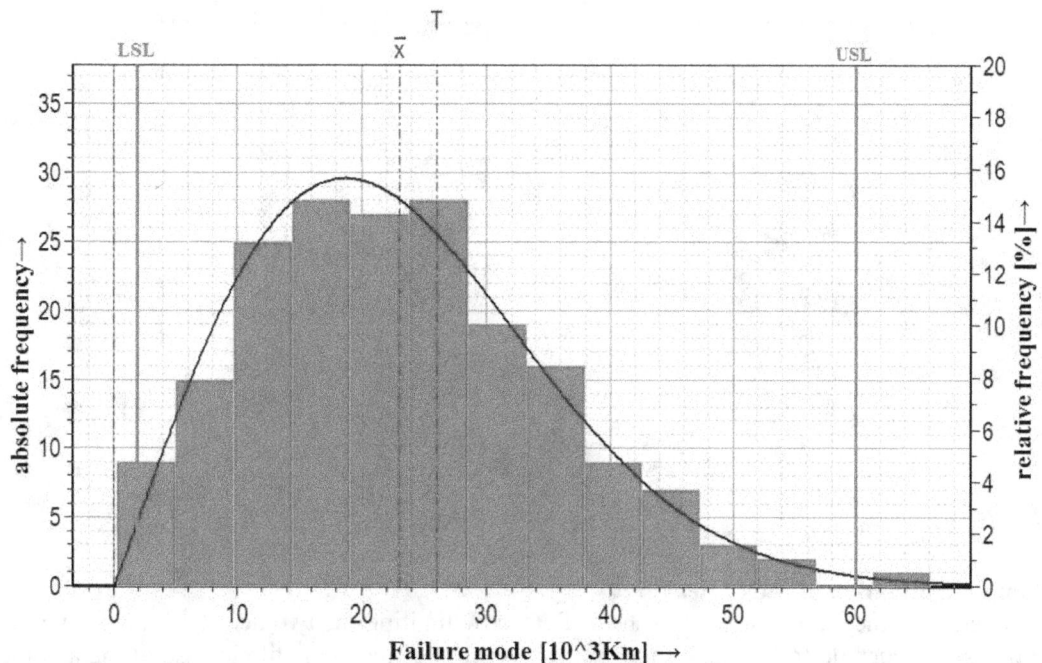

After two years of field observation time, a Weibull distribution model can be fitted based on field data: Figure 3 shows the field data and the fitted Weibull density function $f(x)_{FM}$ after two years of field observation (24 MIS). In this case study a comparison between the predicted parameters (cf. figure 2) and the estimated parameters based on the full data set is feasible and shows similar values. E.g. the location parameter at 24 MIS is predicted to $19*10^3$ km (based on 12 MIS knowledge) and the value of the estimated location parameter is $22.5*10^3$ km (based on 24 MIS knowledge).

5.2 Step B: Mapping field load profile

A kilometrage profile can be generated (cf. table 1) out of the empirically known field data of the product fleet (*PF*). An important criterion is the timeframe between the product approval (e.g.: car registration) and the failure occurrence, the latter describes the risks observation point inside the RAPP method. In many cases distribution functions for mapping a product fleet kilometrage profile can be described with the help of a log-normal distribution (2) [12], normal distribution as well as the Weibull distribution model (1).

$$f(x)_{PF} = \frac{1}{x \cdot \sigma \cdot \sqrt{2 \cdot \pi}} \cdot e^{-\frac{(\lg(x)-\mu)^2}{2\sigma^2}} \tag{2}$$

An example for a approximated density function $f(x)_{PF,2}$ of the product fleet (observation time 2 years) is shown in figure 4. The estimated density function is a general log-normal distribution with two parameters [1].

Figure 4: An example of an empirical kilometrage function $f(x)_{PF}$ (log-normal destiny distribution) of the product fleet based on an observation period of two years (cf. [1])

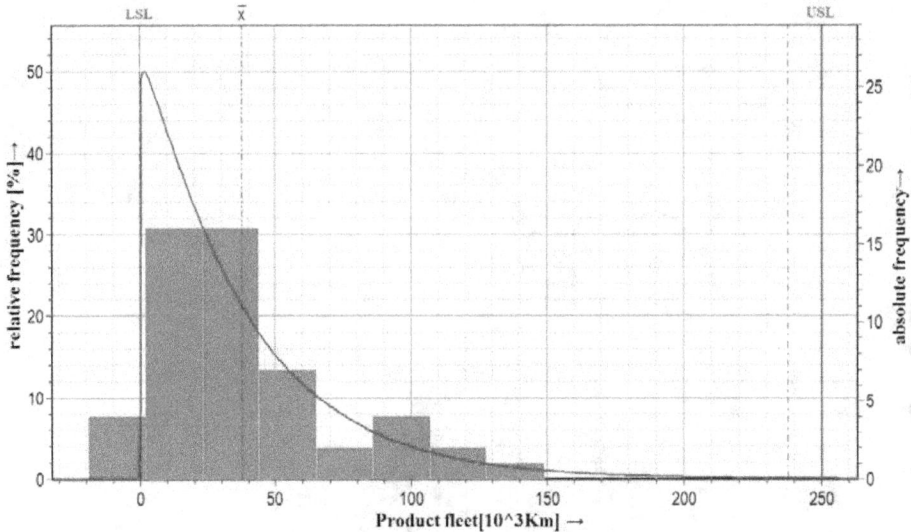

5.3 Step C: Estimation of the critical area

The estimation of the critical area has to be evaluated by limiting the two density functions within the given one-sided quantile (Q) of each function. A typical one-sided quantile in case of the automotive industry (Lower Limit $LL = 0.05$ / Upper Limit $UL = 0.95$) is used in the constituted case study. Therefore, the left-sided failure mode distribution was limited by using a 0.05 quantile (Q_{LL}) and the kilometrage distribution by applying a right-sided 0.95 quantile (Q_{UL}). In case of a right-sided failure mode, the quantile borders have to be switched. The schematic visualisation of the critical area within the oil tube case study is shown in figure 5.

5.4 Step D: Risk analysis and risk prognosis

The RAPP method provides the calculation of three characteristics:
- P_{RPF}: Risk probability regarding the product fleet (Index: RPF),
- $f(x)_{DA}$: Probability function of damage appearance (Index: DA) and
- $P_{RPF\text{-}Prog}$: Prognosis of the future risk probability (Index: $RPF\text{-}Prog$).

Besides the kilometrage function $f(x)_{PF}$ and the failure mode function $f(x)_{FM}$, the introduced characteristics are visualised in figure 6.

Figure 5: Visualisation to determine the critical area [$c_{LL} = Q_{LL;0.05}$; $c_{UL} = Q_{UL;0.95}$] as well as the overlapping of $f(x)_{PF}$ and $f(x)_{FM}$ to determine the risk parameters P_{RPF} and $f(x)_{DA}$ (cf. [1])

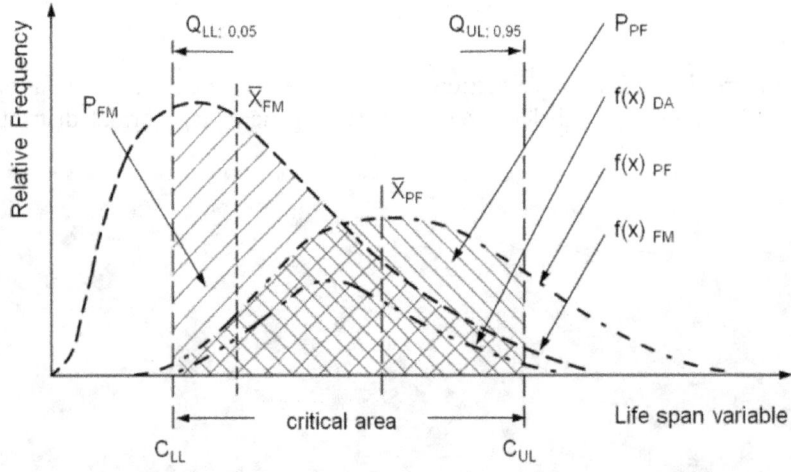

The calculation of the RAPP characteristics is as follows: The risk probability of the critical product fleet P_{PF} can be estimated using the failure occurrence probability P_{FM} (Integral of $f(x)_{FM}$ (1)) regarding to the critical area, cf. Step C) and the product fleet kilometrage probability P_{PF} (Integral of $f(x)_{PF}$ (2) regarding to the critical area, cf. Step C).

$$P_{RPF} = P_{FM} \cdot P_{PF} = \int_{C_{LL}}^{C_{UL}} \frac{b}{T-t_0} \left(\frac{x}{T}\right)^{b-1} \cdot \exp\left[-\left(\frac{x}{T}\right)^b\right] \cdot \int_{C_{LL}}^{C_{UL}} \frac{1}{x \cdot \sigma \cdot \sqrt{2 \cdot \pi}} \cdot e^{\frac{(\lg(x)-\mu)^2}{2\sigma^2}} \quad (3)$$

The function $f(x)_{DA}$ describes the damage appearance probability of the product fleet in dependence on a life span variable (e.g.: kilometrage). With this specific characteristic the damage appearance probability can be determined (4) at any point of observation time with respect to an appropriate life span variable. Furthermore, the probability to find one unit at the point of observation can be easily calculated by finding the maximum of the failure occurrence function (5).

$$f(x)_{DA} = \frac{b}{T}\left(\frac{x}{T}\right)^{b-1} \cdot \exp\left[-\left(\frac{x}{T}\right)^b\right] \cdot \int_{C_{LL}}^{C_{UL}} \frac{1}{x \cdot \sigma \cdot \sqrt{2 \cdot \pi}} \cdot e^{\frac{(\lg(x)-\mu)^2}{2\sigma^2}} \quad (4)$$

$$f'(x)_{DA} = 0 \quad (5)$$

The probability P_{RPF} maps the prognosis in the field of expected damage cases in the future at a specific point of observation time t_{RP}. The general procedure is identical to the determination of the first risk characteristic P_{RPF}. The difference is the adjustment of the product fleet's kilometrage function $f(x)_{PF}$ in relation to the prognosis point of time and thus the renewed determination of the left-sided lower quantile $Q_{LL(Prog);0.01}$. In case of additional changing failure behaviour, the adjusted failure mode function and its appropriate quantile have to be considered.

Referring to equation (3) the probability $P_{RPF\text{-}Prog}$ can be calculated as follows:

$$P_{RPF,\,Prog} = P_{FM,\,Prog} \cdot P_{PF,\,Prog} \quad (6)$$

In the presented oil tube case study the risk analysis and prognosis based on the RAPP method is using the field data set regarding the first eleven MIS. Afterwards, the results of the risk prognosis are compared to the real failure behavior at the points of time 15, 18 and 21 MIS. The results of the RAPP risk analysis and the respective risk prognosis (Step D; cf. Figure 7) are shown in Table 3.

Figure 6: Visualisation of the overlapping f(x)$_{PF}$ and, f(x)$_{FM}$ and the product of both to determine the risk parameters f(x)$_{DA}$ after one MIS (left) and eleven MIS (right)

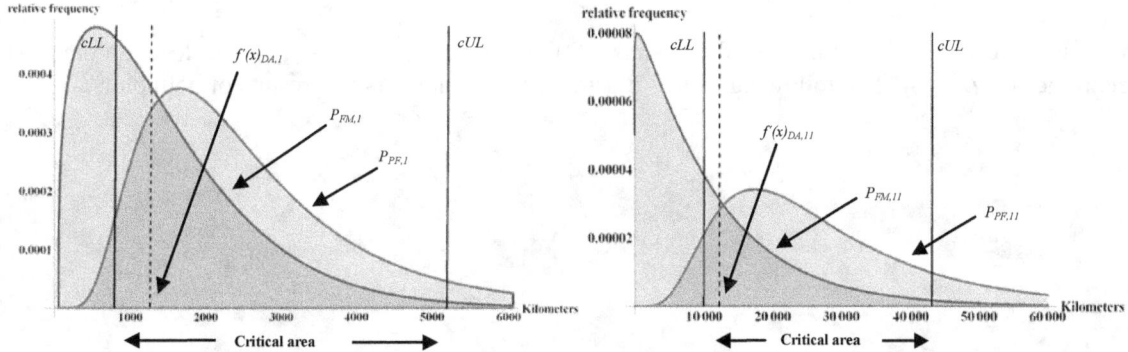

Figure 6 shows the overlapping of the density functions of cars based on the oil tube leak failure mode (Weibull distributed, (1)) and the associated product fleet (Lognormal distributed, (2)) after one and eleven MIS. After the one MIS the major section of the product fleet is moved in the range of cars with oil tube leak failure mode (84.52%). The extremum $f'(x)_{DA,1}$ is at 1,237 kilometres. At the beginning of 12 MIS regarding the product batch the fleet moved to a mean kilometrage of 11,927 kilometers. The risk analysis in product fleet (P_{RPF}) dropped from 44.02% to 15.83% by simultaneous increasing the range of the critical area (4,550 to 33,900 kilometers). Comparison of $f'(x)_{DA,1}$, $f(x)_{PF,1}$ and $f(x)_{FM,1}$ with $f'(x)_{DA,11}$, $f(x)_{PF,11}$ and $f(x)_{FM,11}$ shows a proportional process, therefore it can be assumed, that the failure mode is not completely developed. But it can be received that a short quantity of cars is in the critical area (max 15.83%) of the oil tube leak and could breakdown in higher kilometrage. The following table 2 resumes the results regarding risk analysis at 1 MIS and 11 MIS.

Table 3: Results of the RAPP risk analysis after one and eleven MIS; case study oil tube leakage

Risk characteristics	Results	Note
Results after one month in service		
$P_{PF,1}$	44.02 %	Risk analysis in product fleet; observation point of time: 1 MIS
$P_{PF,1} : cLL / cUL$	0.71 / 5.26 [10^3 km]	Critical area regarding to the life span variable within 1 MIS observation time
$f'(x)_{DA,1} = 0$	2.204 10^3 km	Damage appearance, 1 MIS
Results after eleven months in service		
$P_{PF,11}$	15.83 %	Risk analysis in product fleet; observation point of time: 11 MIS
$P_{PF,11} : c_{LL} / c_{UL}$	9.98 / 43.88 [10^3km]	Critical area regarding to the life span variable within 11 MIS observation time
$f'(x)_{DA,11} = 0$	16.538 10^3 km	Damage appearance, 11 MIS

The extrapolation of the Weibull distribution parameters of $f(x)_{FM,1-11}$ show a prognosis of the oil tube leak failure behaviour after 21 MIS based on the prognosis of the shape and location parameters (cf. figure 1). The following figure 7 shows the verification of the RAPP method: Risk prognosis regarding 21 MIS versus real field data regarding 21 MIS.

Figure 7: Comparison of the destiny functions $f(x)_{FM,Prog\ 21}$ (left) and $f(x)_{FM,21}$ (right)

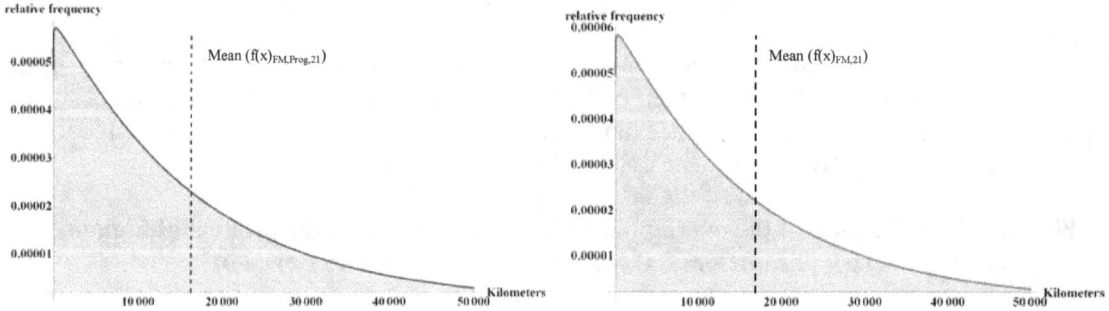

In figure 7 can be observed, that the results of the prognosis $f(x)_{FM,Prog,21}$ are close to the real occurrence of $f(x)_{FM,21}$. The following table 4 illustrates the parameters and results of both plots.

Table 4: Results of the $P_{FM,Prog\,21}$ and $P_{FM,21}$

Risk characteristics	Result	Note
Parameters of the 21 months in service prognosis		
$\beta_{FM,Prog\,21}$	0.988	Shape parameter of Weibull distributed prognosis for 21 MIS
$\alpha_{FM,Prog\,21}$	19,540	Location parameter of Weibull distributed prognosis for 21 MIS
Mean($P_{FM,Prog\,21}$)	19,450	Arithmetic mean of prognosis for 21 MIS
Values of parameters after 21 months in service		
$\beta_{FM,\,21}$	1.0141	Shape parameter of Weibull distributed the real values after 21 MIS
$\alpha_{FM,21}$	19,720	Location parameter of Weibull distributed the real values after 21 MIS
Mean($P_{FM,21}$)	19,500	Arithmetic mean of the real values after 21 MIS

The determination of the function "damage appearance" (4) for each month in service $f(x)_{DA,1-11}$ shows the centered area of the $P_{RPF,1-11}$. After finding the maximum with (5) the x-value of $f'(x)_{DA,1-11} = 0$ expose the focus of kilometrage from the oil tube leak for each month. Figure 8 presents $f(x)_{DA,1}$ till $f(x)_{DA,11}$ and the course of the peaks (black line) $f'(x)_{DA,1} = 0$ till $f'(x)_{DA,11} = 0$

Figure 8: Visualisation of $f(x)_{DA,1-11}$ as well as the course of $f'(x)_{DA,1-11}$ (Damage appearance peaks) from one till eleven months in service

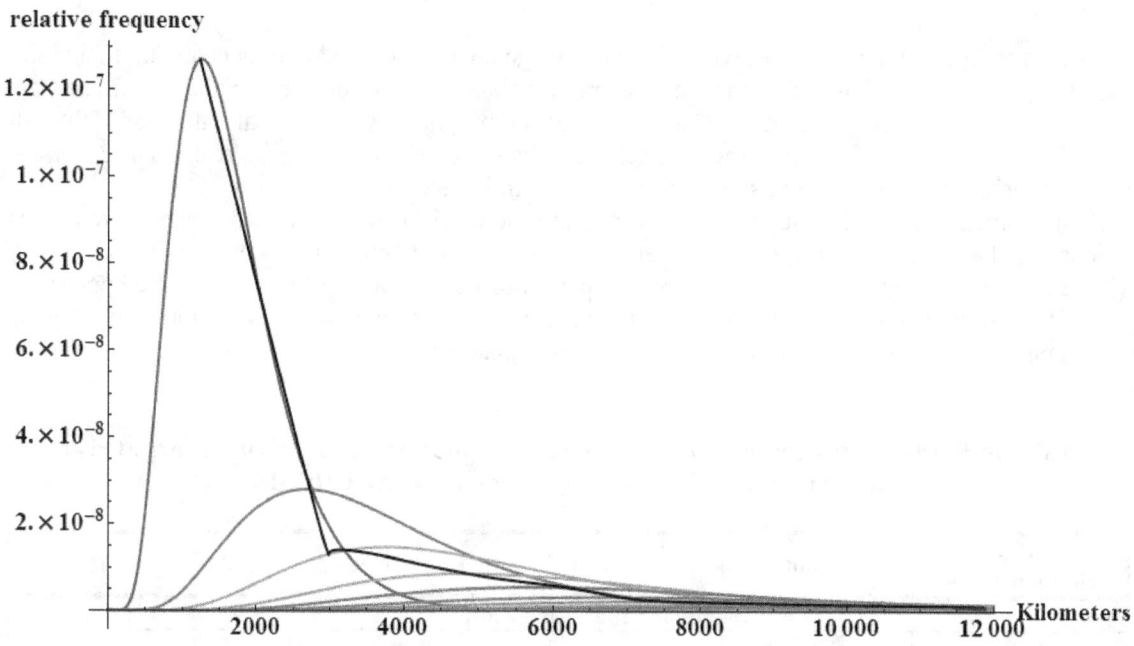

Primarily, it can be observed that the y-value of the $f'(x)_{DA,1-11}$ is very small, therefore it is not possible to show $f(x)_{FM,1-11}$ $f(x)_{PF,1-11}$ and $f(x)_{DA,1-11}$ in one plot. In the first month the focus of the failure mode ($f'(x)_{DA,1}=0$) can be located at 1,237 kilometers. In the following months the dispersion regarding $f(x)_{DA,2-11}$ gets higher. With the increase of the range of $f'(x)_{DA,1-11}$ the validity of focus of the failure mode gets lower. The black line describes $f(x)_{DA,1-11}$. The following figure 9 compares the failure mode focus (damage appearance, cf. (5)) with the proportion of cars from P_{PF} in the critical area of P_{FM} for each month.

Figure 9: Prognosis of the damage appearance focus ($f'(x)_{DA} = 0$) and the proportion of cars from P_{PF} in the critical area P_{FM} based on eleven months in service knowledge compared with the real values after 15, 18 and 21 months in service

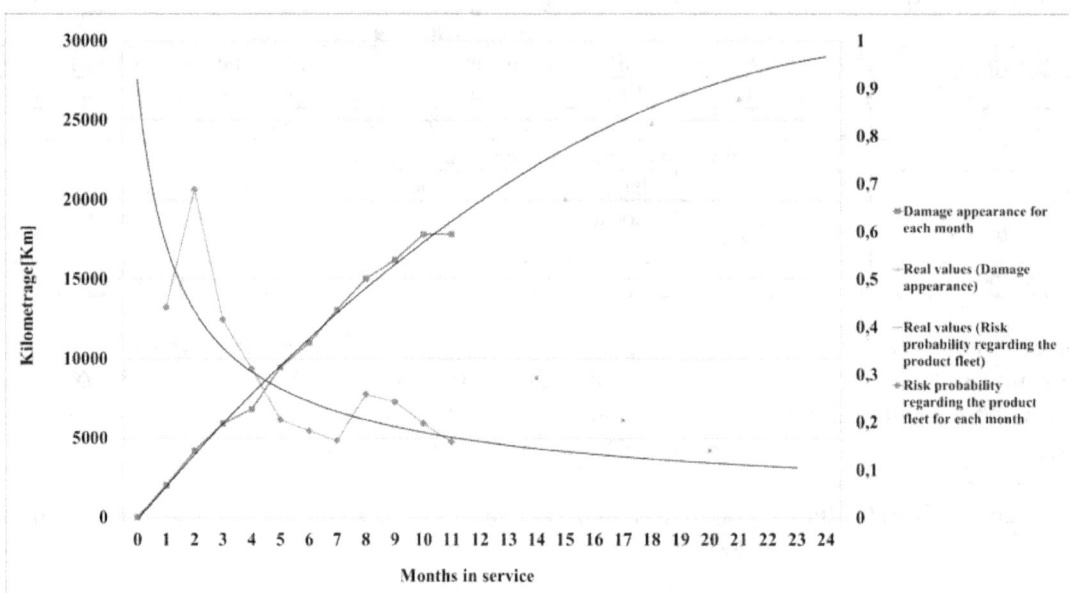

According to figure 9 it can be observed, that the damage appearance (red line) is changing to a higher kilometrage. The prognosis shows the increase trend of failure mode focus regarding the kilometrage for the next 11 MIS (Total: 22 MIS). Comparison of the prognosis with the real values of 15th, 18th and 21st MIS based on field data shows a related behaviour, but overrates the real occurrence. The reason is the increase of the dispersion after eleven month in service.

The critical area of the risk probability regarding the product fleet (blue line) has a regressive course. Comparing the prognosis with real values after 15, 18 and 21 months in service the risk probability will be underrated, that implies outliers after the period of observation. While the damage appearance gets higher simultaneously the proportion of cars at risk gets lower which is an indicator that the overlapping of P_{FM} and P_{PF} gets smaller. The following table summarizes the important values of the figure 9.

Table 5: Results of prognosis from focus of failure mode and proportion of cars at risk compared with the fair values after 15, 18 and 21 MIS

Risk characteristics	Result	Note
Prognosis of the damage appearance		
$f'(x)_{DA,Prog,15}$	16,410 Km	Damage appearance prognosis: 15 MIS
$f'(x)_{DA,Prog,18}$	18,180 Km	Damage appearance prognosis: 18 MIS
$f'(x)_{DA,Prog,21}$	21,040 Km	Damage appearance prognosis: 21 MIS
Real values of Damage appearance		
$f'(x)_{DA,15}$	14,890 Km	Damage appearance 15 MIS
$f'(x)_{DA,18}$	17,280 Km	Damage appearance 18 MIS
$f'(x)_{DA,21}$	20,450 Km	Damage appearance 21 MIS
Risk probability prognosis		
$P_{RPF,Prog,15}$	9.13%	Risk probability prognosis: 15 MIS
$P_{RPF,Prog,18}$	8.80%	Risk probability prognosis: 18 MIS
$P_{RPF,Prog,21}$	8.31%	Risk probability prognosis: 21 MIS
Real values proportion of the risk probability		
$P_{RPF,15}$	17.82%	Risk probability 15 MIS
$P_{RPF,18}$	11.31%	Risk probability 18 MIS
$P_{RPF,21}$	8.52%	Risk probability 21 MIS

6. CONCLUSION

This paper outlines the RAPP method and its application within the oil tube leak case study for determination of risks in product fleets within the usage phase. The focus is the analysis and prognosis of the field failure behaviour of a defect oil tube unit based on product field data. The results lead to a visualised risk analysis and prognosis on the basis of the existing failure mode and the product fleets life span variables at a certain point of observation. Furthermore, the estimation of the risk prognosis with respect to the expected damage cases within the product fleet is feasible. The comparison of the RAPP method and industrial state-of-the-art methods [7] has not been verified yet and is an inherent part of future research work.

The advantages of the present procedure are: The RAPP method, in comparison to the industrial standards of risk prognosis methods, considers a direct link of different dimensions (e.g. operating time, switching cycles) causing a specific damage case and the product's life span variables without a mathematical conversion. Thus, the method generates the possibility of direct visualisation of damage cases (failure behaviour) and life span areas at a specific point of observation time to estimate the existing risk potential. The visualisation includes the critical area between a product's life span variable and a present failure mode, which can be calculated by using p-quantiles of the failure behaviour probability function and the empirical life span variables. Furthermore, the adjusting empirical life span variable as related to the failure behaviour can be visualised properly and analysed subsequently. Finally the developing of the damage appearance regarding the predicted behaviour of the failure mode and the product fleet is calculable.

The disadvantages of the shown procedure are: The fitted density functions of the empirical failure modes and the product fleet's life span variables are fundamental for the presented method. Thus the deviation due to different probability functions and the fitted model itself can lead to different results of the risk analysis. The current research results due to different probability functions show a marginal influence in the result. However, some case studies led to adverse risk prognosis based on small critical areas, which is also a part of the future research work.

References

[1] S. Bracke. *"RAPP: A new approach for risk prognosis on technical complex products in automotive engineering"*, Safety, Reliability and Risk Analysis: Beyond the Horizon. Proceedings: ESREL 2013; European Safety and Reliability Association – ESRA, (2013).

[2] B. Bertsche. *"Reliability in Automotive and Mechanical Engineering. Determination of Component and System Reliability"*, Springer, 2008, Berlin.

[3] A. Birolini. *"Reliability engineering: Theory and Practice"*, Springer, 2007, Berlin.

[4] S. Bracke and S. Haller. *"Field damage analysis (FDA) concept: Contribution to the comprehensive reliability analysis of complex field damage cause"*, Proceedings: RAMS 2011, Annual Reliability and Maintainability Symposium, (2011).

[5] S. Bracke and S. Haller. *"The RAW concept: Early identification and analysis of product failure behaviour in the use phase"*, Proceedings: ESREL 2011, European Safety and Reliability Conference, (2011).

[6] M. Celis, J. E. Dennis, and R. A. Tapia. *"A trust region strategy for nonlinear equality constrained optimization"*, Proceedings: SIAM, (1985).

[7] G. Eckel. *"Determination of the initial course of the reliability function of automotive components"*, Qualität und Zuverlässigkeit 22, pp. 206 – 208, (1977).

[8] G. Linß. *"Quality Management for Engineers"*, Carl Hanser, 2005, München.

[9] A. Meyna and B. Pauli. *"Zuverlässigkeitstechnik – Quantitative Bewertungsverfahren"*, Hanser, 2010, München.

[10] S. Persin, S. Haller and S. Bracke. *"IDREMA-Process: Identification of Reference Market for Defect Parts Routing."* Proceedings: RAMS 2013 - Annual Reliability and Maintainability Symposium, IEEE Reliability Society, (2012).

[11] T. Pfeifer. *"Quality Management"*, Carl Hanser Verlag, 2002, München.

[12] L. Sachs. *"Applied Statistics"*, Springer Verlag, 2002, Berlin.

Stress-Dependent Weibull Shape Parameter Based on Field Data

Jochen Juskowiak[a*] and Bernd Bertsche[a]
[a] University of Stuttgart, Stuttgart, Germany

Abstract: The Weibull shape parameter is often assumed to be constant, with no dependency on stress. However, some cases exist, in which it is a function of stress. If the stress-dependency is not considered, vague assumptions of the Weibull shape parameter may lead to inaccurate results, e. g. for reliability prediction or demonstration testing purposes. Drawbacks in choosing an adequate parameter are e.g. extensive testing at a specific stress level, or insufficiently established mathematical descriptions.

This paper presents an approach which allows a stress-dependent derivation of the Weibull shape parameter based on field data. In order to do so, simulations of the customer behavior and additional information from the customers themselves are used. Linking the occurred failure with the corresponding stress-level is thus possible.

Keywords: Non-constant Weibull Shape Parameter, Stress-dependency, Field Data, Customer Behavior, Automotive Engineering.

1. INTRODUCTION

The Weibull distribution is commonly used in mechanical engineering for the characterization of the failure behavior of specific components and systems. The failure behavior of certain parts in combination with certain failure modes can be described by using a shape parameter. A specific value is often assumed to be constant, e.g. in the case of roller bearings. However, for other components such as gears and shafts, the shape parameter is non-constant, as it also depends considerably on the load [1]; in this case, the values are within a specific range, e.g. known from literature. Yet the dependency often cannot be described by a function. Hence, regarding a reliability prediction for future products under consideration of these dependencies, certain challenges need to be faced:

Firstly, to describe the dependencies by a statistically representative function, extensive testing would be required for each failure mode of a given component, which leads to considerable costs. The correlation between the component's behavior under testing conditions and field use has to be known. This is to ensure that the shape parameter is the same under both conditions. Secondly, gaining relevant customer data for automotive applications in the private sector is generally difficult, as strict legal constraints apply. In case of a failure, conclusions on the actual load history, i.e. the stress endured, can hardly be drawn systematically. Therefore, the failure of interest is allocated to the set of failures under field conditions, resulting in a single shape parameter for failures actually stemming from different load scenarios.

The goal is to find a method which allows for describing the shape parameter more realistically based on the load history actually experienced in the field and declared by the user. The field data on hand could then be used more efficiently and effectively; expenditures for testing could be minimized even further. The stress-dependent determination of the shape parameter will lead to a more realistic reliability prediction for the given failure mode to be fed back as input for future applications or developments. It can also be used for reliability demonstration testing. Its benefits are clarified by a brief example regarding the success run (see [2]): Assuming a reliability of required lifetime $R = 0.9$, a confidence level $P_A = 0.9$, a lifetime ratio $L_V = 0.6$, an acceleration factor $r = 3$ and a "constant" shape parameter $b = 1.5$, results in a required sample size $n = 12$. If the shape parameter can be described more precisely concerning lower stress in the field ($b_{field} = 1.3$) and higher stress in test ($b_{test} = 1.7$) the sample size will be reduced to $n = 8$.

---

* jochen.juskowiak@ima.uni-stuttgart.de

In order to make field data usable for efficient analysis, case-dependent characteristics need to be gathered, e.g. driving style and driver type [3]. In case of an occurred failure these data are gathered by collecting information directly from the customer. By means of a sum of square error approach this information is matched with statistical data. In order to classify the user data, simulations for different load scenarios are implemented. The simulated load time functions are used to calculate damage values by means of the damage accumulation hypothesis [4]. Based on these results, the required rules for correctly combining the customer data are derived. The lifetime characterizing unit (e.g. miles, hours or load cycles) is then linked with the stress-intensity. The data used are multiple censored data. A Weibull analysis of the stress-specific data using Sudden-Death approaches [1] leads to stress-specific shape parameters derived from field data.

The Section below discusses briefly the shape parameter and the assumption of its stress-dependency. The third Section describes the customer's role and the way how inference from the customer's behavior about the applied stress can be drawn. Section 4 introduces the entire approach step-by-step whereas Section 5 and 6 provide a short example and conclusion.

2. STRESS-DEPENDENT SHAPE PARAMETER

The Weibull distribution is determined by its parameters: the shape parameter b, the characteristic life (or scale parameter) T and – in case of a three parameter Weibull – the failure free time t_0. All these parameters are dependent upon geometry, material, machining and stress [1]. In this paper, the focus is on the stress-dependency of the shape parameter.

Experiments with some components such as gears and shafts have shown that the shape parameter depends significantly on the load [1]. In many cases, a higher stress yields a higher shape parameter. However, a steep Weibull does not immediately go along with high stress. A large shape parameter corresponds with a small variation in the times to failure. This fact is for instance used to control the quality of turbine blades made from metal as purer metal results in steeper shape parameter than dirtier metals [5]. The stress-dependency is shown by several conducted experiments:

Maenning [6] demonstrated a stress-dependency for shafts made from C35. The observed failure mode is crack due to fatigue. He proved his findings by experiments on 19 stress levels from 295 up to 385 MPa with at least 20 specimens in each level. A higher stress results in a larger shape parameter.

Figure 1: Stress-Dependent Weibull Shape Parameter as well as the Characteristic Life Including their Confidence Intervals; a) Brodbeck and b) Groß

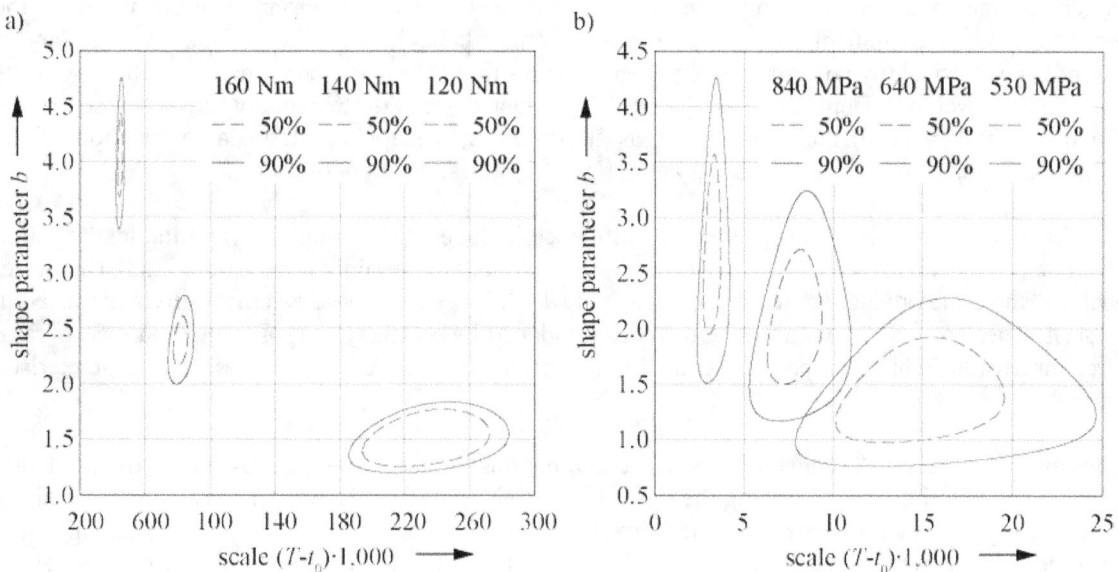

The same dependency is determined for gears with failure mode crack. Groß [7] tested three samples with 12 specimens each at 530, 640 and 840 MPa. He used gears made from 42CrMo4V. Brodbeck [8] considered carburized steel 16MnCr5 gears and carried out extensive tests at 120, 140 and 160 Nm with each 100 specimens. Results from Groß and Brodbeck are summarized in Figure 1. Here, a three parameter Weibull analysis is applied.

Beier [9] revealed a similar positive correlation for gear pairs made from plastic. In this case the observed failure modes were crack due to fatigue and additional wear. He investigated at least 8 specimens in different variants. The occurred mechanisms are discussed separately.

Nelson [10] states that experience confirms the dependency between the shape parameter and stress levels in terms of metal fatigue and even roller bearings. Bergling on the other hand [11] assumes a constant shape parameter for roller bearings. Additionally, for some electrical insulation Nelson suggests a negative correlation, i.e. a higher shape parameter at lower stress.

3. APPLIED STRESS BY THE CUSTOMER

Field data is generally one of the most important resources that can be used by any reliability program. No data will better demonstrate the true reliability of a product, nor identify the failure modes that exist in the field. Thus, a more comprehensive reliability database can be created by incorporating field usage data. Particularly, information about customer behavior helps to differentiate the individually occurring stress that leads to failure. For instance, Lucas et al. [12] introduced a complementary FRACAS approach, which focuses on differentiated usage conditions linked to failed and non-failed units, in terms of the oil and gas industry.

Normally, in automotive engineering, important data on operating and environmental conditions are missed to draw interference about the applied stress leading to the specific failure. On the other hand, in case of leasing models, e.g. car-sharing or car rental, gathering stress-related data should be possible. Leopold [13] analyzed several sources of reliability data during operating time. A common drawback is the missing link between the applied stress and an occurred failure, especially for non-commercial passenger cars. For this reason, the influences on stress from customer behavior are discussed first, followed by a proposed adequate procedure which is implemented in the entire approach in Section 4 to challenge this fact.

3.1. Influences on Stress from Customer Behavior

In general, the stress on a certain part depends on the operating and environmental conditions of the product. Both are immediately affected by customer behavior. In the context of automotive engineering, particularly powertrain components, the resulting stress is a combination of the driver, the road and the vehicle. Depending on the actual product design, the externally applied load on the product is broken down into single parts as stress. Thus, the applied stress, as a function of the individual load, ultimately leads to wear and fatigue failures at different times [3].

The driver's - respectively the customer's - influence includes the driving style and the loading. In a wider sense the road type can be inferred as well, as the driver certainly chooses the road. The driver decides where the product is used and the ways in which it is used. These external criteria are listed in Table 1. These are in line with the 3F method introduced by Kücükay [14]. The method incorporates three dimensions which can be represented by the criteria. Each criterion is expressed in characteristic attributes.

Consequent, there are 48 (equal to 4 x 4 x 3) combinations of customer types. By means of simulation of these combinations, incorporating the results from field experiments, Müller-Kose [15] pointed out some examples which indicate different damage values for several combinations. The ranking of the customer type combinations regarding the damage values depends on the focused part and failure mode as well.

Table 1: Stress Influencing Criteria and Characteristic Attributes, c.f. [3,15]

Criteria	Description	Characteristic attributes
Road type	Proportion of total mileage on various types of roads such as motorway, rural road, urban traffic or mountain road	Motorway, rural road, urban traffic or mountain road
Loading	Percentage distribution of journeys with numbers of passengers, cargo and trailer weight	Light, average, heavy or extreme (trailer)
Driving style	Shifting frequency, gearshift engine speed, acceleration habits in town (moving-off from traffic lights), on rural roads (when leaving built up areas) and on the motorway (overtaking)	Sporting, average or moderate

As mentioned above, different vehicle configurations result in a different stress related to a certain part. In other words, a theoretical identical customer behavior leads to different stress due to different vehicle configurations. For instance, there are 64 different vehicle configurations when assuming each two car types, engines, moving-off elements, hybrid elements, main gearboxes and rear axle drives [16]. However, a certain part, e.g. a definite gear, is assembled in all configurations. This fact can be used to increase the database in reference to the field data approach.

3.2. Identification of the Customer Type Combination of Interest

Identification of the appropriate customer type combination enables inference about the applied stress. Required information is: a representative mapping of the behavior of all customers and on the other hand an assessment of the single customer whose product failed.

Müller-Kose collected a representative data set by means of comprehensive field experiments. The results given in Table 2 represent exemplary for each criterion a percentage allocation of the characteristic attributes. With this 11 customer types can be defined:

Table 2: Percentage Allocation of 11 Customer Types, Following [15]

[%]	Road type				[%]	Loading				[%]	Driving style		
Customer type	Urban traffic	Rural road	Mountain road	Motorway	Customer type	Light	Average	Heavy	Extreme (trailer)	Customer type	Sporting	Average	Moderate
Urban traffic driver	93	7	0	0	Light loading	94	6	0	0	Sporting driver	46	54	0
Rural road driver	4	92	0	4	Average loading	0	100	0	0	Average driver	6	86	8
Mountain road driver	10	50	30	10	Heavy loading	0	20	80	0	Moderate driver	5	10	85
Motorway driver	5	5	0	90	Extreme loading (trailer)	0	10	80	10				

If such a percentage allocation is known, this allocation can be matched with the customer behavior. In order to do this, the customer is encouraged to provide information if a failure occurred; e.g. to complete a questionnaire. The customer is then asked to estimate their preferred road type, loading and driving style. This results in an individual percentage allocation of the customer. For instance, customer "C" states, that the proportion of their total mileage is 20 % in urban traffic, 60 % on rural roads and 20 % on motorways. In the next step, based on this information the most proper customer type (see Table 2) is to be figured out. For this purpose, the sums of square errors are calculated (see Section 4). The customer type with the least square error represents the appropriate actual customer type. Repeating this analogously for loading as well as driving style yields the appropriate customer type combination of interest.

4. APPROACH

The following procedure is generally applicable for several parts and different failure modes. However, for the sake of simplicity it is exemplary introduced based on a gear pair of a rear axle drive. The observed failure mode is pitting due to fatigue. To transfer this on other parts and mechanisms, lifetime model and material data must be adapted among others. Here, the lifetime model is assumed to be valid for 50 % failure probability and the procedure is presented for identical products, i.e. an identical vehicle configuration.

Using this procedure, the following requirements must be taken as given: First, measurements are needed to establish a validated powertrain simulation for the vehicle longitudinal dynamics. Second, a geometry data set of the considered part is necessary. If the simulation does not provide the data directly, e.g. gears in a gearbox with just a simulated input or output rotation speed, this can be done indirectly by additional use of gear ratios. Third, material data and the underlying lifetime model are required, i.e. the Wöhler curve and its defining parameters respectively.

There are two main paths in the depicted algorithm (Figure 2):

- The determination of load values for each type. Lifetime model and material data is needed and damage values are determined (see I. simulation path).

- On the other hand, the given field data, its gathering procedure and Weibull analysis (see II. field path).

Both paths are combined in the algorithm. With this, the estimation of a stress-dependent Weibull shape parameter can be done and the optimization of the initial assumed lifetime exponent is enabled.

To simplify in the following remarks, "one" out of 48 customer type combinations is called "type i". The implemented steps in the algorithm are as follows:

I. Simulation path

   a) Simulate the load time function for each type i: The simulated variables are torque T_i and rotation speed n_i over time. The simulation is done for a given reference route. The reference route consists of a type-specific mixture of a representative road type course. The simulation ends, if the reference route length w_i is reached. In the automotive industry the use of shortened reference routes are standard in order to reduce simulation time significantly [3].

   b) Transform the simulated torque T_i to stress σ_i by means of geometry data [3]: The stress of interest in case of pitting is the existing Hertzian stress. It can be calculated by using the standard ISO 6336 part 2 [17].

   c) Apply the two parametric level distribution classification [1]: For each type i the results of this step are the dwell time $t_{i,j,v}$ in stress class j and rotation speed class v. The dwell times are a function of the reference route length w_i.

   d) Transform the rotation speed $n_{i,j,v}$ [rpm] in revolutions of the wheel $r_{i,j}$ [absolute frequency], i.e. load cycles, per each stress class j by using equation

$$r_{i,j} = \sum_{v=1}^{n_v} \frac{n_{i,j,v}}{60 \text{ s} \cdot \text{rpm}} \cdot t_{i,j,v} \qquad (1)$$

with dwell time $t_{i,j,v}$ [s] for type i, stress class j and rotation speed class v [16].

Figure 2: Algorithm of the Approach

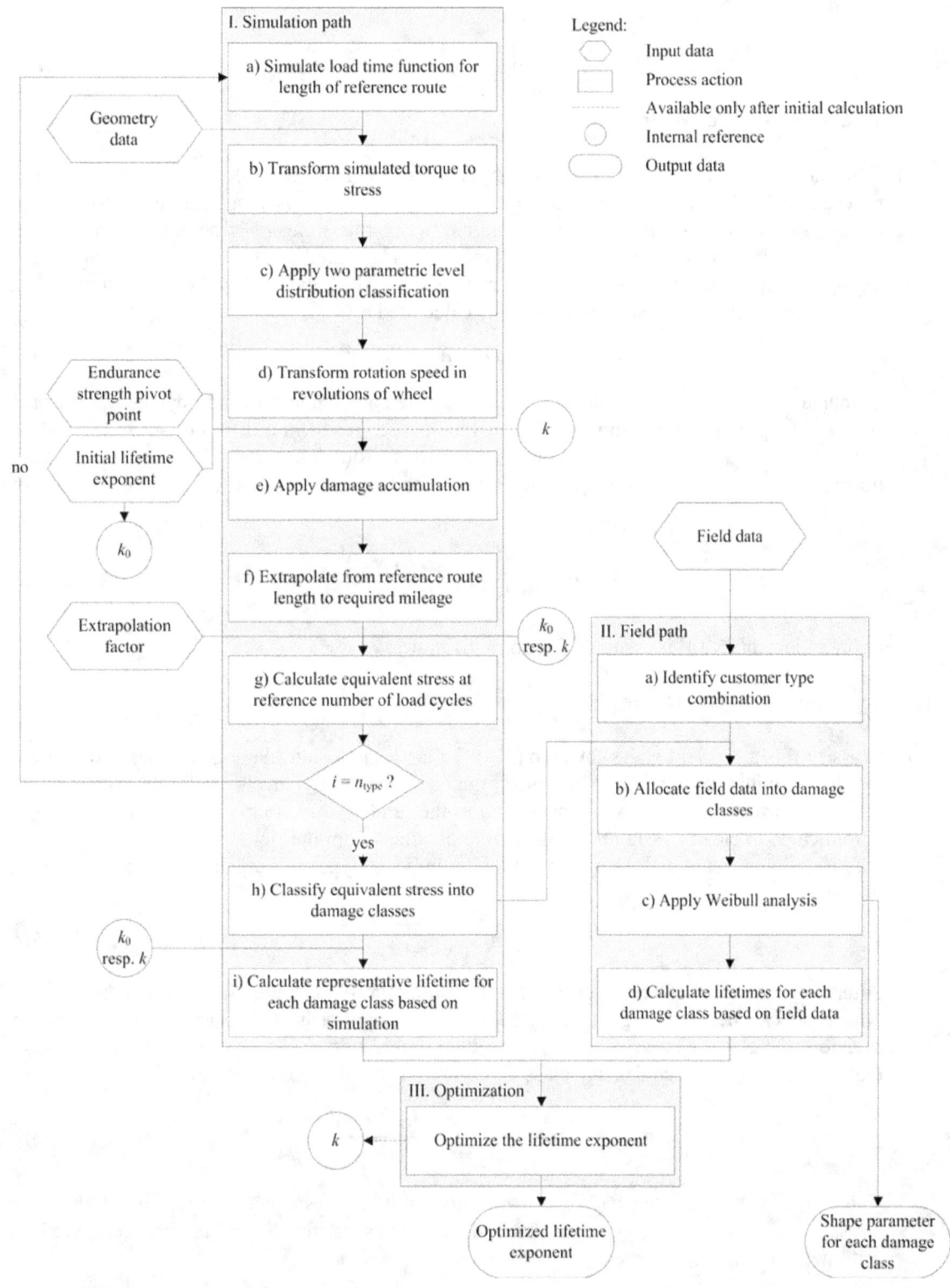

e) Apply the damage accumulation: In this case the Miner-Haibach damage accumulation hypothesis is assumed, due to experience in practice [3]. The specific Wöhler curve must be known. The slope of the fatigue strength zone is represented by the lifetime exponent k and the endurance strength pivot point is defined by the assumed number of endurance strength N_D

and the material specific endurance strength σ_D itself. First, an initial lifetime exponent k_0 is assumed. Afterward, for least square estimation purpose an optimized lifetime exponent k is used (see step III.). With this, the damage sum of the simulated reference route length $D_{i,w}$ is calculated by

$$D_{i,w} = \sum_{j=1}^{u} \frac{r_{i,j}}{N_D} \cdot \left(\frac{\sigma_{i,j}}{\sigma_D}\right)^k + \sum_{j=u+1}^{n_j} \frac{r_{i,j}}{N_D} \cdot \left(\frac{\sigma_{i,j}}{\sigma_D}\right)^{2k-1} \quad (2)$$

with $\sigma_{i,u} \geq \sigma_D$ and $\sigma_{i,u+1} < \sigma_D$.

f) Extrapolate from the reference route length w_i to the required mileage w_{req}: The required mileage is for instance assumed to 200,000 km. It is obvious that various customer type combinations lead to different damage values when this mileage is achieved. To compare these damage values from simulation, they are extrapolated to the required mileage. Each extrapolation factor EF_i is derived by the ratio w_{req}/w_i. Thus, the expected damage sum of type i, extrapolated up to the required mileage, is calculated by

$$D_i = EF_i \cdot D_{i,w}. \quad (3)$$

g) Calculate an equivalent stress level $\sigma_i(N_{ref})$ at a reference number of load cycles N_{ref} for each type i: Along with the assumption, that different types result in a different stress level about the required mileage, the underlying stress varies around a certain level if the stress is normalized at a definite reference number of load cycles. This normalization is done by equation

$$\sigma_i(N_{ref}) = \sigma_D \cdot \left(\frac{D_i \cdot N_D}{N_{ref}}\right)^{1/k} \quad (4)$$

(Miner elementary [4] is assumed for normalization).

Repeat steps a)-g) for all types "type i".

h) Classify the equivalent stress levels $\sigma_i(N_{ref})$ for $i = 1(1)n_{type}$ into damage classes: To increase the database it is recommended to classify the equivalent stress levels of all types into damage classes. Furthermore, the vagueness about the underlying stress in the field makes it complicated to clearly assign the failure time obtained from the field to a certain type. Thus a classification is beneficial. Depending on the number of types, a statistically proper amount of damage classes n_d is

$$n_d = \sqrt{n_{type}}. \quad (5)$$

Alternative approaches are given in [1]. If the amount of damage classes is determined the allocation can be done, e.g. by means of a linear approach. The class range is given by division of the stress range of all types by the number of damage classes n_d. With this upper class limits $\sigma_{UL,d}$ for each damage class $d = 1(1)(n_d-1)$ are defined by equation

$$\sigma_{UL,d} = \min(\sigma) + d \cdot \left(\frac{\max(\sigma) - \min(\sigma)}{n_d}\right). \quad (6)$$

The equivalent stress levels $\sigma_i(N_{ref})$ are allocated to the classes consequently. Finally the mean stress values $\sigma_{m,d}$ for all types in each damage class d for $d = 1(1)n_d$ are computed by arithmetic mean.

i) Calculate a representative lifetime for each damage class: At first identify an average damage class d_{ave}. Select the type i which is close to the damage sum $D_i \approx 1$ and set its mileage equal to B_q-lifetime (here: $q = 50\%$), i.e. the required mileage in case of $D_i = 1$. The selected type indicates the average damage class d_ave. With this the representative simulated lifetimes $B_{q,d,\text{sim}}$ for each damage class d is given by equation

$$B_{q,d,\text{sim}} = B_{q,d_\text{ave}} \cdot \left(\frac{\sigma_{d_\text{ave}}}{\sigma_{m,d}}\right)^k. \qquad (7)$$

II. Field path

a) Identify the customer type combination: If a failure occurs the associated type must be determined. The proposed method is a sum of square procedure that requires input from both statistics and the customer directly. More theoretical explanations are given in Section 3.2. With statistical percentage amount $p_{c,a,\text{stat}}$ of customer type c and characteristic attribute a (c.f. Table 2) as well as customer information $p_{c,a,\text{customer}}$ of customer type c and characteristic attribute a, the sum of square errors Φ regarding customer type c results in equation

$$\Phi_c = \sum_{a=1}^{n_a} \left(p_{c,a,\text{stat}} - p_{c,a,\text{customer}}\right)^2. \qquad (8)$$

The minimum of Φ_c indicates the appropriate customer type of each criterion. Replicate for all three criteria determines the appropriate 3F customer type combination of interest:

$$\min(\Phi_\text{road type}) \wedge \min(\Phi_\text{loading}) \wedge \min(\Phi_\text{driving style}) \Rightarrow \text{combination of interest}. \qquad (9)$$

b) Allocate field data into damage classes: After the determination of the associated type, the failure can be allocated into the corresponding damage class d by using the results of step I. h). This is enabled, because of the link between the unknown customer stress causing the failure and the stress quantified by the simulation path. In other words, matching the determined type and the simulated type empowers the allocation. As a result of this, field failure data sets for n_D damage classes are obtained.

c) Apply Weibull analysis for each damage class: For analyzing field data the Sudden Death assessment is an appropriate method. Field data are usually multiple censored data. Thus additional information, as the delivered output and the amount of intact products in each damage class, is needed. To approximate the amount of intact products in each damage class d a normally distributed customer stress σ_customer is assumed. That means, the majority of customer represent types, which exhibit a stress allocated in a middle damage class. On the other hand, only a small proportion of customer represent types which exhibit a stress allocated in a lower or upper damage class. This assumption is qualitatively illustrated in Figure 3 exemplary for three damage classes. Hence, the amount of intact products $n_{s,d}$ in damage class d is approximated by equation

$$n_{s,d} = P(\sigma_{\text{UL},d-1} < \sigma_\text{customer} \leq \sigma_{\text{UL},d}) \cdot n_s = \int_{\sigma_{\text{UL},d-1}}^{\sigma_{\text{UL},d}} f(\sigma_\text{customer}) d\sigma \cdot n_s \qquad (10)$$

with the overall number of intact products n_s. For more explanation in Sudden Death, the reader is referred to [1]. Finally, this step leads directly to stress-dependent Weibull shape parameters.

d) Calculate the lifetimes for each damage class: Convert the characteristic life (or scale parameter) T to the $B_{q,d,\text{field}}$-lifetime based on field data, c.f. [1].

III. Optimize the lifetime exponent: The initial assumed lifetime exponent k_0 can be updated by using pre-processed field data (see II. above), e.g. by means of least square estimation (LSE) [18]. Finally, the lifetime exponent k is optimized by minimizing the sum of square errors Φ with

$$\Phi = \sum_{d=1}^{n_d} \left(B_{q,d,\text{sim}} - B_{q,d,\text{field}} \right)^2. \qquad (11)$$

Figure 3: Assumed Probability Density Function of the Customer Stress and Allocation in Damage Classes

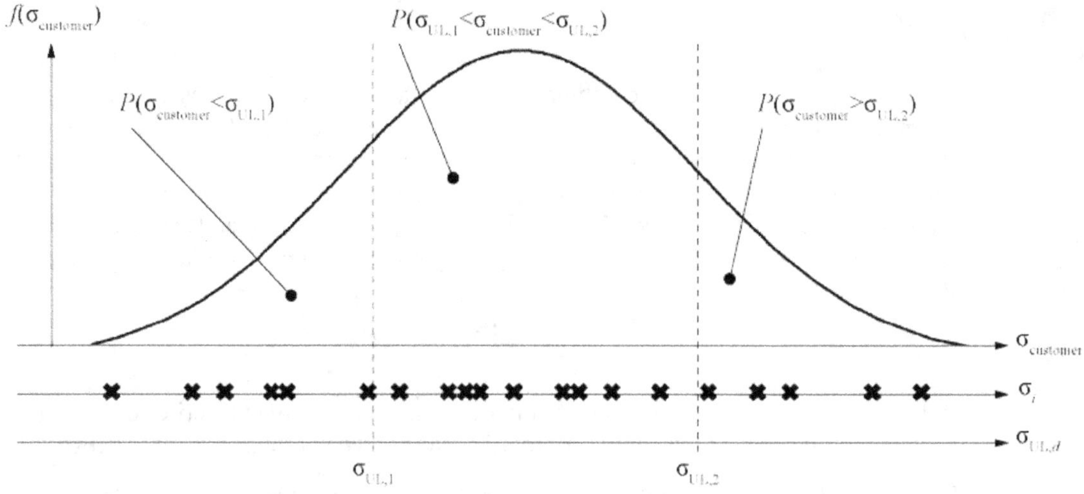

Remark: The exact choice of the reference number of load cycles N_{ref} in step I. g) does not matter regarding the allocation in step II. b) as only relative ratios are considered.

5. EXAMPLE

The approach is illustrated by a simplified synthetic calculation example which is in line with the steps in Section 4. Some steps are omitted due to missing data, such as simulations and both identification and allocation of the customer type combinations. Table 3 depicts the obtained values after the optimization.

Eight customer type combinations are assumed. Each type i is simulated. The results of the two parametric level distribution classifications and the transformations in revolutions of the wheel $r_{i,j}$ per stress class j, are listed in the columns stress class 1 and 2. The next columns depict results of the steps I. e)-g). The mean stresses $\sigma_{m,d}$ are calculated for each damage class in step I. h). The next step identifies the average class d_{ave} and calculates representative lifetimes $B_{50,d,\text{sim}}$ based on the simulation path. Next, field data are gathered and analyzed analogously to steps II. a)-c). The resulting Weibull shape parameters b_d as well as the characteristic lifetimes T_d are stated. Thus, the $B_{50,d,\text{field}}$-lifetimes based on field data are computed. Finally, a least square estimation regarding the lifetime exponent k is conducted (step III.).

Figure 4 depicts the Weibull shape parameter b as a function of an applied stress. By fitting the data, a convenient shape parameter can be derived from a specific stress level.

6. CONCLUSION

Products in the field provide an enormous amount of data. By means of linking the occurred failure event with a certain stress level, a stress-dependent analysis can be executed. The introduced approach empowers the decision-maker to gain a sharper understanding of a stress-dependent Weibull shape parameter based on field data. Using a more precisely determined stress-dependent Weibull shape parameter, results in a more realistic reliability prediction for future products. Thereby, drawbacks from product similarity related assumptions are avoided, as individual components' failure modes are the focal point. Better known Weibull shape parameters, e.g. stemming from both field and test stress levels, might be used to reduce the necessary sample size regarding reliability demonstration testing.

In addition, initially assumed lifetime model parameters can be optimized. Also, analyzing field data using this approach allows for identifying of stress-dependent components and failure modes, which are still unknown.

Table 3: Calculation Example

				stress class 1		stress class 2					
	k_0	5	Initially assumed								
	k	4.884	After optimization (step III.)								
	σ_D	400									
	N_D	5.0E+07								N_{ref}	5.0E+07
	type i	w_i		$\sigma_{i,1}$	$r_{i,1}$	$\sigma_{i,2}$	$r_{i,2}$	$D_{i,w}$	EF_i	D_i	$\sigma_i(N_{ref})$
I. e)-g)	1	8,000		420	300,000	500	250,000	0.022	25	0.562	355.5
	2	10,500		450	300,000	480	300,000	0.025	19	0.482	344.4
	3	10,000		420	300,000	450	350,000	0.020	20	0.401	331.8
	4	5,000		420	350,000	500	325,000	0.028	40	1.129	410.0
	5	5,500		450	350,000	480	300,000	0.027	36	0.984	398.7
	6	7,500		450	350,000	500	325,000	0.032	27	0.847	386.7
	7	6,000		500	300,000	550	400,000	0.056	33	1.858	454.1
	8	7,000		450	350,000	550	350,000	0.046	29	1.303	422.3
I. h)	n_d	3									
	d	1	2	3							
	$\sigma_{UL,d}$	372.5	413.3								
	$\sigma_{m,d}$	343.9	398.5	438.2							
I. i)	d_{ave}		2								
	$B_{50,d,\text{sim}}$	417,218	200,000	127,770							
II. c)	T_d	589,363	264,408	150,324	Estimated based on assumed field data						
	b_d	1.135	1.243	1.573							
II. d)	$B_{50,d,\text{field}}$	426,718	196,887	119,079							
III.	Φ		2.1E+08		Least square estimation						

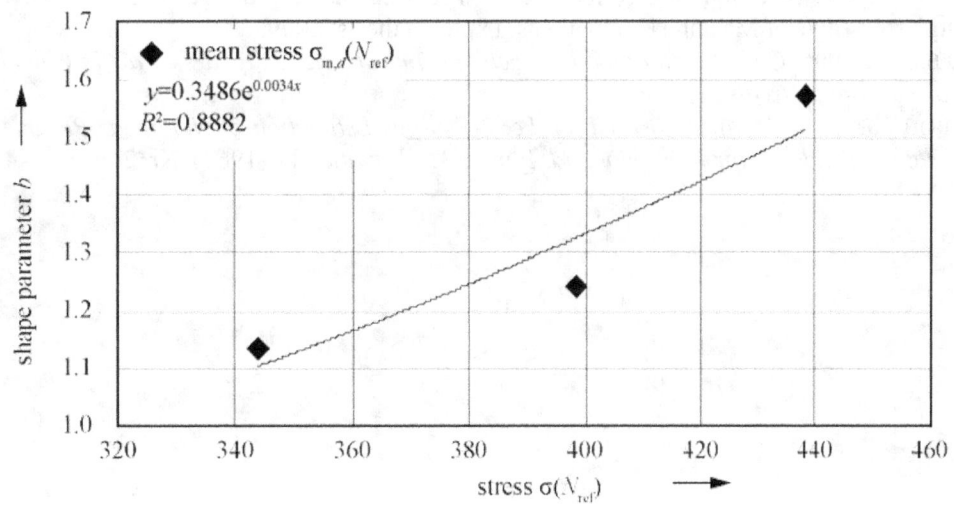

Figure 4: Weibull Shape Parameter as a Function of Stress

References

[1] BERTSCHE, B.: *Reliability in automotive and mechanical engineering: Determination of component and system reliability.* Berlin, Heidelberg: Springer, 2008.

[2] KROLO, A.: *Planning of Reliability Tests Considering Prior Information.* University of Stuttgart, Institute of Machine Components. PhD Thesis. 2004.

[3] NAUNHEIMER, H.; BERTSCHE, B.; RYBORZ, J.; NOVAK, W.: *Automotive transmissions: Fundamentals, selection, design and application (2th ed.).* Berlin, Heidelberg: Springer, 2011.

[4] HAIBACH, E.: *Betriebsfestigkeit: Verfahren und Daten zur Bauteilberechnung (3th ed.).* Berlin, Heidelberg: Springer, 2006.

[5] ABERNETHY, R. B.: *The new Weibull handbook : Reliability & statistical analysis for predicting life, safety, risk, support costs, failures, and forecasting warranty claims (5th ed.).* North Palm Beach, FL: R.B. Abernethy, 2006.

[6] MAENNING, W. W.: *Untersuchungen zur Planung und Auswertung von Dauerschwingversuchen an Stahl in den Bereichen der Zeit- und Dauerfestigkeit.* Düsseldorf: VDI-Verlag, 1967 (Fortschrittsbericht 5).

[7] GROß, R.W.H: *Beitrag zur Lebensdauerabschätzung von Stirnrädern bei Zahnkraftkollektiven mit geringem Völligkeitsgrad.* Aachen, RWTH. PhD Thesis. 1974.

[8] BRODBECK, P.: *Experimentelle und theoretische Untersuchungen zur Bauteilzuverlässigkeit und zur Systemberechnung nach dem Booleschen Modell.* University of Stuttgart, Institute of Machine Components. PhD Thesis. 1995.

[9] BEIER, M.: *Lebensdaueruntersuchungen an feinwerktechnischen Planetenradgetrieben mit Kunststoffverzahnung.* University of Stuttgart, Institute of Design and Production in Precision Engineering. PhD Thesis. 2010.

[10] NELSON, W.: *Accelerated testing: Statistical models, test plans and data analyses (pbk. ed.).* Hoboken, N.J.: Wiley-Interscience, 2004.

[11] BERGLING, G.: *Betriebszuverlässigkeit von Wälzlagern.* In: *SKF Kugellager* 51 (1976), Nr. 188, p. 1–10.

[12] LUCAS, J.; BROCKMAN, E.; THIRAVIAM, A.: Beyond FRACAS: Integrating field failures and field data. In: *2012 Proceedings Annual Reliability and Maintainability Symposium:* IEEE, 2012, p. 1–5.

[13] LEOPOLD, T.: *Holistic data collection for improved reliability analysis.* University of Stuttgart, Institute of Machine Components. PhD Thesis. 2012.

[14] KÜCÜKAY, F.: *Repräsentative Erprobungsmethoden bei der Pkw-Getriebeentwicklung.* In: *Getriebe in Fahrzeugen '95: Tagung.* Düsseldorf: VDI-Verlag, 1995 (VDI-Berichte, 1175), p. 49–65.

[15] MÜLLER-KOSE, J.-P.: *Repräsentative Lastkollektive für Fahrzeuggetriebe.* Technische Universität Braunschweig, Institute of Automotive Engineering. PhD Thesis. 2002.

[16] JUSKOWIAK, J.: *Determination of load profiles for automatic gearbox concepts.* University of Stuttgart, Institute of Machine Components. Diploma thesis. 2009.

[17] ISO 6336-2:2006: *Calculation of load capacity of spur and helical gears - Part 2: Calculation of surface durability (pitting).* 2006.

[18] MARQUARDT, D. W.: *An Algorithm for Least-Squares Estimation of Nonlinear Parameters.* In: *Journal of the Society for Industrial and Applied Mathematics* 11 (1963), Nr. 2, p. 431-441.

APTA approach: Analysis of accelerated prototype test data based on small data volumes within a car door system case study

Marcin Hinz[a*], Philipp Temminghoff[a], and Stefan Bracke[a]

[a] University of Wuppertal, Chair of Safety Engineering and Risk Management, Wuppertal, Germany

Abstract: Knowledge of failure behavior and failure modes regarding the component´s complete life cycle is fundamental within the early development phases of technical and complex products. Here, an overview of the design of prototype test procedures as well as the transformation of expected field failure behavior in prototype test characteristics is described. This provides the required knowledge for the understanding of accelerated testing and is the basis for understanding of the developed "Accelerated Prototype Test data Analysis" (APTA) approach. The APTA approach is demonstrated with the help of a case study with regard to a car door system. The analysis of the design principles, expected impacts in the usage phase and car door prototype test procedure is discussed. With the use of nonparametric as well as parametric statistical methods, the wearing and ageing of specific door mechanism characteristics (e.g. forces or displacements) in relation to life span variables are analyzed. Furthermore a method for the comparison of qualitative and quantitative characteristics and their impact on the door system is described. Finally the interpretation of the results and deduction of general issues and recommendations regarding to the design of prototype test procedures are presented.

Keywords: Accelerated Life Test, (Non-) parametric statistics, product reliability, prototype test data analysis

1. INTRODUCTION

Continuously increasing functionality of components and products as well as their complexity lead to complex failure modes and failure behavior. This applies particularly to the automotive industry: Car systems and components of the technical disciplines like electronic/electric, powertrain, auto body (interior/exterior) and chassis which are characterized by growing complexity and functionality. The goal is the detection of product risks and the analysis of the product and component reliability. Hence, the challenge is the analysis of the potential prospective customer usage conditions and their transformation in prototype test procedures during the product development process. Furthermore, the low number of prototypes - mainly due to financial limits - causes a major problem: From a statistical point of view, the validity of the test results based on a low test data volume is considerable. E.g. parametrical distribution models are unverifiable and subsequently parametric significance tests are not applicable.

The estimation of the failure behavior and failure prediction of highly reliable technical products is a very complex and difficult process. New, modern products are developed with a service life of many years, in plenty of cases in permanent running. Competitive national as well as international markets demand the reduction of the time-to-market point, which results in shorter product development and testing times. Within shorter innovation cycles and high customer requirements it is hardly possible to test the products under the normal conditions in a feasible time. The usage of Accelerated Tests (ATs) reduces the testing times and offers a solution for the detection of failure modes. Understanding of the AT models and its applications as well as adequate analysis of the product and selection of the right statistical methods in the design phase combined with the knowledge of testing methods is essential for the development of the right testing strategy. Choosing the proper test strategy will cause in the reduction of development time, number of prototypes and cost.

This paper shows the analysis of life span variables of a car fleet during the usage phase. This is the base of operations for the correlation analysis between prototype accelerated testing and usage phase. Furthermore the limits of parametric and applications/advantages of nonparametric statistics are discussed. Lastly, the interpretation of the results and deduction of general issues and recommendations regarding to the design of prototype test procedures are presented.

*) m.hinz@uni-wuppertal.de

2. PRINCIPALS OF ACCELERATED TESTING

This chapter focuses on theoretical principals and application areas of accelerated testing procedures. Based on these theoretical aspects the analysis of the main characteristics regarding the test cycle within the case study "car door system" is performed. Both areas, accelerated testing theory and case study test method, are the fundamentals of operations with respect to the APTA approach in chapter 3.

2.1. Comparison of testing methods

According to [3], "Accelerated tests" (ATs) is a term used for two completely different kinds of tests which both have different purposes, "qualitative ATs" and "quantitative ATs". The key differences are presented briefly in figure 1.

Figure 1: Accelerated testing methods

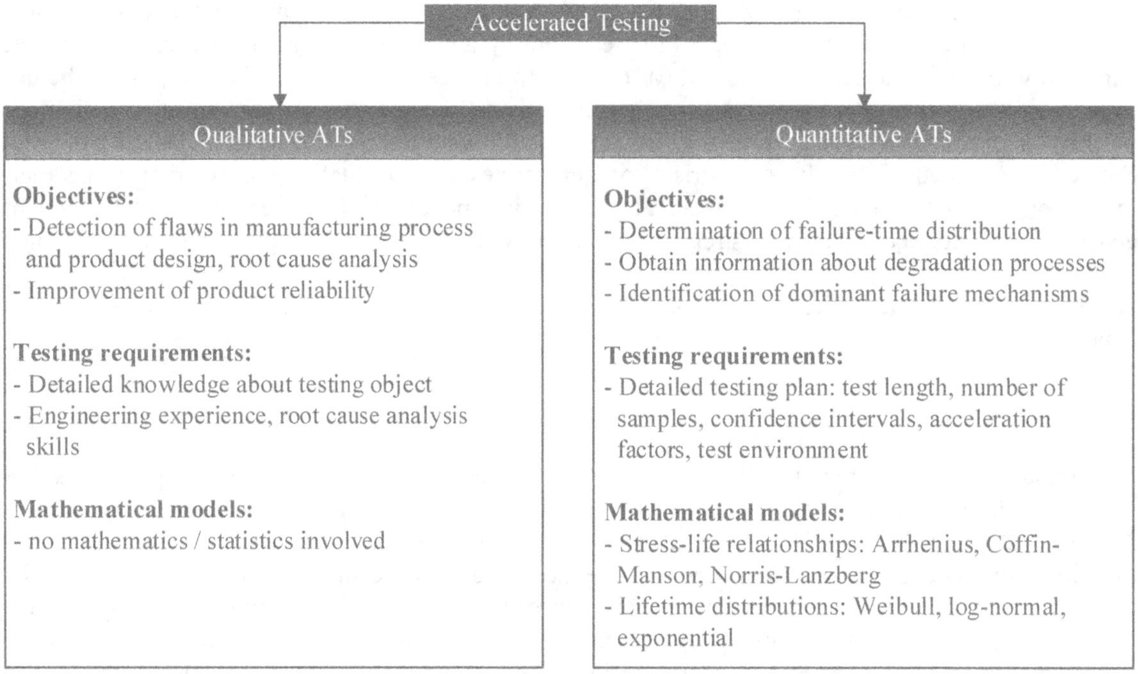

Qualitative ATs	Quantitative ATs
Objectives: - Detection of flaws in manufacturing process and product design, root cause analysis - Improvement of product reliability	**Objectives:** - Determination of failure-time distribution - Obtain information about degradation processes - Identification of dominant failure mechanisms
Testing requirements: - Detailed knowledge about testing object - Engineering experience, root cause analysis skills	**Testing requirements:** - Detailed testing plan: test length, number of samples, confidence intervals, acceleration factors, test environment
Mathematical models: - no mathematics / statistics involved	**Mathematical models:** - Stress-life relationships: Arrhenius, Coffin-Manson, Norris-Lanzberg - Lifetime distributions: Weibull, log-normal, exponential

Qualitative ATs are used to identify product weaknesses caused by flaws in the product's design or manufacturing process. This is done by gradually increasing stresses until the product fails. The aim is to improve the product reliability in a very short period of time, usually hours or days. They are mostly performed on entire systems but can be performed on individual assemblies as well. Common names for this type of tests are HALT (Highly Accelerated Life Test), STRIFE (Stress-Life) and EST (Environmental Stress Testing) [11].

An essential component of qualitative ATs is root cause analysis. It needs to be determined whether the identified potential flaws can happen in the field or if they are caused by testing a product above / below its specification levels and changing the failure mechanism. Based on this analysis corrective action is eventually implemented to improve the product reliability. Tests of this type do not involve any statistical or mathematical methods. They are only used to make the product more robust, not to gain any knowledge about the product lifetime or the degradation process over time.

Quantitative ATs are used to obtain information about the failure-time distribution and degradation in a relatively short period of time (usually weeks or months) by accelerating the use environment. In most cases a model to describe the relationship between failure mechanism and accelerating variables already exists. They are also well-suited for finding dominant failure mechanisms and are usually performed on individual assemblies rather than full systems. In order to set up a quantitative AT, several different parameters must be known, for example test duration, number of samples, desired confidence intervals, field and test environment, stress-life relationship and distribution model.

2.2 Acceleration models

One key factor is to determine the acceleration factor. It can be obtained by two different methods, either by using existing physical / chemical acceleration models or by determining empirical acceleration models by experimentation.

Physical / chemical acceleration models
Various physical / chemical models for well-understood failure mechanisms are already available. They describe the failure-causing process and allow extrapolation to use conditions for specific environment variables. The relationship between accelerating variables and the actual failure mechanism is usually extremely complicated. However simple models which adequately describe the process already exist. Examples for some acceleration models can be found in [3].

Empirical acceleration models

If there are no chemical / physical models to describe the failure-causing processes adequately it may be necessary to develop an own empirical model by experimentation. An empirical model usually provides an excellent fit to the available data, but still can provide false extrapolations for working conditions. Extensive empirical research regarding possible failure mechanisms and different stress variable combinations is needed to justify the needed extrapolation.

2.3. Acceleration methods

Increasing the use rate: Products, that are not in continuous use, can be accelerated by increasing the usage rate. This means that the time between the load phases is reduced. It has to be considered whether the increased use rate changes the cycles to failure distribution. A toaster for example could heat up increasingly cycle after cycle until it reaches temperatures which would not appear under normal use conditions. A case study for this purpose was performed by [4].

Increasing the aging-rate: By increasing experimental variables like temperature, humidity or radiation, chemical processes that lead to certain failure modes can be accelerated (e. g. the chemical degradation of a PV module can be accelerated by inducing an electric potential).

Increasing the level of stress: If the environmental stress exceeds the strength of the testing object, the unit (prototype) will fail. This means that a unit operating at higher stress levels will generally fail after a shorter space of time than a unit at lower stress levels (e.g., amplitude in temperature cycling, voltage, or pressure).

2.4 Types of responses

Accelerated tests can be distinguished based on the type of response as following:

- **Accelerated Binary Tests (ABTs):** The response of an ABT is binary, so the only reliability information obtained from each unit is whether it has survived the test or not, cf. [3].

- **Accelerated Life Tests (ALTs):** The response of an ALT is directly related to the lifetime of the product at use conditions. In most cases ALT data are right censored because the test is stopped before all units fail. If failures are only discovered during periodic inspection times the ALT response is interval censored, cf. [6] for a comprehensive overview of ALTs.

- **Accelerated Repeated Measures Degradation Tests (ARMDTs):** In an ARMDT the degradation of a unit is measured at several different points in time. This information is used to extrapolate the lifetime of the testing object. The degradation response could be actual chemical or physical degradation or performance degradation. An example for ARMDT modelling and analysis can be found in [7].

- **Accelerated Destructive Degradation Tests (ADDTs):** ADDTs are similar to ARMDTs, except that the measurements are destructive, so only one observation per test unit can be obtained. A discussion of the ADDT method and a detailed case study can be found in [8].

These different kinds of ATs are often closely related because they can involve the same failure mechanisms and physical / chemical model assumptions. However they are different regarding the statistical models and analyses which are performed because of the different kinds of observed responses. An important attribute of all quantitative ATs is the necessity to extrapolate to lower stress levels. Tests are performed in an environment with accelerated conditions, but estimates are needed at use conditions.

The test used in the case study "car door system" is a collection of worst-case scenarios of different use types and markets. Because of the irregular test plan it is not possible to perform a profound prognosis by using existing acceleration models. The purpose of the test procedure is only to ensure that the product does not fail at higher stress levels. For now the data is not used to extrapolate to normal use conditions. To do this, it would be advised to rearrange the test procedure to follow some of the more commonly used stress profiles like constant-stress or step-stress.
Studies to create appropriate testing plans were already done in the past, e. g. [9]. Because of that, the statistical analysis of the prototype data is only used to compare the different prototypes themselves as well as the two test facilities. In addition, test characteristics can be correlated and the impact of simulated environment influences can be evaluated quantitatively.

3. APTA APPROACH

In this chapter, the theory and application of the developed APTA approach is shown. First, in chapter 3.1 the authors outline an overview regarding the essential APTA procedure. Subsequently, the five APTA approach steps are explained in detail by an application within the case study with regard to a car door system (cf. chapter 3.2 – 3.6).

Figure 2: APTA approach

3.1. Overview of fundamental steps of the APTA approach

The "Accelerated Prototype Test data Analysis (APTA)" approach is a tool for definition, analysis, evaluation and optimization of existing test procedures and its application for the common as well as next generation products in all application areas, were testing plays a major role in the development process. It is a five step bottom-up algorithm with the main focus on:

- (Non-) parametric analysis of the test data regarding similarities/differences in the tested products and life span variables
- Analysis of the test procedure in order to detect its weak and strong points
- Optimization of the test procedure and optionally a prognosis of the system performance

With regard to the definition of APTA (figure 1) the single steps are to be characterized as follows.

Step 1: Analysis of the test procedure should answer the following questions: What is the testing product? Where is it tested? What are the relevant environmental influences? How many tested objects exist? How many testing machines are available? Are there similarities between the products (e.g. left-right variants or symmetrical design)? Are the samples independent? Answers to these questions are fundamental for the choice of proper statistical methods.

Step 2: This step includes a definition and allocation of life span variables with at least more than 10 measurements. Furthermore the boundary conditions which influence the measured product should be defined and classified.

Step 3: According to step 1, the statistical methods have to be chosen in a proper way. For this purpose distribution tests (e.g. Kolmogorov-Smirnov, Shapiro-Wilk, Anderson-Darling or Chi-squared test) [1] are the base of operations in order to choose either parametric or nonparametric methods. Normal distribution of all life span variables is the requirement for the use of many parametric tests whereas nonparametric statistics can be used without any knowledge of the distribution model regarding the data base. In spite of the wide application possibility of nonparametric methods which are more conservative, there is an information loss of about 15% [12].

Step 4: Based on step 1 and 2, the test procedure can be evaluated. Hence, a system of equations with BCs defined as unknowns and measurements as right-hand side has to be solved. Solution of this problem provides the impact of every single BC on each life span variable.

Step 5: With the solution of step 4 the test procedure can be optimized. A further development of the system behavior according to the test plan can be predicted.

Table 1: Boundary conditions and results of the tests

Boundary condition	Unknown	Result Force	Result displacement
Temperature T1, Humidity H1	x1	15.3	1.3
Temperature T2, Humidity H1	x2	-803.4	-82.4
Temperature T3, Humidity H1	x3	736.1	76.7
Temperature T4, Humidity H1	x4	-72.7	-16.7
Temperature T5, Humidity H2	x5	75.5	8.8
Temperature T6, Humidity H1	x6	2.8	-0.2
Temperature T7, Humidity H2	x7	31.2	-4.4
System adjustment	x8	-18.4	-4.2
Particle 1	x9	-601.9	-70.5
Erasure of particle 1	x10	534.5	64.7
System wash	x11	318.6	28.8
Particle 2	x12	-181.6	-14.3
Erasure of particle 2	x13	-46.3	-5.6
Additional stress of the system	x14	-1744	-174.3
Particle 3	x15	1655.7	162

The introduced APTA approach will be explained with the help of a case study involving a car door mechanism which is an assembly of multiple separate components. The used data in form of measurements is anonymized. Nevertheless, the presented findings correspond to the real, physical performance effects and failure behavior. The applied boundary conditions are an example of a composite of qualitative and quantitative parameters which can stress a door mechanism and are representative loads of an automotive usage phase. However, the focus of this paper is the optimized analysis of small data and information volumes. The optimization of the test procedure and its prognosis doesn't play the key role of this research. The mechanism will be explained on the basis of the five steps of the approach in the following chapters 3.2 to 3.6.

3.2. Analysis of the test procedure (APTA Step 1)

The tests within the case study "car door system" were performed in a climate chamber with the possibility of simulation of different temperatures and humidities. The mechanisms were tested in two separate and identically constructed testing facilities. Every test facility consists of two symmetrical doors which are driven by one power train. The symmetrical setup reflects the buildup of a car (left and right door). The power train opens and closes both doors within one test facility at the same time. Both test rigs (placed in the same climatic chamber) are driven separately. Every mechanism can be defined as one sample.

3.3. Definition of the life span variables and boundary conditions (APTA Step 2)

All mechanisms were tested under the influence of different boundary conditions (cf. table 1) which can be split into two groups. The first group (quantitative boundaries) is represented by seven combinations of different temperatures and humidity. The second one (qualitative boundaries) consist of eight factors with the objective of additional influence on the system (e.g. particles, additional stress or system adjustment). Particles simulate foreign objects and represent a typical, additional impact affecting the car door within the usage phase (e.g. sand or dust). Boundaries listed in the table 1 are anonymized and correspond to the real application conditions.

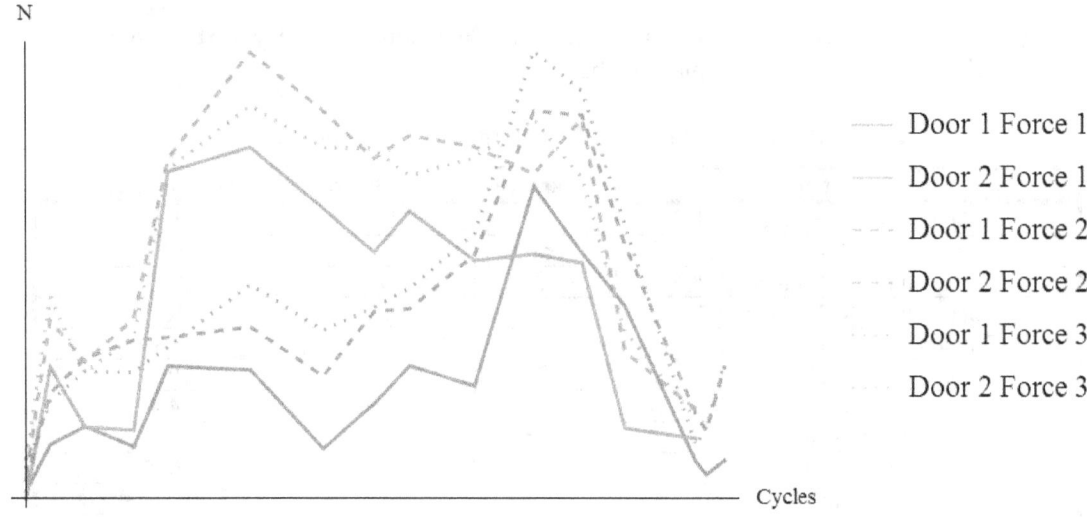

Figure 3: Force distribution of door 1 and 2

All doors were tested for several hundred thousand cycles. In the meantime, three forces and three displacements were measured within the mechanisms in irregular cycle intervals. The analyzed forces and displacements are attributes with an influence on the performance of the products in the whole product lifecycle. Due to the confidentiality of the data the accurate description of all functions of measured quantities as well as the testing program has to be disclaimed. Nevertheless it doesn't have any influence on the reliability analysis of prototypes.

Every series of measurement consist of 16 gaging points. An example of the distribution of the points is shown in the figure 2 and 3 and represents the features of all measurements.
A plot of the force distribution along the whole measurement series for two doors with three forces each is illustrated in figure 2. Cycles are plotted on the abscissa and forces in newton in the ordinate. Colors of the lines differentiate between door 1 (green) and door 2 (orange). Line types characterize the different forces, whereas the same line types indicate the same forces in different doors. Hence the green and orange continuous lines represent the same forces of door 1 and 2 respectively. By the examination of the line profiles some similarities can be identified and shall be discussed.

A systematic offset of force 1 and the other ones can be observed in both cases. Force 2 and 3 in both doors are nearly congruent. All forces within every door show a very similar development which means a similar reaction of the technical system with regard to the test program. Thus, with respect to door 1 after a rather constant trend followed by an increase of the line occurs a fall into nearly the same level as at the beginning of the test. This means, that the system is recovered after the whole test cycle and the amount of force before and after the test is nearly the same.

Differences in the reactions of the door systems (e.g. trends) between the doors can be explained by the manual production and assembly and as an outcome of both of them the dispersion effects of the prototypes.

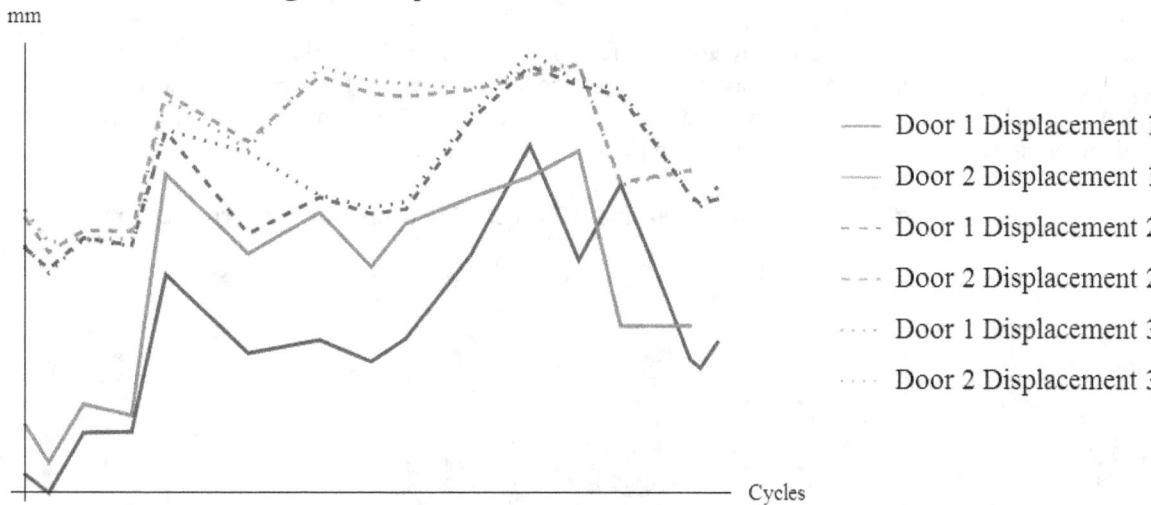

Figure 4: Displacement distribution of door 1 and 2

Figure 3 shows the displacement characteristics of the same doors as in figure 2. Door 1 is marked with the blue, door 2 with the red line. Line styles correspond to the figure 2. Cycles are plotted on the abscissa and displacements in millimeter in the ordinate. Systematic analysis of the lines distribution provides similar results (offset, congruence, development regarding the testing program) as in figure 2. Contrary to the force, after the whole test, the system doesn't recover in terms of the displacement, which means that the displacement at the end is remarkably higher than at the beginning.

A comparison of force and displacement yields a correlation in the trends for every single door. An increase of values followed by a constant distribution and a fall in door 2 as well as constant, increase, and fall distribution in door 1 show small differences in mechanisms and a similar reaction to the environment boundaries. This leads to the assumption of statistical similarities of the prototypes.

3.4. Statistical analysis of the test data (APTA Step 3)

Systematic analysis of the graphs provides many optical similarities (e.g. trends, peaks, offsets). In order to perform a statistical analysis of the test data all facts will be summarized. Overall there are 24 samples (three forces, three displacements in four doors) with 16 values available. In case of all doors the samples are statistically independent.
The statistical test of Shapiro-Wilk and Anderson-Darling [1] with the null (and alternative) hypothesis:

$$H_0 : F = F_0 \ (and \ H_1 : F \neq F_0) \tag{1}$$

The hypothesis that the sample data are normally distributed by a significance level of $\alpha=5\%$ can't be rejected in case of all samples. Nevertheless the amount of the data within a sample as well as the number of prototypes stay at a low level, therefore - depending on the sample size - parametric and nonparametric test will be applied [2].

For the comparison of samples, different significance tests according to different applications are available. In this case nonparametric tests will be described. Parametric test were also undertaken and provide the same conclusions. The null (and alternative) hypothesis of Mann-Whitney U test [10]

$$H_0 : F(z) = G(z) \ (and \ H_1 : F(z) \neq G(z - \theta) \ ) \tag{2}$$

that the samples have the same focus and the null (and alternative) hypothesis of Levene's test [10]

$$H_0 : F(z) = G(z) \ (and \ H_1 : F(z) \neq G(\theta z) \ ) \tag{3}$$

that the variances of the samples are equal weren't rejected in any case. Hence it can't be proved, that all forces and displacements are significantly different in terms of the focus and the variance. The same conclusions can be drawn by the analysis of the boxplots. An example is shown in figure 4 and demonstrates the comparison of force 2 (on the left) and force 3 (on the right) of prototype door 1.

Figure 5: Comparison of the samples (force 2, blue color; force 3, green color)

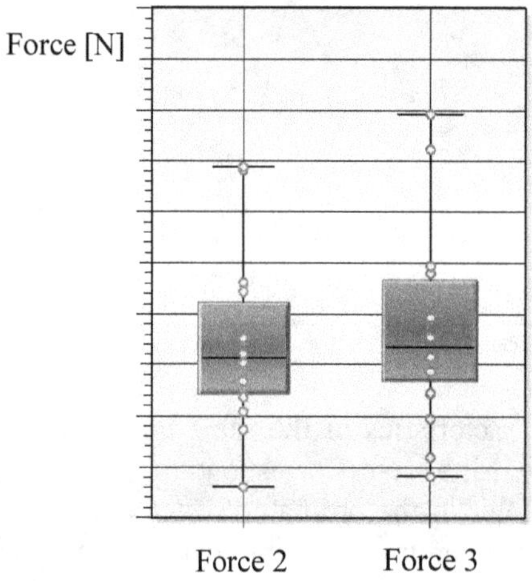

After application of the significance tests regarding the comparison of the focuses and variations of the samples, a regression analysis of all forces and displacements shall be executed. For the reason that the probes are normally distributed but the data volumes are small, the correlation analysis will be performed parametrically as well as nonparametrically.

Figure 5 shows the results of the correlation analysis with the assumption of normally distributed data with the use of Pearson's product-moment correlation coefficient [1]. In this case, all attributes of left door of both test facilities have been correlated with each other. The result of every single correlation is shown in the corresponding combination of the attributes in the matrix.

All Pearson coefficients within the first door are greater than 0.81, whereupon by a comparison of the same attributes (force with force and displacement with displacement), the coefficient increases to $r \geq 0.94$. In the second door the coefficients are slightly lower with $r_{min} = 0.62$. However by an analog comparison of the same characteristics the value increases to $r \geq 0.69$. In the case of both single doors it is a high correlation. Therefore, an assumption can be met, that the attributes of the mechanisms react in a similar way to the external influences. It is the same assumption as in case of the interpretation of values characteristics shown in figure 2 and 3.

The correlation of two separate doors (upper right and lower left corner of the correlation matrix) yields weak to very weak solution *(0.67 ≤ r ≤ 0.04)*. This can be traced back to the manual manufacture of the prototypes.

Figure 6: Pearson correlation matrix of the analyzed door prototype characteristics

	Door 1 Disp. 1	Door 1 Force 1	Door 1 Disp. 2	Door 1 Force 2	Door 1 Disp. 3	Door 1 Force 3	Door 2 Disp. 1	Door 2 Force 1	Door 2 Disp. 2	Door 2 Force 2	Door 2 Disp. 3	Door 2 Force 3
Door 1 Disp. 1	1.	0.85	0.95	0.81	0.96	0.82	0.67	0.46	0.42	0.58	0.48	0.54
Door 1 Force 1	0.85	1.	0.84	0.94	0.87	0.95	0.39	0.31	0.19	0.47	0.24	0.47
Door 1 Disp. 2	0.95	0.84	1.	0.82	0.95	0.82	0.62	0.41	0.4	0.45	0.47	0.45
Door 1 Force 2	0.81	0.94	0.82	1.	0.84	0.98	0.32	0.18	0.067	0.42	0.12	0.43
Door 1 Disp. 3	0.96	0.87	0.95	0.84	1.	0.85	0.64	0.46	0.42	0.54	0.47	0.48
Door 1 Force 3	0.82	0.95	0.82	0.98	0.85	1.	0.31	0.17	0.04	0.4	0.11	0.39
Door 2 Disp. 1	0.67	0.39	0.62	0.32	0.64	0.31	1.	0.92	0.89	0.79	0.88	0.71
Door 2 Force 1	0.46	0.31	0.41	0.18	0.46	0.17	0.92	1.	0.93	0.78	0.89	0.69
Door 2 Disp. 2	0.42	0.19	0.4	0.067	0.42	0.04	0.89	0.93	1.	0.75	0.98	0.64
Door 2 Force 2	0.58	0.47	0.45	0.42	0.54	0.4	0.79	0.78	0.75	1.	0.73	0.94
Door 2 Disp. 3	0.48	0.24	0.47	0.12	0.47	0.11	0.88	0.89	0.98	0.73	1.	0.62
Door 2 Force 3	0.54	0.47	0.45	0.43	0.48	0.39	0.71	0.69	0.64	0.94	0.62	1.

3.5. Evaluation of the test procedure (APTA Step 4)

According to the assumption of similar reaction of the mechanisms on external influences, the system with its boundary conditions can be mapped onto a mathematical system of equations, which can be easily solved. All external effects and changes of the mechanisms during the test phase have been defined as boundary conditions and are listed in table 1. Furthermore, an unknown has been allocated to every boundary condition. Between every measurement of the attributes the number and combination of the boundaries has been varied. The lowest number of driven cycles by given temperature and humidity was 500, in many cases multiples of it. Between two measurements multiple changes of the temperature occurred many times. By a definition of the variables the number of cycles was implemented with a factor of 1/1000. Hence 2000 cycles at temperature T2 correspond to the expression $2*x2$. In addition to the variation of the temperature and humidity, the mechanisms were stressed with a set of boundary conditions which are not quantitatively measurable in terms of time and amount. Because of the fact that it is not possible to define how long the qualitative factors stress the mechanisms, their number of cycles were not considered and implemented by the definition of the equations. A realistic example provides a better understanding of the procedure. After the first measurement (m1) following steps have been performed:

- 1000 cycles at temperature T1
- 2000 cycles at temperature T4
- Application of particle 2
- Additional stress of the system
- 3000 cycles at temperature T7

Subsequently a second measurement (m2) took place. In this case 6000 cycles were performed between the two measurements. These steps can be described as follows:

$$m1 + 1*x1 + 2*x4 + x12 + x14 + 3*x7 = m2 \qquad (4)$$

and rearranged to:
$$x1 + 2*x4 + 3*x7 + x12 + x14 = m2 - m1 \tag{5}$$

All in all 16 measurements and 15 unknowns are available. With 16 values, 15 equations can be defined. Therefore the system of equations can be solved.

The interpretation of the results (cf. table 1) is shown by using an example of a measurement of one force and one displacement of the same door. The factors have no physical meaning and serve only the interpretation of the weighting of the influence of the boundary conditions on the system behavior. Following statements can be made (the notation Tx correspond to 1000 cycles at temperature x):

- T6 has the smallest and particle 3 the highest influence on both attributes
- T2 has the highest negative and T3 the highest positive impact on the system
- T3 has 10-times higher effect than T5
- T2 and T4 influence the force in contrary to all others temperatures negative
- T1, T3 and T5 influence the displacement in contrary to all others temperatures positive
- Particle 1 and particle 2 have, in contrast to particle 3, a negative impact on the mechanism
- System adjustment has the smallest effect out of all qualitative boundary conditions

Furthermore, the solution of this mathematical procedure allows a direct comparison of qualitative and quantitative boundary conditions.

3.6. Optimization of the test procedure (APTA Step 5)

Because of the low number of measurements and missing individual measurements (those after the application of every single boundary condition) it is not possible to verify the factors. Such a validation would not only confirm or reject the method but would also deliver new and important information (e.g. functional relationship between temperature and force/displacement or the size and shape of particle and force/displacement).

With the help of the factors it would be possible to provide a prognosis of the further performance of the system by a further development of the testing plan. This would reduce the amount of the measurements and costs.

Furthermore it is possible to define the most disadvantageous composite of the BCs (worst case scenario). An example for it would be the configuration of temperature T3 (x3), erasure of particle 1 (x10), system wash (x11) and particle 3 (x15). This compound would heavily (positively) influence the force and could exceed the limits prescribed by the requirements or could lead to system breakdown.

4. CONCLUSIONS

Development of technical complex products is affected by the challenge to identify and eliminate the potential weak points in early phases of product and components design. Here, the increasing complexity and functionality of the products are the key factors also because they can cause a wide range of failure spectrum. Accelerated tests offer the possibility to map the user behavior and, as its consequence, the potential expected failure spectrum over the load spectrum. The great challenge is the design of the test plans. The aim is to test the prototypes with regard to a wide failure spectrum in a short time and, as far as possible, with less test objects.

This paper presents on one hand the basic possibilities of prototype testing. On the other hand a case study with regard to a car door system outlines the statistical analysis of the test data based on small data volumes. The systematic application of nonparametric statistical models in combination with parametric methods enables the comparison of the quality and reliability of the products and test facilities already in the early phases of the design phase.

Furthermore, the impacts of the simulated environmental effects on the prototypes can be described by a mathematical model which is the base of operations for further design of product improvement activities. A prediction of the expected life time by using the existing data is not possible, which is caused by the miscellaneous impacts of different markets (worst-case scenario). An adaption of the test plan for specific target markets would increase the significance of the statistical analysis and reliability.

References

[1] L. Sachs and J. Hedderich. *"Angewandte Statistik Methodensammlung mit R."*, 13. Edition, Springer, 2009, Berlin.

[2] S. Bracke and R. Breede. *"Design of Prototype test Procedures (DPP concept) for complex damage causes in automotive engineering"*, Reliability, Risk and Safety: Back to the Future, London, Taylor & Francis Group, (2010).

[3] L. Escobar and W. Meeker. *"A Review of Accelerated Test Models"*, Statistical Science 21, No. 4, pp. 552–577, (2006).

[4] Y. Guangbin and Z. Zaghati. *"Accelerated life tests at higher usage rates: a case study"*, Annual Reliability and Maintainability Symposium (RAMS), pp. 313–317, (2006).

[5] W. Meeker and G. Hahn. *"Asymptotically Optimum Over-Stress Tests to Estimate the Survival Probability at a Condition with a Low Expected Failure Probability"*, Technometrics 19, No. 4, pp. 381–399, (1977).

[6] W. Nelson. *"Accelerated testing: Statistical models, test plans and data analyses"*, Wiley, 1990, New York.

[7] W. Meeker and L. Escobar. *"Accelerated Degradation Tests: Modeling and Analysis"* Technometrics 40, No. 2, pp. 89–99, (1998).

[8] L. Escobar and W. Meeker. *"Accelerated destructive degradation tests: Data, models and analysis"* Mathematical and Statistical Methods in Reliability, World Scientific, River Edge, pp. 319–337, (2003).

[9] M. Haiming and W. Meeker. *"Strategy for Planning Accelerated Life Tests with Small Sample Sizes"*. IEEE Transactions on Reliability 59, No. 4, pp. 610–619, (2010).

[10] J. Hartung and B. Elpelt. *„Lehr- und Handbuch der angewandten Statistik; mit zahlreichen, vollständig durchgerechneten Beispielen. 11. Edition."*, Oldenbourg, 1998, München, Wien.

[11] H. Chan and P. Englert. *"Accelerated stress testing handbook: Guide for achieving quality products"*, IEEE Press, 2001, New York.

[12] S. Siegel. *"Nonparametric statistical methods."* Klotz, 2001, Eschborn bei Frankfurt a. M.

www.ingramcontent.com/pod-product-compliance
Lightning Source LLC
Chambersburg PA
CBHW081141180526
45170CB00006B/1878